CLIMATE-CHANGE MITIGATION AND EUROPEAN LAND-USE POLICIES

Climate-change Mitigation and European Land-use Policies

Edited by

W. Neil Adger
University of East Anglia, UK

Davide Pettenella
University of Padova, Italy

and

Martin Whitby
University of Newcastle upon Tyne, UK

CAB INTERNATIONAL

CAB INTERNATIONAL
Wallingford
Oxon OX10 8DE
UK

CAB INTERNATIONAL
198 Madison Avenue
New York, NY 10016-4314
USA

Tel: +44 (0)1491 832111
Fax: +44 (0)1491 833508
E-mail: cabi@cabi.org

Tel: +1 212 726 6490
Fax: +1 212 686 7993
E-mail: cabi-nao@cabi.org

©CAB INTERNATIONAL 1997. All rights reserved. No part of this publication may be reproduced in any form or by any means, electronically, mechanically, by photocopying, recording or otherwise, without the prior permission of the copyright owners.

A catalogue record for this book is available from the British Library, London, UK.

Library of Congress Cataloging-in-Publication Data
Climate-change mitigation and European land-use policies / edited by
  W. Neil Adger, Davide Pettenella, and Martin Whitby.
     p. cm.
  Includes bibliographical references and index.
  ISBN 0-85199-185-8
    1. Climatic changes--Environmental aspects--Europe. 2. Greenhouse gases--Environmental aspects--Europe. 3. Land use--Environmental aspects--Europe. 4. Agriculture and state--Europe. 5. Forest policy--Europe. I. Adger, W. Neil. II. Pettenella, Davide.
III. Whitby, Martin Charles.
QC981.8.C5C595   1997
363.738'747--dc21   97-39882
                                                      CIP

ISBN 0 85199 185 8

Typeset in Garamond by Advance Typesetting Ltd, Oxford
Printed and bound in the UK by Biddles Ltd, Guildford and King's Lynn

Contents

Contributors ix

Preface xv

Part I: Agriculture, Forestry and the Policy Issues

1 Land Use in Europe and the Reduction of
   Greenhouse-gas Emissions 1
   W. Neil Adger, Davide Pettenella and Martin Whitby

2 Agricultural Policy Reform and Climate-change Mitigation in
   Organization for Economic Cooperation and Development
   Countries 23
   Mark Storey and Merylyn McKenzie-Hedger

3 Forestry Options for Mitigating Predicted Climate Change 35
   William M. Ciesla

4 A Critical Review of the Scientific Basis of Projected
   Global Warming 49
   Antonio Zecca and Roberto S. Brusa

5 The Economic Costs of Climate Change and Implications for
   Land-use Change 59
   Samuel Fankhauser

Part II: Issues in the Analysis of Greenhouse-gas Mitigation

6 Analysis of Time Profiles of Climate Change 71
   Colin Price

7 Public Policies and Incentives to Accelerate Irreversible
   Green Investments 89
   Cesare Dosi and Michele Moretto

8 Human Adaptation in Ameliorating the Impact of Climate
   Change on Global Timber Markets 103
   *Brent Sohngen, Roger A. Sedjo, Robert Mendelsohn
   and Kenneth S. Lyon*

9 Economic Instruments and the Pasture–Crop–Forest Interface 113
   G. Cornelis van Kooten and Henk Folmer

Part III: Mitigation Options and Policies in Agriculture

10 Agricultural Policy Impacts on United Kingdom Carbon Fluxes 129
    *Susan Armstrong Brown, Mark D.A. Rounsevell, James D. Annan,
    V. Roger Phillips and Eric Audsley*

11 Full Cycle Emissions from Extensive and Intensive Beef
    Production in Europe 145
    Susan Subak

12 Reduction of Emissions in Farming Systems in Germany 159
    Konrad Löthe, Clemens Fuchs and Jürgen Zeddies

13 The Effects of the Dutch 1996 Energy Tax on Agriculture 171
    Marinus H.C. Komen and Jack H.M. Peerlings

Part IV: Mitigation Options and Policies in Forestry

14 Policy Instruments for Environmental Forestry:
    Carbon Retention in Farm Woodlands 187
    Robert Crabtree

15 The Role of the Common Agricultural Policy in Inhibiting
    Afforestation: the Example of Saxony 199
    Christian Lippert and Michael Rittershofer

16 Forest Management and Policy Options for Emission
    Mitigation in Finland 215
    Heikki Seppälä and Kim Pingoud

17 German Forests in the National Carbon Budget:
an Overview and Regional Case-studies   227
Klaus Böswald

18 Carbon Fixation in Swedish Forests in the Context of
Environmental National Accounts   239
Peter Eliasson

Part V: Joint Implementation in Forestry and International Perspectives

19 The Potential Role of Large-scale Forestry in Argentina   255
Roger A. Sedjo and Eduardo Ley

20 Forestry and Agroforestry Land-use Systems for Carbon
Mitigation: an Analysis in Chiapas, Mexico   269
*Ben H.J. De Jong, Lorena Soto-Pinto, Guillermo Montoya-Gómez,
Kristen Nelson, John Taylor and Richard Tipper*

21 Institutional Premises for the Fulfilment of Carbon-credit
Requirements by Russia   285
Andrei A. Gusev and Nina L. Korobova

Part VI: Summary

22 Issues and Implications for Agriculture and Forestry:
a Focus on Policy Instruments   295
Paola Gatto and Maurizio Merlo

References   313

Index   339

Contributors

W. Neil Adger is a Lecturer in Environmental Economics, School of Environmental Sciences, University of East Anglia, Norwich NR4 7TJ, and Senior Research Fellow, CSERGE, University of East Anglia and University College London, UK.

James D. Annan is a Research Scientist, Silsoe Research Institute, Wrest Park, Silsoe, Bedfordshire MK45 4HS, UK.

Susan Armstrong Brown is a Research Scientist with the Soil Survey and Land Research Centre, Cranfield University, Silsoe, Bedfordshire MK45 4DT, UK.

Eric Audsley is Principal Research Scientist and Head of the Mathematics and Decision Support Group, Silsoe Research Institute, Wrest Park, Silsoe, Bedfordshire MK45 4HS, UK.

Klaus Böswald is Junior Research Scientist in Forest Economics, Albert-Ludwigs-Universität Freiberg, Institut für Fortspolitik und Raumordnung, Beroldstrasse 17, D-79085 Freiberg, Germany.

Roberto S. Brusa is a Research Scientist in the Atomic and Positron Physics Laboratory at Trento University, via Sommarive 12, 38050 Povo (Tn), Italy.

William M. Ciesla is an international consultant in forest protection and is based in Fort Collins, Colorado. He was formerly Forest Protection Officer, Food and Agriculture Organization, Rome, Italy.

Robert Crabtree is Head of the Environmental and Socio-economics Group at the Macaulay Land Use Research Institute, Craigebuckler, Aberdeen AB15 8QH, UK.

Ben H.J. De Jong is a Senior Researcher in Tropical Silviculture at El Colegio de la Frontera Sur, San Cristóbal de las Casas, Chiapas, Mexico.

Cesare Dosi is a Lecturer in Public Finance, Department of Economics, University of Padova, via del Santo 28, I-35100 Padova, Italy, and Research Fellow, Fondazione Eni Enrico Mattei, Milano, Italy.

Peter Eliasson is a doctoral student in Forestry at the Department of Forest Economics, Faculty of Forestry, Swedish University of Agricultural Sciences, S-901 83 Umeå, Sweden.

Samuel Fankhauser is with the World Bank, Washington, DC, and previously with the Centre for Social and Economic Research on the Global Environment, University College London and University of East Anglia, UK.

Henk Folmer is Professor in the Department of Economics, Wageningen Agricultural University, Hollandseweg 1, 6706, KN Wageningen, the Netherlands.

Clemens Fuchs is an Agricultural Economist and a Professor of Farm Management at the Fachhochschule in Neubrandenburg, Brodaer Strasse 2, 17009 Neubrandenburg, Germany.

Paola Gatto is a Research Associate in the Department of Land and Agroforestry Systems, University of Padova, AGRIPOLIS, via Romea, 35020 Legnaro (PD), Italy.

Andrei A. Gusev is Professor in Economics and Director of Scientific Direction at the Institute for Market Economy Problems, Krasikova Street 32, Moscow, Russia.

Marinus H.C. Komen is a doctoral student in the Department of Agricultural Economics and Policy, Wageningen Agricultural University, Hollandseweg 1, 6706, KN Wageningen, the Netherlands.

Nina L. Korobova is a Senior Researcher in Economics and Environmental Economics at the Institute for Market Economy Problems, Krasikova Street 32, Moscow, Russia.

Eduardo Ley is a Research Fellow, Resources for the Future, Washington, DC, USA.

Christian Lippert is a doctoral student in the Department of Agricultural Economics at the University of Halle-Wittenberg, Germany.

Konrad Löthe is an Agricultural Economist in the Department of Farm Management, University of Hohenheim, 70593 Stuttgart, Germany.

Kenneth S. Lyon is a Professor of Economics at Utah State University, Logan, Utah, USA.

Merylyn McKenzie-Hedger is the Climate Change Policy Coordinator, World-Wide Fund for Nature (WWF) International, based at WWF UK, Panda House, Weyside Park, Godalming, Surrey GU7 1XR, UK.

Robert Mendelsohn is a Professor in the School of Forestry and Environmental Studies, Yale University, New Haven, Connecticut, USA.

Maurizio Merlo is Professor of Forestry Economics and Policy in the Department of Land and Agro-forestry Systems, University of Padova, AGRIPOLIS, via Romea, 35020 Legnaro (PD), Italy.

Guillermo Montoya-Gómez is an Associate Researcher specializing in agroforestry at El Colegio de la Frontera Sur, San Cristóbal de las Casas, Chiapas, Mexico.

Michele Moretto is Lecturer in Economics, Department of Economics, University of Padua, via Del Santo 22, I-35100 Padua, Italy.

Kristen Nelson is a Senior Researcher specializing in rural sociology at El Colegio de la Frontera Sur, San Cristóbal de las Casas, Chiapas, Mexico.

Jack H.M. Peerlings is Assistant Professor of Agricultural Economics and Policy in the Department of Agricultural Economics and Policy, Wageningen Agricultural University, Hollandseweg 1, 6706, KN Wageningen, the Netherlands.

Davide Pettenella is a Lecturer in Forestry Economics, Dipartamento Territorio e Sistemi Agro-forestali, University of Padova, AGRIPOLIS, I-35020 Legnaro (PD), Italy.

V. Roger Phillips is a Principal Research Scientist in the Environment Group, Silsoe Research Institute, Wrest Park, Silsoe, Bedfordshire MK45 4HS, UK.

Kim Pingoud is Senior Research Scientist, VTT Energy, PO Box 1606, FIN-02044 VTT, Finland.

Colin Price is a Professor of Environmental and Forestry Economics at the School of Agricultural and Forest Sciences, University of Wales, Bangor, Gwynedd LL57 2UW, UK.

Michael Rittershofer is a doctoral student in the Department of Agricultural Economics at the University of Halle-Wittenberg, Germany.

Mark D.A. Rounsevell is a Senior Research Scientist and Head of the Climate Change Unit, Soil Survey and Land Research Centre, Cranfield University, Silsoe, Bedfordshire MK45 4DT, UK.

Roger A. Sedjo is a Senior Fellow and Director of the Forest Economics and Policy Program at Resources for the Future, Washington, DC, USA.

Heikki Seppälä is Senior Research Economist, Finnish Forest Research Institute (METLA), Unioninkatu, Finland.

Brent Sohngen is Assistant Professor, Department of Agricultural Economics, Ohio State University, Columbus, Ohio, USA.

Lorena Soto-Pinto is a Senior Researcher in Natural Resource Economics at El Colegio de la Frontera Sur, San Cristóbal de las Casas, Chiapas, Mexico.

Mark Storey is a Climate Change Analyst in the Pollution Prevention and Control Division of the OECD Environment Directorate, 2 rue Andre-Pascal, 75775 Paris, Cedex 16, France.

Susan Subak is a Senior Research Associate in the Centre for Social and Economic Research on the Global Environment, University of East Anglia, Norwich NR4 7TJ and University College London, UK.

John Taylor is a Forestry Adviser to the Union de Crédito Ya Kac'Tic SA de CV, San Cristóbal de las Casas, Chiapas, Mexico.

Richard Tipper is a consultant with the Institute of Ecology and Resource Management, University of Edinburgh, Darwin Building, Mayfield Road, Edinburgh EH9 3JU, UK.

G. Cornelius van Kooten is Professor in the Department of Agricultural Economics, University of British Columbia, No. 303, 2053 East Mall, Vancouver V6T 1Z2, Canada.

Martin Whitby is Professor of Countryside Management and Director of the Centre for Rural Economy, University of Newcastle upon Tyne, Newcastle NE1 7RU, UK.

Antonio Zecca is Professsor of Atomic Physics and the Head of the Atomic and Positron Physics Laboratory at Trento University, via Sommarive 12, 38050 Povo (Tn), Italy.

Jürgen Zeddies is Professor of Farm Management at the University of Hohenheim, 70593 Stuttgart, Germany.

Preface

This book forms one element of a European Union (EU) Concerted Action on Policy Measures to Control Environmental Impacts from Agriculture. It results from a workshop held at Monte Bondone, Italy, in May 1996, which brought together academics and policy-makers to examine the critical issues of greenhouse-gas emissions from agriculture and forestry. There are direct linkages and correlation between diverse environmental issues examined under other parts of the Concerted Action: pesticides, mineral flows and landscape conservation with fluxes of greenhouse gases. This is particularly so in the forestry sector, where increases in forest area, as well as the lengthening of rotations and the management of 'old growth' and natural forests, bring about the joint benefits of carbon sequestration in soil and vegetation biomass, as well as enhancing landscape and conserving habitats. Some of the chapters in this book address this theme and the related issue of the interface between agriculture and forestry, where the largest carbon fluxes associated with land use actually occur.

Part of the explanation for inertia on tackling both climate-change mitigation, and environmental problems in general, is the separation of the causes of human-induced climate change (emissions of greenhouse gases) and their effects, both in space and in time. From an economic perspective, climate change is a classic externality, where abatement of greenhouse-gas emissions must take place against a background of uncertain future potential climate changes, some of which may even benefit parts of the economy in some areas in Europe. The precautionary principle would suggest that the uncertainty of impact is reason enough for present mitigation, and indeed this principle is enshrined in the Framework Convention on Climate Change (see Bodansky, 1993, for an in-depth legal review). The policy impetus for mitigation is undoubtedly strengthened, however, by an appraisal of the economic costs of the impact of climate change, demonstrating the efficiency, as well as the equity and sustainability rationales, for proactive abatement policies. Indeed, the analysis of several chapters effectively stresses the need for much greater international action than is presently observed.

The case for concerted international action and the relationship between European land-use policies and those in other countries are presented in various ways in the following chapters. Economic theory suggests that, if the costs of climate-change mitigation are to be minimized in an optimizing world, opportunities for carbon sequestration or emission reduction at low or negative marginal cost should be undertaken before those with higher marginal costs. Yet this principle is controversial when it translates into enhancing carbon sinks in one part of the world, while allowing continued emissions in another unabated. This issue is relevant to actions within Europe, where differential responsibility is implicitly accepted between the climate policy 'leaders' and the states within the EU where 'catch-up' economic growth is the dominant policy objective. The cost-minimization principle is also pertinent in the relationship between European land use and the rest of the world. The series of chapters in this book dealing with the experience of land use and greenhouse-gas fluxes on other continents illustrates these themes.

The editors wish to thank the EU for funding the workshop, its participants, and the organization under the Agro Industrial Research Programme (AIR3-CT93-1164). In particular, we are grateful to Arie Oskam, Maurizio Merlo, Paola Gatto, Luca Cesaro, the Universities of Padova and Wageningen, the Centro di Ecolgia Alpina, Monte Bondone and the Centre for Rural Economy, where Richard Hill prepared the text for final publication. We thank Philip Judge at the University of East Anglia for preparing the diagrams.

The timing of the workshop coincided with the completion and publication of the second assessment report of the Intergovernmental Panel on Climate Change (Bruce *et al.*, 1996; Houghton *et al.*, 1996; Watson *et al.*, 1996), thereby grounding the analysis presented in this book on the latest scientific knowledge and policy appraisal on the subject. The greater challenge taken up in these pages is to integrate analysis of climate-change mitigation in an interdisciplinary analysis of the relationship of agriculture and forestry to its environment.

W. Neil Adger
Davide Pettenella
Martin Whitby

Land Use in Europe and the Reduction of Greenhouse-gas Emissions

W. Neil Adger, Davide Pettenella and Martin Whitby

INTRODUCTION: THE POLICY CONTEXT

Climate change induced by the enhanced greenhouse effect waxes and wanes as an issue in European public attention, with hot summers, droughts and snowless winters. Climate change, a phenomenon now confirmed as being human-induced, will have important and significant ramifications for global resource use well into the next century, particularly in the agricultural sectors of the world's economies, but also more generally in the biotic world. Current projections of the magnitude of changes suggest that, if a 'business as usual' path of emissions increase is followed, mean temperature will rise by approximately 2.0°C (range 1.0–3.5°C) by 2100, with an associated rise in mean sea-levels of 0.5 m (range 0.15–0.95 m) (IPCC, 1996b).

In the period since the United Nations (UN) Conference on Environment and Development in Rio de Janeiro in 1992, national governments have been obliged to consider the global environmental impacts of their economies and land uses. The Earth Summit culminated in the opening for signature of Conventions on Biodiversity and Climate Change, as well as setting out Agenda 21 on sustainable development and producing principles for forest management. Progress on global environmental protection since then has been fitful. The 5 years since Rio have shown that such agreements only become meaningful when they are implemented at the national and regional level and where there already exist political will and motivation to do so. This book mainly considers the European actions in complying with one of the agreements, namely the UN Framework Convention on Climate Change (FCCC). Specifically, we focus on land-use-related greenhouse-gas fluxes, recognizing that the prospects of success for the global environmental agreements of Rio are fundamentally influenced by trends in global land use. The importance of the issue of climate change is underlined by the existence of FCCC, ratified in 1993 by the countries of the European Union (EU) (O'Riordan and Jäger, 1996). The policy commitments of the EU governments, as well as other signatory parties to the FCCC, are to limit overall emissions and to enhance sinks of greenhouse gases, with the target of limiting net emissions in 2000 to those in 1990. The EU Member States have heterogeneous policy positions within this framework, with some countries, such as the Netherlands and Germany, setting self-imposed targets for overall reductions by 2000 or 2005. Others, such as Ireland, Spain and Portugal have stated their intention to increase emissions and further stated that it is the responsibility of Europe as a whole to meet the external target of the FCCC.

EUROPEAN CLIMATE POLICIES

Does the Climate Change Convention have the political impetus in Europe to make an effective difference to the prospect for climatic changes over the next century? The countries of the EU, on which the majority of the chapters in this book concentrate, will most probably meet their commitments under the Convention to stabilize overall emissions at 1990 levels by 2000. This might suggest the existence of some political will. However, there is a remarkable heterogeneity of policy positions within the EU. Some countries are undertaking unilateral carbon and energy taxes, with the policy objective to reduce overall emissions significantly. In stark contrast, other countries have maintained their right to increase emissions. In that policy context, this book seeks to provide the analytical tools for policy design and analysis and to appraise the national and international situation with regard to greenhouse-gas fluxes associated with agriculture and forestry.

In the global context, Europe has been, and continues to be, important in the formulation, negotiation and implementation of the FCCC. This importance may be greater than its relative contribution to emissions of greenhouse gases, although this in itself is significant from energy use and industry. The 'block vote' nature of the EU and its proactive role in many aspects of the debate on climate contrast with other industrialized regions of the world, such as the USA and Australia.

As demonstrated by the diverse national policies, however, Europe is also a microcosm of global climate policy, in that the differentials in income levels and economic structure between European countries also lead to internal conflicts when a multilateral European policy on energy use or carbon tax is proposed. For the study here, the differential impacts of climate change on agriculture and forestry and the diverse assignment of property rights governing their management in Europe are additional policy factors.

Outside the climate policy field, there is momentum for policy reform in European agriculture. The reforms begun in earnest in 1992 are not, however, related to the international climate debate and only tangentially to the evolving environmental agenda. The economic driving forces for change of price support to agriculture throughout the industrialized world are undoubtedly the large budgetary cost of such support and the incorporation of agriculture into the multilateral trade agreements and talks of the General Agreement on Tariffs and Trade (GATT) and now the World Trade Organization (WTO) (Josling, 1994; Tangermann, 1996). Some agricultural policy reform may well have positive effects on reducing emissions of greenhouse gases, and this is explored by several chapters in this book. But, if climate-change mitigation and the other environmental issues were fully incorporated in multiple objective policy reforms, many aspects of the natural environment could be enhanced to a greater degree than is at present expected.

Table 1.1 documents the reported climate policy of European countries, as at late 1994. The table demonstrates the heterogeneity of the European situation with regard both to overall targets and to the means of achieving them, either through a comprehensive policy or by treating each gas separately. The implications of comprehensive, in contrast to gas-by-gas, approaches to climate policy are demonstrated within the EU by various countries with different profiles of emissions. Reaching a carbon dioxide (CO_2)-equivalent target (counting all gases as their CO_2 equivalents) could be easier for the Netherlands through large reductions in agriculture-related non-CO_2 emissions than achieving the same reduction targets for each gas independently. The Netherlands has, however, set targets for each gas, though at different levels, over and above its requirements under the FCCC.

The so-called 'Cohesion States' (Greece, Ireland, Portugal and Spain) intend to increase their overall emissions. Their National Communications under the Climate Change Convention are simply a reporting of their national energy plans, which are based on the development of energy resources to meet planned economic growth as their economies converge with the rest of the EU (Haigh, 1996).

There is also significant divergence of stated policy from actual implementation. In the first 3 years since ratification of the Climate Change Convention, some of the member states of the EU have significantly missed their stated 'targets', as illustrated in Fig. 1.1. This shows the per cent change only in CO_2 emissions from energy sources (although this makes up 90% of total emissions in Europe) from 1990 to 1994.

The slow or even negative growth in economic activity and the economic restructuring associated with the reunification of Germany have ensured that the EU total emissions have fallen by around 2.7% over the period. Germany

Table 1.1. Summaries of European national policies on greenhouse-gas emissions (late 1994) (based on OECD, 1994a; Grubb, 1995a).

Country/region	CO_2 reduction target (%)*	Details of greenhouse-gas emission policy
Belgium	−5 (2005, 1990)	A 5% reduction in all CO_2 emissions by 2005 based on 1990 levels
Denmark	−20 (2005, 1988)	Transport emissions stabilized by 2005 and a 25% reduction by 2030 NO_x and halocarbons reduced by at least 40% by 2000
Finland	0 (2000, 1990)	Although signed up to the stabilization 2000 scenario, unlikely to be able to achieve this. Aims to stop energy-related growth by the end of the 1990s
France	Stabilization at 2 t C per capita	Due to significant reductions in emissions during the 1980s (25%), aim is to limit per capita emissions to 7.3 tonnes CO_2 per year. This will result in an approximately 10% increase in emissions by 2000
Germany	−25 (2005, 1987)	A 25–30% reduction in energy CO_2 emissions by 2005 based on 1987 figures. This applies to gross emissions and therefore does not include sinks. No fixed targets on other gases but aims to reduce overall impact by 50% by 2005
Greece	+25 (2000, 1990)	Believes climate change should be dealt with in terms of the whole EU
Ireland	+20 (2000, 1990)	Aiming to stabilize emissions by 2000 but unlikely to do so. Limitation of gross emissions to 20% above 1990 levels by 2000
Italy	0 (2000, 1990)	Stabilization of emissions at 1990 levels by 2000. Figure based on net emissions but plan mainly related to reduction in energy-related emissions
Luxemburg	0 (2000, 1990)	Stabilization of emissions at 1990 levels by 2000 followed by a 20% reduction by 2005

Netherlands	−3 to −5 (2000, 1989)	Has a binding target on reductions: if not achieved then more vigorous measures will be taken. Target of 3–5% reduction in CO_2 emissions from 1990 by 2000. Eventual target of total reduction in CO_2-equivalent emissions of 20–25%, based on gas-by-gas strategy: CFC and halons phased out; NO_x reduced by 53%; VOC reduced by 60% from 1988 by 2000; N_2O stabilized; CH_4 reduced by 10%; CO reduced from 1990 by 2000
Norway	0 (2000, 1989)	Aims to stabilize all emissions of CO_2 from 1989 by 2000
Portugal	+29 to +39 (2000, 1990)	Due to low energy per capita emissions, environmental standards should not constrain economic growth. Has no CO_2 target
Spain	+25 (2000, 1990)	Aims to limit energy-related CO_2 emissions to an increase of 25% by 2000. This is 20% below what would otherwise have been expected
Sweden	0 (2000, 1990)	Aims to stabilize CO_2 emissions from 1990 by 2000. Reduction after 2000. CH_4 in refuse will be reduced by 30% from 1990 by 2000
Switzerland	0 (2000, 1990)	Aims for at least a stabilization of CO_2 emissions from 1990 by 2000, followed by a reduction after 2000. National targets for other gases
UK	0 (2000, 1990)	Aims to stabilize emissions of CO_2, methane and other gases from 1990 by 2000 in a gas-by-gas approach
EU	0 (2000, 1990)	Aims for a stabilization of gross CO_2 emissions from 1990 by 2000. This target will be met with some countries actually increasing emissions, with others having to reduce to balance this

* Target year and base year in parentheses.
CFC, chlorofluorocarbon; CO, carbon monoxide; NO_x, nitrous oxide; VOC, volatile organic compounds.

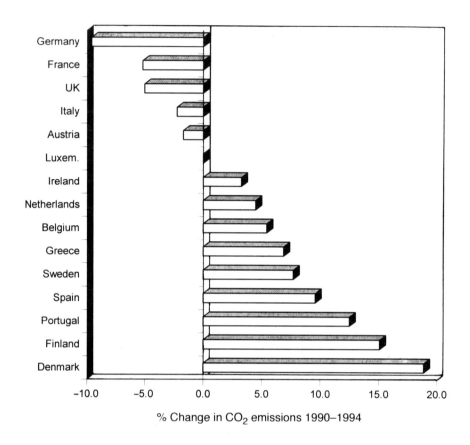

Fig. 1.1. Change in total CO_2 emissions of EU Member States, 1990–1994.

is the largest single emitter of CO_2 in the EU and has a per capita emission rate double the EU average as a whole, with the former West Germany increasing its overall emissions between 1987 and 1992. The observed reduction of almost 10% between 1990 and 1994, in Fig. 1.1, is almost solely due to fuel switching away from lignite in the former East Germany and the reduction of economic activity there (Beuermann and Jäger, 1996). But, with economic recovery in Germany and in the EU in general, the Commission of the European Community is projecting, in 1996, a 3% rise in total emissions of CO_2 in the EU between 1990 and 2000 (European Commission, COM(96)91).

The issue of individual countries of the EU not complying with the FCCC remains relatively unimportant until, as projected, the EU as a whole exceeds its emissions and fails to comply with the international agreement. Then, as pointed out by Haigh (1996) among others, the issue of formal or informal allocation of the burden to be shared among the Member States will re-emerge as a significant political issue.

In the area of mitigation of greenhouse-gas emissions, there are large uncertainties in estimating emission sources and sinks, particularly those associated with the land-use sector. Uncertainty surrounds the magnitude of carbon sinks in Europe's forests and the emissions of the important non-CO_2 greenhouse gases, methane (CH_4) and nitrous oxide (N_2O), from various agricultural activities, as well as monitoring actual land use and land-use change itself. The uncertainty in sources and sinks fundamentally affects mitigation strategies, in that cost-effectiveness or desirability of policies to reduce emissions must be based on information on direct emissions, as well as on knock-on effects, such as on energy use for farming or forestry practices, for the policy measures to be comparable (Adger and Brown, 1994). These issues are discussed in many of the chapters in this volume, both in the European context and elsewhere. The issues of the magnitude of the sources and sinks of greenhouse gases and the implications and approaches to agricultural and forest policy in Europe are outlined in this chapter to set the scene for subsequent chapters on policy measures.

THE MAGNITUDE AND IMPLICATIONS OF EMISSIONS FROM LAND USE IN EUROPE

This section reviews the emissions of greenhouse gases from land-use sources and the relative magnitude of carbon sinks. The major sources of greenhouse-gas emissions are from loss of carbon stored in soils and in above-ground vegetation through conversion of forest land or other land to intensive agricultural uses; from livestock in the form of CH_4, both through enteric fermentation and through emissions from stored slurry; from fossil-fuel energy use in agriculture; and from N_2O emissions associated with fertilizer use.

It is necessary when estimating such emissions to recognize that the global carbon cycle occurs through the processes of photosynthesis and respiration and accumulation in vegetation and soils, as well as through the interaction of the oceans and atmosphere. So, even without human intervention in the terrestrial systems of the world, the processes of the global carbon cycle persist. In considering the emissions of greenhouse gases associated with land use, it is necessary to assign emissions as either being natural or anthropogenic. Virtually all energy use and current land-use changes are anthropogenic activities, hence the scale of land conversion is important.

Consideration of historical trends shows that, in the last 300 years, Europe in general has reduced its forest cover by around 10% net overall, despite a 6% increase in forest cover in the second half of this century (Richards, 1990). This has occurred mainly due to the doubling of crop land

and increased urbanization, at the expense of both forest and grassland. Houghton and Skole (1990) demonstrate that these land-use changes ensured that, globally, annual fluxes of carbon to the atmosphere from land-use change were only surpassed by those from fossil-fuel use in the early decades of this century. The aerial extent of land-use changes in Europe underplays the implications for greenhouse-gas fluxes, due to both the energy intensity of the agricultural systems and the related increase in livestock numbers, as greater quantities of feed have become available from crop land. The following sections review the different sources of greenhouse-gas emissions from agriculture and the critical role of temperate forests in this system.

Agricultural Emission Sources

The relative contribution of land-use activities to total greenhouse-gas emissions is significant, possibly 25% of current greenhouse-gas emissions, with their climate-forcing impact. Adams *et al.* (1992), for example, estimate the global contribution of agriculture, forestry and land-use emissions to total greenhouse-gas emissions, weighting the emissions of the main greenhouse gases as CO_2 equivalents (the other greenhouse gases exhibit greater radiative forcing), and show the relative contribution as a proportion of the overall total. Their results show that land-use-related activities contribute 25% to the global total. If a shorter time horizon is used to calculate the relative contribution of the gases, this results in the contribution of land-use activities being estimated as higher, due to the predominance of CH_4 and N_2O emissions in the total.

While many of the anthropogenic and natural sources of emissions of the main greenhouse gases have been identified, the magnitude of emissions from these sources is not, in all cases, well established. There is greater knowledge about human-induced emissions than about some natural processes, which constitute both sources and sinks. Particular uncertainty surrounds sources and sinks of CH_4, for example (Cicerone and Oremland, 1988). Releases of CO_2 from biomass burning, such as occurs during deforestation, are also inexact. Current estimates give the large range of 0.4–2.9 billion (thousand million) tonnes per year in the 1980s. This range results from discrepancies in the estimates of the area deforested and in the emissions per unit area from changing land use. Similarly, with regard to CH_4 emissions from soils, wetlands and rice cultivation, large ranges have been estimated. Direct *in situ* measurement of CH_4 fluxes from wetlands or even from cultivated experimental rice plots is highly site-specific, and the range of climatic and soil conditions give the extrapolated global estimates wide confidence intervals.

The principal present sources of greenhouse-gas emissions from European agriculture and land use include carbon fluxes associated with land-use

Table 1.2. Major anthropogenic methane-emission sources: EU in 1990 (from European Commission (COM(66) 557), Brussels, 19 November 1996).

Source	CH_4 (million tonnes)	CH_4 as % of total emissions
Agriculture		44.7
Enteric fermentation of ruminant livestock	6.8	
Livestock manure	3.6	
Waste		31.5
Landfills	7.0	
Waste-water treatment	0.2	
Energy		23.0
Coal-mining	2.6	
Gas production and distribution	2.0	
Combustion	0.5	
Transport	0.2	
Total	22.8	

change and with energy use in both the agriculture and forestry sectors, CH_4 associated with ruminants, storage of animal wastes, biomass burning and irrigated rice under anaerobic conditions and N_2O from fertilizer applications on cultivated soils. In addition, there are 'natural' fluxes associated with land use, such as CH_4 from wetland areas, as well as other land-related emissions generally attributed to agriculture or forestry, such as CH_4 from waste disposed in landfills. For the EU states, CH_4 currently contributes almost 45% of total emissions, as shown in Table 1.2, referring to 1990. The European Commission has set out plans to reduce these emissions, particularly from agricultural sources, which it claims could lead to a reduction in emissions of 30% by 2005 and 40% by 2010 (European Commission, COM(96)557).

The greatest carbon fluxes from land-use change are associated with existing forests and with afforestation. Existing forests are a net sink of carbon, whereas land-use changes associated with forestry may be either a source or a sink. For example, it has been shown that historical afforestation in the UK in the last half-century has led to a net flux of carbon to the atmosphere rather than net sequestration, primarily because of the location of new afforestation on former peat land and on the sites of old-growth forest (Adger *et al.*, 1992).

The Role of the European Forestry Sector in the Global Carbon Cycle

The forests and other wooded land of Western Europe cover about 195 million hectares (FAO, 1995), most of which are actively managed. Both exploited and unmanaged forests play an important role in the carbon-emission budget of some European countries (Tables 1.A1 and 1.A2). European forests are generally estimated to be a net sink of carbon: currently 34 billion tonnes of carbon are fixed in forest ecosystems, 9 billion tonnes of carbon in standing vegetation and 25 billion tonnes of carbon in soils (Kauppi et al., 1992; Dixon et al., 1994b). The annual carbon flux associated with this stock is estimated to be sequestration in the range of 0.09–0.12 billion tonnes of carbon. In global terms, European forests are of relatively minor significance. European carbon pools in forest vegetation and soils are respectively 2.7 and 3.8% of the world total (Brown et al., 1996). However, the fluxes associated with this stock play a significant role in overall emissions for particular European countries. The proportion of total fossil-fuel emissions offset by forest-carbon sequestration varies from 1–2% for countries with low forest cover (the Netherlands, UK, Germany) to 5–10% for Poland and Italy; to 28% in Finland and 90% in Sweden (Cannell et al., 1993; Karjalainen and Kellomäki, 1993; Kauppi and Tomppo, 1993; Federal Ministry of Environment, Germany, 1994; Galinski and Kuppers, 1994; Cannell and Dewar, 1995; Ministero del'Ambiente, Italia, 1995).

In the evaluation of the carbon budget for the forestry sector, more than for other sectors, there are large areas of uncertainty. Uncertainties depend not on statistical procedures (methods of inventory and assessment of mitigation options have been agreed within the work carried out by the Intergovernmental Panel on Climate Change (IPCC)) but on lack of data and biases in the existing data. Relevant uncertainties are also associated with estimating the carbon-fixing function in wood products, such as paper, lumber and furniture (Harmon et al., 1990; Melillo et al., 1996).

Land-use trends and forest practices in Europe show the increasing positive role of the forest ecosystems in global-warming mitigation. Three categories of forest-management practice help to reduce carbon emissions in the atmosphere (Brown et al., 1996), namely, through increasing forest area, changing forest management to minimize carbon loss and exploring options for fossil-fuel substitution.

Carbon storage in forest biomass could be increased through the expansion of the forest area as a consequence of the abandonment of marginal agricultural land and its natural conversion to forest. The potential area for afforestation in Europe has been estimated as 44 million hectares (Health et al., 1993). Afforestation programmes in intensive agricultural areas are at present partly financed by the EU through Regulation 2080/92. Forest expansion and the increased density of natural and plantation forests are generally measures with short-term effects on the carbon budget.

Net losses of carbon from forests can be minimized through a range of active management for conservation; by controlling anthropogenic disturbances, such as fires, pest attacks and air- and water-pollution impacts; and by changing harvesting regimes through, for example, selective cutting; and by protecting forest in parks and reserves. The impacts of these measures on the carbon budget would be immediate. An increase of approximately 30% in carbon storage through net ecosystem production has been estimated for the two decades up to the late 1980s for European temperate forests (Kauppi *et al.*, 1992). This net sink may be expanding, however, through trends such as importation of timber products (Solomon *et al.*, 1996), which offset net national emissions. Future changes in costs of imported roundwood could have a relevant impact on carbon sequestration by European forests, if they provoke increased rates of felling within the EU.

Management for substitution of fossil-fuel energy and products, cement-based products and other building materials is a third measure to curb the rate of increase in CO_2 in the long term. Also in this case, at the EU level, there is financial support for research and development activities.

In this context, European forest ecosystems of the temperate zone may have an increasingly important role for global-warming mitigation. Young and middle-aged understocked forests of the Mediterranean area can also easily accumulate carbon, while old-growth forests prevailing in Central and Northern Europe tend to accumulate carbon at a much lower rate through retaining a higher stock of carbon. Temperate forests are recovering from past disturbances, such as overexploitation, illegal grazing and deforestation (Kauppi *et al.*, 1992; Wofsy *et al.*, 1993), and a large area of marginal agricultural land, mainly in mountainous regions, is now potentially viable as a site for natural expansion of the forest.

Forest fires and pest outbreaks are serious impediments to the process of increasing carbon sequestration by temperate forests. Most forest fires in Mediterranean areas are anthropogenic in origin. Their effects on net emissions are linked to the reduction of the above-ground carbon sink. In addition to CO_2, biomass burning releases non-CO_2 gases as a result of incomplete combustion (CH_4, carbon monoxide, N_2O and nitrogen oxide) (Crutzen and Andreae, 1990); it also reduces litter and soil organic matter, with resulting decreases in soil fertility, undermining the long-term potentials of biomass increment. The potential and problems with forest expansion to reduce net emissions for temperate forests reflect those of tropical regions; their possible contribution to global-warming mitigation is theoretically relevant, but risk and uncertainties, in terms of costs and benefits of future development, are extremely high.

For the future, the potential role of the European forestry sector for mitigation of greenhouse-gas emission will be constrained by social and economic factors, such as the rate of economic growth, technological changes, and land-use and energy policies. It will also be directly affected by future changes in climate (Melillo *et al.*, 1996). Increased levels of atmospheric CO_2 and nitrogen (fertilization effects) and changes in temperature, moisture regime and growing-season length will certainly have an impact on any proposed strategy in the forestry sector (Apps *et al.*, 1993). As a consequence of global warming, temperate forests may show little change in net carbon storage, because increases in net primary productivity may be offset by increased soil respiration due to higher temperatures and other factors, such as water and nutrient availability or leaf-size increases (e.g. Kerstiens *et al.*, 1995; Kirschbaum *et al.*, 1996).

A key issue in the potential future role of temperate and boreal forests as mitigation instruments is the cost of forest management. In the assessment by Dixon and colleagues (1993b), cost per unit of carbon sequestered or conserved generally increases from low- to high-latitude nations, implying that temperate forest management is not such a cost-effective mitigation option as in the tropics. However, the estimates relate only to the national direct costs, which would be higher if the opportunity cost of land (embodying agricultural protection) and the costs of establishing protective infrastructures were included, together with the transaction cost of bringing about this change in land use. On the other hand, costs would be offset by social benefits from non-timber products and services. The carbon sink is only one of the large sets of benefits that can derive from forest investments; even in the absence of climate-change considerations, many forest-sector investments that promote carbon sequestration can be socially, economically and ecologically beneficial. In particular, the net benefits of European temperate forests could be significantly enhanced by the inclusion of the benefits deriving from the maintenance of biodiversity, watershed protection and other environmental functions.

Protection of temperate forests could be, in many cases, a 'no regret' investment that can be realized at low positive or even negative costs, similarly for some options in energy conservation and efficiency improvement (Hourcade *et al.*, 1996). No complete cost estimate for carbon sequestration in Mediterranean forests is currently available. However, from the magnitude of the issues such as forest fires, the demand for tourism, recreation and countryside protection, the economic role of non-wood forest products in some areas (cork, chestnut, resin, mushrooms and truffles, for example) and the low opportunity cost of land, it would mean that any investment aimed to control anthropogenic disturbances to the carbon balance could potentially harmonize with other social objectives.

As observed by Brown *et al.* (1996), Hourcade *et al.* (1996) and Jepma *et al.* (1996) in the IPCC Second Assessment reports, only recently have cost estimates been improved to evaluate individual countries rather than regions, to include opportunity costs of land and maintenance costs or to develop cost functions. This improvement is despite the obvious importance of mitigation costs of some forest practices in considering the rise in costs associated with large-scale forest investments (Cline, 1992a; Turner *et al.*, 1995).

Arrow *et al.* (1996) and Pearce *et al.* (1996) have underlined the sensitivity of marginal mitigation costs to the choice of appropriate social discount rate. The discount rate fundamentally affects some long-term forestry mitigation investments in comparison with other options. Inside the forestry sector, discounting tends to favour short-term forest investments (such as active management for conservation) against long-term options (expansion of forest areas). So the choice of an appropriate discount rate for global climate analysis involves ethical and normative as well as positive economic issues. The debate on these questions raises complex questions of intertemporal equity and efficiency, as well as irreversibility and risk evaluation (Cline, 1992a; Nordhaus, 1993a; see also Price, Chapter 6, this volume).

The EU policy-making process seems to be more influenced by short-term problems, such as the need to reduce the Common Agricultural Policy (CAP) costs; this reform will probably be a more important driving force for future changes in land use (and the consequent carbon-storage function in forest ecosystems) than changes induced by the decision-making process to address global-warming problems. Since the current carbon-sink function in European forests could potentially disappear within 50–100 years (Kauppi *et al.*, 1992) as the present forests mature, so the role of the forestry sector in mitigation of global warming can only be that of an intermediate response policy and a temporally limited instrument to complement other actions.

EUROPEAN EMISSIONS OF GREENHOUSE GASES FROM AGRICULTURAL AND FORESTRY SOURCES

The previous sections have outlined the relative importance of greenhouse-gas emissions from agricultural, forestry and land-use sources and sinks. As has been discussed, Europe is probably more important for carbon fluxes from agriculture than its land area would suggest: its forest cover has been massively reduced over the last five or more centuries, but the forest area is now increasing again, and emissions of non-CO_2 gases are high, as with energy-related emissions from agriculture, because of high input intensity. Detailed estimates of land-use emissions reported in National Communications under the FCCC and by a separate European land-use assessment (CORINAIR) for the EU and other European countries are presented in Appendix 1.

One major conclusion from the estimates presented is that there are a number of notable discrepancies between the figures from the National Communications and those from the CORINAIR analysis. The estimates for forest carbon storage for both Italy and Portugal from the National Communications, for example, are very large. They imply either that different data sources were used in compiling the estimates or, more probably, that the estimates reported are gross rather than net carbon flux. In this context, gross accumulation is the carbon locked up as a result of tree growth, where net carbon flux accounts for soil dynamics and timber use. The Portuguese National Communication, for example, states that its 3.2 million hectares of forests on average sequester 22 tonnes of carbon per hectare per year, compared with 3–5 tonnes of carbon net sequestration for many European forests. This undoubtedly excludes forest felling and litter and soil dynamics.

At the same time, the Italian national plan (Ministero del'Ambiente, Italia, 1995) notes that a high proportion of the forested land is neglected coppice of low productivity and with little private or public planting taking place. Sweden has a large difference between the total CO_2 measured by CORINAIR and its National Communication and Spain has also included its national sink of forest carbon in the net emissions. Other countries that have measured the forest sink have included this in the emissions table but have not included it in the overall emissions.

Discrepancies for the CH_4 emissions are significant for most countries. Some, such as those for Sweden, may be due to measurement of emissions from land-use change and other natural sources, as has occurred in the CO_2 accounting. Unlike the overall CH_4 emissions, those for agriculture are more consistent. Exceptions are the UK, Finland, Austria and the Czech Republic. For the N_2O emissions, the UK agricultural output is again vastly different from the UK total figure.

How can the differences in country estimates of land-use sources and sinks be explained? There are three related factors. Firstly, National Communications submitted in 1994 do not follow the same protocols. On the recommendation of the FCCC, they should follow the IPCC/Organization for Economic Cooperation and Development (OECD) Emissions Inventory Guidelines, but these simply allow comparison of estimates across different methods, rather than advocating a single consistent method. A more directed approach and a single method, it is argued, would have pulled countries down to a lowest common denominator of sophistication defined by data availability (Subak, 1996).

This leads to the second factor, which is the pervasive uncertainty surrounding both land-use data and many land-use-related emissions. Accurate factors for agricultural emissions of CH_4, for example, require experimental monitoring to generate even first-order estimates for some processes. In forestry, as discussed in the section above, large-scale models of land cover and forest change, incorporating historical changes, are required if accurate emission factors are to be derived (for UK, see, for example, Cannell and Dewar, 1995; Cannell and Milne, 1995).

The third factor in uncertain estimates of greenhouse-gas emissions is the undoubted political nature of the process by which the estimates are arrived at and published. In independent reviews of National Communications, it has been shown that government assessments usually inflate the 1990 baseline level of emissions, thereby making the task of meeting future targets easier and 'deflating the value of the stabilization target' (Subak, 1996, p. 62). For example, estimates of CH_4 from landfills in the final National Communications of EU states of the 1990 emissions were higher than in earlier draft reports. No significant lobby argued for emission estimates to be reduced for them, in contrast with coal-mines and other sources.

The differences in the treatment of CH_4 from landfills and CH_4 from coal, in terms of the revision of initial estimates, can therefore be explained by the existence or absence of influential lobbies realizing the significance of being perceived as a large emitter which requires reduction. Has the land-use sector realized the significance of estimates of greenhouse-gas fluxes? The strongest grasp of this issue has undoubtedly been in the forestry sector, where carbon sequestration through afforestation is seen by state forestry authorities as an emerging policy issue and a contribution of forestry to national sustainability (for UK forestry, see, for example, Pearce, 1991a; although disputed by Price, Chapter 6, this volume).

The national-level inventories of emissions of greenhouse gases for EU states and their respective policy positions give the context for policy prescriptions within the agriculture and forestry sectors. The discrepancies and difficulties in identifying the goalposts at which to aim have been set out, though without any prospect of resolution. Many commentators see a positive role for the European Community in the monitoring of emissions (Brown and Adger, 1993; Grubb, 1995a), with the 'monitoring mechanism' for reporting annual emissions setting an international lead in this area. Despite technical advances in quantifying particular emissions, emission monitoring remains a contested and uncertain area which is difficult to resolve. As Grubb puts it:

> It remains one of the quaint and troubling facts of the European situation that the Commission's projections frequently vary from those of Member States, and there are no institutional mechanisms for exploring the reasons for the differences and achieving consistency (Grubb, 1995a, p. 50).

COMMON AGRICULTURAL POLICY REFORM AND GREENHOUSE-GAS EMISSIONS

The continuous process of CAP reform was given particular impetus both by the Uruguay Round agreement under GATT and by the EU response to those events initiated by the MacSharry proposals of 1992. The resulting policy was to move the prices farmers receive towards world price levels and to compensate them for the resulting loss of income through direct payments decoupled from production decisions. While it is possible to detail what has been implemented, it remains to be seen to what extent these policy instruments will achieve the objectives set for them. In particular, for commodities other than cereals, it is as yet not known whether movement towards world prices will prove sufficient to reduce production and allow the reduction of subsidized exporting of food surpluses.

Implementation of these proposals can now be seen in many aspects of agricultural policy. The decoupling of policies from production incentives is now being sought through measures such as the Arable Area Payments Scheme (set-aside), which requires farmers to reduce their area cultivated for arable production, first by 15% and now by 5%, if they wish to receive the compensation payments. The environmental impact of set-aside is assessed by Brouwer and van Berkum (1996), who point out that the incentives operating will have the important effect of freezing the area of arable land at the base level (average of 1989–1991). This has the substantial environmental impact of checking the conversion of extensive grassland to more intensive cropping systems. For those areas remaining in arable production, the payments, however, provide a flow of funds to increase intensification on the remaining area of land. The area prescribed for set-aside has partly been reduced as the budgetary cost of price support has been substantially lower than expected, throughout the mid-1990s, due to a global shortfall of supply in the major cereal crops, hence reducing the discrepancy between support prices and the world price. These authors report total set-aside in 1994/5 at 7.3 million hectares, compared with a base area of 49 million hectares.

Similar measures in the livestock sector limit the headage payments paid to a stated number of stock per farm, hence reducing the incentive to produce beyond a certain level of intensity. One impact of these crop- and livestock-reduction measures is to freeze the relationship between crops and livestock on farms by effectively specifying a maximum area of crops and numbers of stock on the individual farm (Saunders, 1994). Subsequent adjustments to the regulations have allowed individual countries to vary the set-aside mechanisms used, and a full analysis of the reforms is not yet available.

Measures accompanying CAP reform were also adopted by the EU, through three Regulations. Regulation 2078/92 on agrienvironment measures, 2080/92 on forestry measures in agriculture and 2079/92 on early retirement in agriculture. A recent German Ministry of Agriculture study reported under the headline 'Germany nets 20% of agrienvironmental funds'. For the EU15, some 13% of utilized agricultural area (UAA) is entered into various programmes, but the range of variation is literally from zero for Greece to 100% of UAA for Austria. Germany had 29% of its UAA entered, France 17% and the UK 8%. These variations reflect a complex of agronomic, legal, technical and policy-derived factors and they have yet to be explored on an EU basis. There is similar diversity within individual states in the requirements for participation in schemes (Whitby, 1996). Many impose stocking limitations, some apply constraints on the use of fertilizers and manures, some measures carry incentives for positive developments in the maintenance or regeneration of natural capital and others promote the conversion from conventional to organic farming. The general tenor of the schemes is towards reducing the intensity of agricultural production and, as such, they would be expected to make some modest net contribution to the reduction in greenhouse-gas emissions from agriculture.

In a detailed analysis of the implementation of the Uruguay Round, Tangermann (1996) describes the likely impact of the Agreement on Agriculture. He explains how implementation has produced a situation in which, for the first time, participant countries 'now have Schedule commitments in quantitative terms which define what they can and cannot do, in the areas of market access, export competition and domestic support' (p. 334). He nevertheless recognizes significant variation from country to country in the way in which the quantitative dimensions of the schedules were defined, which has introduced some slack into them. Apart from subsidized exports, therefore, the schedules do not bind individual policies very much.

Various elements of the agreement may turn out to be less effective than was thought. The benefits Tangermann (1996) identifies, in comparison with the situation before the Uruguay Round, are that now countries' agricultural and trade policies may be judged against defined criteria. He suggests that, as it proceeds, the current implementation period will begin to squeeze current levels of protection; that the miniround of agricultural negotiations to begin in 1999 will bring opportunities for further reduction in protection; and that the WTO disciplines for agriculture are already influencing agricultural policies in ways intended to reduce overall production and potentially production intensity.

It must be concluded from the studies so far available that CAP reform may well have some impact on the net emissions of greenhouse gases from European agriculture in the future, when the recent trends in agricultural production and land use combine with some of the issues raised in specific case-studies later in this book. Thus, trends such as the reduction in livestock intensity and increased on-farm afforestation, through the recent changes in set-aside, are likely to bring benefits in terms of reduced emissions. The likelihood that reduced levels of input use, especially chemicals, and lower stocking levels, especially of ruminant stock, will reduce emissions is reasonably clear, but the scale of the effects must await the availability of evidence of their impact before it can be assessed.

The observed trajectories of overall greenhouse-gas emissions from the European economies, highlighted for example in Fig. 1.1, show that extra-sectoral influences, such as the state of the macroeconomy, are in some circumstances more important than agriculture and forestry policy changes themselves. They are also difficult to predict and analyse. The impact of the EU's ban on beef imports from the UK in 1996, for example, will contribute to a reduction in CH_4 emissions from the UK livestock sector in the short run. This will follow an initial boost to emissions from the slaughter policy to the extent that lost UK output is supplied by production elsewhere. However, there will not necessarily be a change in the global carbon balance. Indeed, by diverting potential production into other member states the net carbon flux to the from EU policy must be passive in the short term. But the long-term impact on land use and on the size of the European beef herd depends on shifts in the long-run elasticity of demand for beef in the light of changing tastes. Evidence from statistical analysis of the impact of publicity of bovine spongiform encephalopathy (BSE) on the UK beef market from the early 1990s onwards indicates a related long run reduction in beef market share of the magnitude of 4.5% in the 3 years to 1993 (Burton and Young, 1996). Only a brave agricultural economist would have predicted the 1996 'beef crisis' and only a brave politician would point out the benefits of CH_4 reduction from the BSE crisis to the agricultural lobby.

CONCLUSION

This chapter has introduced many of the central issues taken up in subsequent analyses. Firstly, there is great diversity within Europe in terms of policy relating both to climate-change mitigation and to agriculture and forestry and in the precision with which it is monitored. The physical environment and history of land use, resulting in the present land cover, also differentially affects the constraints and opportunities for climate mitigation in the land-use sector throughout Europe.

Secondly, this chapter has shown that the quality and quantity of information relating to land use and greenhouse-gas emissions are variable within the EU. This occurs despite the information-provision regulations it operates, which are more stringent than those under the FCCC. Meanwhile, information provision for the rest of Europe under the Climate Change Convention is less fully implemented. The data reported here show that the emissions of non-CO_2 greenhouse gases are particularly significant for agriculture. Forestry acts as a net carbon sink in Europe, a sink which could be enhanced through appropriate management and through novel uses of marginal agricultural land.

The impact of the reforms of both agricultural and forestry policy in the early 1990s has yet to be fully observed in terms of land-use change and agriculture and forestry practices. The refocusing and 'greening' of agricultural and forestry policy in Europe have led to novel policies and approaches (for reviews, see, for example, Whitby, 1996; Winter, 1996) to incorporate wildlife conservation, protect landscapes and reduce agricultural pollution. Yet there is little evidence, except perhaps within forestry circles, of any consideration of policies which reduce greenhouse-gas emissions or which have the primary objective of enhancing carbon sinks. This omission occurs fundamentally because of the reliance of agriculture and forestry, as with the major part of all European economies, on carbon-based fossil fuels and intensive production systems. As has been illustrated in this chapter, extrasectoral influences on emissions associated with land use are also critical in this arena.

The following sections in the book detail the central analytical issues surrounding both the scientific and policy uncertainties in this area. The temporal and spatial dimensions of climate impacts and of mitigation policies are highlighted in these sections. The theme of uncertainty also emerges strongly from the empirical investigations of the interaction of land use with greenhouse-gas emissions.

The global nature of the issues raised here is further emphasized in the penultimate section of the book, which discusses how land-use issues affect the global carbon cycle outside Europe. The Climate Change Convention hints at possibilities for offsetting emissions beyond countries' domestic borders. The chapters on Argentina, Mexico and Russia demonstrate that despite the physical potential for carbon sequestration, the social acceptability, economic reality and the absence of institutional infrastructure for such action, constrain international offsets to a major degree. These themes are common to all the empirical research presented in this book: land-use policies need to be considered in an integrated manner in their social and economic context, not merely as carbon sinks to offset continuing emissions from other sectors of Europe's industrialized economies.

APPENDIX 1: GREENHOUSE-GAS EMISSIONS FROM AGRICULTURE FORESTRY AND IN TOTAL FOR EUROPEAN STATES

This appendix presents information on the land-use-related and overall emissions of greenhouse gases of the EU and other European countries. This information is presented in Tables 1.A1 and 1.A2. The information comes from three main sources.

National Communications to the Framework Convention on Climate Change

Under the terms of the FCCC, the developed-country parties to the Convention were obliged to report their greenhouse emissions at 1990 during 1994 before the first conference of the parties. The FCCC highlights all greenhouse gases from anthropogenic sources, although it excludes natural sources, such as CH_4 flux from lakes, oceans and natural wetlands. The EU

Table 1.A1. Greenhouse-gas emissions (agricultural and total) for EU and selected European countries, as reported in National Communications. (From National Communications to the Climate Change Convention; Eurostat (1995) for land use; and World Bank (1994) for GDP estimates.)

	Land area (kha)	Agricultural area (kha)	Forest area (kha)	Agriculture as % GDP	Total emissions				Agricultural			Forestry balance CO_2 C (kt)
					CO_2 C (kt)	CH_4 (kt)	N_2O (kt)		CO_2 C (kt)	CH_4 (kt)	N_2O (kt)	
EU15 total	323,946	148,319.4	112,060.6	4.35								
Austria	8,385	3,500	3,227	3.15	16,145	603	4		n/r	259	2	n/r
Belgium	3,052	1,362	617	1.87	n/a	n/a	n/a		n/a	n/a	n/a	n/a
Denmark	4,309	2,788	493	3.79	15,914	406	11		220	262	9	n/r
Finland	33,813	2,558	23,222	5.63	15,818	252	23		573	94	12	8,500
France	55,150	30,581	14,811	3.46	n/a	n/a	n/a		n/a	n/a	n/a	n/a
Germany	35,691	18,032	10,393	1.28	276,000	6,218	223		1,364	2,043	80	5,400
Greece	13,199	9,160	2,620	14.91	n/a	n/a	n/a		n/a	n/a	n/a	n/a
Ireland	7,028	5,635	343	7.37	8,900	n/r	n/r		n/r	n/r	n/r	n/r
Italy	30,127	16,850	6,751	3.21	117,985	3,901	120		2,212	1,860	59	11,018
Luxembourg	259	126.4	88.6	1.93	n/a	n/a	n/a		n/a	n/a	n/a	n/a
Netherlands	3,733	2,006	300	4.04	45,709	1,067	60		2,345.4	508	22	327
Portugal	9,239	4,011	2,968	5.77	11,495	227	11		1,338	176	4	19,200
Spain	50,478	30,472	15,807	4.65	69,948	2,143	94.70		7,889	932	63	1,139
Sweden	44,996	3,401	28,020	2.56	16,706	329	15		147	196	8	9,373
UK	24,488	17,837	2,400	1.62	158,255	4,844	109		733	1,562	18	2,500
Other												
CSFR (former)	n/a	n/a	n/a	n/a	45,609	880	41		1,349	175	2	n/r
Estonia	4,523	1,417.3	1,869.2	15.62	10,761	n/a	n/a		n/a	n/a	n/a	630
Hungary	93,030	6,474	1,695	12.50	19,547	544.6	11.35		671	173	4.1	n/r
Norway	32,390	976	8,330	3.08	9,709	389	16		167	91	6	1,220
Poland	31,268	18,793	8,754	8.42	113,163	6,100	156		n/r	1,860	94	4,090
Russia	1,707,540	212,800	771,000	15.47	n/a	n/a	n/a		n/a	n/a	n/a	4,990
Switzerland	4,129	2,021	1,052	n/r	11,045	274	29		n/r	215	26.7	1,430

CSFR, Czech and Slovak former republic; GDP, gross domestic product; n/a, not available; n/r, not reported.

Table 1.A2. Agricultural and total greenhouse-gas emissions and forestry flux for EU and selected countries as estimated by Dobris Assessment and other sources. (From Eurostat (1995), and based on Kauppi and Tommpo (1993) for forestry estimates.)

	Total emissions			Agricultural emissions				Forest sequestration (mt C)
	C (kt)	CH_4 (kt)	N_2O (kt)	CH_4 (kt)	CH_4 (% total)	N_2O (kt)	N_2O (% total)	
EU15 total	991,709	32,316	1,396.60	10,390	32	468.666	34	
Austria	14,399	855	8.02	356	42	3.031	38	3.4–6.8
Belgium	28,197	370	26.17	270	73	6.044	23	0.6–1.3
Denmark	15,281	761	16.42	263	35	8.539	52	0.6–1.2
Finland	15,554	990	31.68	160	16	10.281	32	10.1–20.3
France	131,382	3,038	222.67	1,611	53	60.763	27	9.4–19.3
Germany	275,783	6,067	214.84	2,063	34	77.499	36	10.3–20.2
Greece	22,310	5,508	205.23	363	7	12.716	6	0.7–1.4
Ireland	8,630	850	45.00	643	76	39.509	88	0.6–1.1
Italy	134,192	3,928	141.69	1,764	45	57.267	40	3.8–6.5
Luxemburg	3,067	25	0.69	18	72	0.475	69	0.14–0.24
Netherlands	43,446	1,040	24.80	520	50	8.695	35	0.4–0.8
Portugal	15,655	391	54.70	204	52	30.912	57	1.1–2.8
Spain	79,036	2,998	201.88	874	29	63.011	31	6.4–11.4
Sweden	46,471	2,106	32.91	205	10	7.924	24	16.7–30.3
UK	158,305	3,389	169.90	1,076	32	82	48	1.6–3.2
Other								
CSFR (former)	42,962	1,552	62.26	508	33	25.898	42	n/a
Estonia	8,104	165	4.45	78	47	3.758	84	n/a
Hungary	16,596	612	21.55	324	53	17.149	80	n/a
Norway	9,391	282	15.50	91	32	6.451	42	3.2–6.0
Poland	113,160	6,107	155.06	1,861	30	93.6	60	n/a
Russia	n/a	n/a	n/a	n/a		n/a		n/a
Switzerland	n/a	n/a	n/a	n/a		n/a		0.9–1.6

CSFR, Czech and Slovak former republic; n/a, not available.

countries also have direct obligations to provide more information than necessary under the FCCC to the EU under the March 1993 'EU Monitoring Decision'. This includes information on strategies to reduce emissions, as well as annual updates on CO_2 emissions. The National Communications tend not to follow a single methodology, style of presentation or unit of measurement, even within the EU. Data from these are presented in Table 1.A1, with the omission of Belgium, France, Greece and Luxemburg, which were unavailable.

The CORINAIR Assessment of Greenhouse-gas Emissions

These data are reported as draft results from the emissions inventory undertaken within CORINAIR, a project originally established in the mid-1980s in Europe to monitor emissions associated with the Long Range Transboundary Air Pollution Convention. The land-use part of CORINAIR uses data from the CORINE project. CORINE attempts to provide comparable databases and classifications of European land use for all countries at a fine resolution from satellite imagery. The CORINAIR draft results, published in the Dobris Assessment (Eurostat, 1995), included estimates based on land-use change, as well as point-source emissions based on other information. The CORINAIR results and methodologies are generally unrelated to individual country contributions and are reported in Table 1.A2.

Other Databases and Sources of Information

Further databases and sources of information on greenhouse-gas emissions can be used to compare these two sets of 'official' estimates. In Table 1.A2, the results of a model of European forest areas, based on United Nations Economic Commission for Europe (UNECE) data, are reported (Kauppi and Tomppo, 1993). Forestry fluxes are not reported separately by CORINAIR.

2 Agricultural Policy Reform and Climate-change Mitigation in Organization for Economic Cooperation and Development Countries

Mark Storey and Merylyn McKenzie-Hedger[1]

INTRODUCTION

Agricultural policy reform is under way in Organization for Economic Cooperation and Development (OECD) countries at different speeds and from very different starting-points. In some countries reform has been aimed at reducing the overall level of support to agriculture, while in others support has remained high but there has been a shift away from market-price support to more direct payments. Objectives of climate-change policy have received little attention in the agricultural policy-reform debate. Rather, the main driving forces have been budgetary and trade-related concerns and, to a lesser extent, other environmental issues related to resource degradation and environmental pollution. Despite this, these policy reforms offer the potential to produce results which are beneficial to reducing greenhouse-gas (GHG) emissions from agriculture.

This chapter reviews recent trends in agricultural policy reforms in OECD countries, both in Europe and outside Europe, and assesses their implications from a climate-protection policy viewpoint. There is also a brief overview of other policy instruments and measures that are being used or considered to varying degrees which also have the potential to reduce GHG emissions. A significant feature of many of these measures is that their primary objective is unrelated to climate-change issues. There is, therefore, a clear need for climate-change issues to be better integrated into mainstream agricultural policy. The current agricultural policy-reform debate and the search for policy instruments and measures to promote more sustainable agriculture systems present an important window of opportunity for this to occur.

THE RELATIVE CONTRIBUTION OF THE AGRICULTURAL SECTOR TO TOTAL EMISSIONS

Agriculture is a significant source of GHG emissions, particularly of methane (CH_4) and nitrous oxide (N_2O). On a global scale, it has been estimated that agriculture currently contributes about 21–25%, 57% and 65–80% of the total anthropogenic emissions of carbon dioxide (CO_2), CH_4 and N_2O, respectively (Kaiser and Drennen, 1993). Overall, it accounts for one-fifth of the annual increase in anthropogenic emissions of GHGs. For individual OECD countries, the agricultural sector, excluding food processing, packaging and distribution, is likely to contribute between 6 and 15% of total national anthropogenic GHG emissions.

The main agricultural sources of GHG emissions include: CH_4 emissions from livestock (CH_4 is produced as part of the normal digestive process of ruminant animals and from the anaerobic decay of livestock waste) and from paddy-rice cultivation; N_2O emissions from the application of synthetic and organic nitrogenous fertilizers to agricultural soils; and CO_2 emissions resulting from both direct and indirect energy use in agriculture. The burning of agricultural wastes is also a significant source of both CH_4 and N_2O. In addition, modern industrial agriculture is a significant source of CO_2, as intensive annual crop production depletes soils of organic matter.

Comparatively, fossil-fuel use by agriculture is small, about 3–4.5% of the total energy consumption for the developed countries (Enquete Commission, 1995). However, fuel requirements by the food sector as a whole (including processing, preservation, storage and distribution) account for 10–20% of total fossil-energy consumption (Pimentel *et al.*, 1990). High-intensity animal production has become the biggest consumer of fossil energy in modern agriculture (Enquete Commission, 1995).

The significance of the agricultural sector as a major source of GHG emissions is closely related to the level and intensity of agricultural production in most OECD countries. Over the past 50 years, agriculture within the OECD has become increasingly dependent upon external inputs of fertilizers, pesticides and mechanization. Dramatic increases in productivity have masked a progressive decline in energy efficiency and an increased reliance and dependence upon fossil fuels. It is this specialization and intensification of agricultural production – often involving the spatial separation of livestock and arable enterprises – that has resulted in the sector becoming an important source of GHG emissions. Where extensive agricultural systems exist, such as in Australia, other factors, such as land clearing, assume more importance as sources of GHGs from the agriculture sector.

AGRICULTURAL POLICY TRENDS IN ORGANIZATION FOR ECONOMIC COOPERATION AND DEVELOPMENT COUNTRIES

In most OECD countries, the agricultural sector has had very high levels of support. Price-support policies have accounted for the largest share of this support. These high levels of support, linked largely to production, have contributed to the intensification of agricultural practices throughout most OECD countries. These measures have been extremely successful in contributing to a marked rise in the agricultural productivity of farm labour and land. On the negative side, however, commodity-related assistance to the agricultural sector has led to significant budgetary outlays, higher prices being paid by consumers of agricultural products and agricultural commodity surpluses, leading in turn to overall economic inefficiencies and to trade frictions. In addition, the intensification of agriculture is increasingly linked with concerns about resource degradation, such as soil erosion, and environmental pollution, such as nitrate leaching.

The need for agricultural support policies to be reformed to address these issues is now widely acknowledged by policy-makers. The 1987 OECD Council at ministerial level specified that agricultural policy reform should aim to strengthen the role of market signals and decrease total support to production (OECD, 1994a). Agricultural policy reform of this nature is being undertaken in several OECD countries, including the European Union (EU), and has been given added impetus by the General Agreement on Tariffs and Trade (GATT) Uruguay Round. However, the total level of support to agriculture in OECD countries, as measured by the Producer Subsidy Equivalent indicator (PSE),[2] has remained at very high levels since 1986–1988. The percentage PSE, which expresses the total level of support in relation to the value of agricultural production, has fallen only slightly, from 45% in 1986–1988 to 42% in 1995 (OECD, 1996).

The major trends which can be identified in the evolution of agricultural support policies in OECD countries since 1986–1988 (OECD, 1996) can be summarized as some reduction in the degree of insulation of domestic producers and consumers from world markets for a number of countries and a shift from market-price support to more direct payment measures.

While the overall level of OECD support remains high, there are wide variations in the levels of support and its composition in OECD countries. As measured by the percentage PSE, total support has declined significantly in Canada, Iceland (although support still remains at high levels), the USA and New Zealand (which now has a very low level) and markedly increased in Turkey. For most other OECD countries, except Australia, initial support has remained high and there has been little change in the PSE (OECD, 1996).

An important development within the OECD countries has been a reduced emphasis on market-price support measures and an increase in budgetary support. Traditionally, the major components of the support policies have been measures aimed at maintaining high agriculture-producer prices. Market-price support measures provide assistance to agricultural producers through consumers paying higher prices for agricultural products in relation to world market prices. Budgetary support mainly comprises direct payments (which include deficiency payments) and other forms of support, such as subsidies for inputs.

Since 1986–1988, market-price support as a share of total OECD support has decreased from 42 to 35%, while the share of budget-financed support has shown a corresponding increase (OECD, 1996). In general, a shift to increased direct payments (such as headage and area payments) is favourable compared with market-price support measures, as direct payments are more transparent, reduce the implicit tax on consumers and allow for better targeting of support.

Agricultural Reform in the European Union

The EU introduced a series of reforms to its Common Agricultural Policy (CAP) in 1993. The main feature of these reforms has been a reduction in support prices, compensated by direct payments in conjunction with production-limiting measures (European Commission, 1994b). While the CAP reform has led to an increase in overall budgetary payments, the shift towards an increased emphasis on direct payments provides scope for environmental benefits to be better targeted. However, from a climate-change perspective, some important sectors have been hardly affected by CAP reform; in particular, the dairy sector remains largely unchanged and surplus beef production has continued to be encouraged by export subsidies in the livestock sector (OECD, 1996).

IMPLICATIONS FOR GREENHOUSE-GAS REDUCTION OBJECTIVES

There are two principal ways in which current trends in agricultural support polices have relevance for climate-change policy objectives. First, the increased emphasis towards direct payments provides the potential for support payments to be better targeted to meet environmental, including climate-change, objectives. Second, net national reductions in GHG emissions may result from a move to less intensive agricultural systems. Whether global emissions would decline as a result is a question that can only be answered through empirical analysis.

The trend to redirect support in the form of direct payments rather than price guarantees or other measures directly linked to production or factors of production is seen as potentially providing more leverage to target environmental objectives and to introduce new environmentally friendly forms of land management, such as farm forestry (OECD, 1995a). There is, however, a long way to go on this: in 1993, the OECD average for direct payments to farmers was only 17% of total assistance to agricultural producers.

A more direct influence on GHG emissions will come about if agricultural reform measures succeed in encouraging the adoption of less intensive farming systems or, in other words, some 'extensification' of agricultural practices. As noted above, the specialization and intensification of agricultural production in many OECD countries in postwar years (encouraged by high levels of production-related support) has had important consequences in terms of net GHG emissions associated with land use. One of these has been the conversion of forests and woodland areas to agricultural use in many OECD countries, which has resulted in large net emissions of carbon (C). For livestock production in general, the more intensive the management regime in terms of feed and nutrient supplements and in terms of stored slurry, the greater the related emissions (Adger and Brown, 1994). Intensive systems are also more likely to result in the use of higher levels of inorganic fertilizers on grasslands, which will result in N_2O emissions (Adger and Brown, 1994). Consequently, there is a strong argument that current trends to reduce total levels of support and to redirect remaining support away from production have significant potential to reduce GHG emissions. Some of these potential benefits will now be discussed.

Fewer Livestock Leads to Reduced Methane Emissions at the National Level

The removal of agricultural subsidies in New Zealand, combined with a downturn in world livestock commodity prices at the time, resulted in a significant reduction in livestock units in New Zealand between 1984 and 1992 (Ministry of Agriculture and Fisheries (NZ), 1994). Recent reforms to both the CAP and the US Farm Bill, which have sought to shift the emphasis of support away from production, have also resulted in a decline in overall livestock numbers. Several parties to the Framework Convention on Climate Change (FCCC) refer to CAP reform and continued reform of the CAP as the mechanism enabling them to reduce agricultural CH_4 emissions by reducing livestock numbers. The UK, for example, has estimated that CAP reform will contribute to a 15% reduction in the number of dairy cattle (UK Department of Environment, 1994).

Using Nitrogenous Fertilizer Inputs More Efficiently Leads to Reduced Nitrous Oxide Emissions

A study by Harold and Runge (1993), examining linkages between the percentage PSE and fertilizer use, found a clear derived linkage from high levels of support to heavy applications of fertilizer across countries. A reduction in production-linked support is likely to have positive impacts in terms of reducing the derived demand for inorganic fertilizers, which in turn may have a beneficial effect in reducing N_2O emissions.

Letting Agricultural Land Revert to Forest or Wooded Land can Help Increase Carbon Sequestration

The reversion of marginal agricultural land to forest, grasslands and wetlands represents an important potential for C sequestration. If soils are left uncultivated, C contents in upper horizons could eventually reach levels comparable to the precultivation situation (IPCC, 1996a). The reduction of agricultural subsidies may potentially result in the reversion of some agricultural land to more natural ecosystems and also the conversion to other land uses, such as forestry. This has been the experience in New Zealand, where the removal of subsidies has removed an incentive to farm marginal land and resulted in a gradual afforestation of hill-country land. The same effect may be predicted elsewhere when support levels are effectively decoupled from production levels, allowing land values to fall.

The capacity for land to be converted from agricultural to non-agricultural land uses depends, however, on the extent to which agricultural land is 'surplus' to needs for food production. In regions with current food surpluses, such as the USA, Canada and Western Europe, the agricultural land base is currently able to be reduced (IPCC, 1996a). A similar situation may occur in the longer term for countries in Eastern Europe and the former Soviet Union as productivity increases. However, climate-induced changes in agricultural production, such as a projected increase in the incidence of drought in key food-producing areas of midcontinental USA, Canada and Australia, could have serious long-term impacts on patterns of food distribution and land availability.

It should be stressed that the implications of an 'extensification' of agricultural practices in terms of GHG emissions are not clear-cut and may not all be positive. For example, CH_4 emissions from dairy cows reared on extensive pastoral-based farming systems are considerably higher than for dairy cows reared in more intensive systems feeding high-quality forage and grain (IPCC, 1996a). More general questions also need to be addressed, such as whether reduced agricultural production (and reduced GHG emissions) resulting from reduced subsidies in one country is only likely to be replaced by increased production (and increased GHG emissions) in another country.

Further research and analysis of the implications of reform in agricultural support policies for GHG emissions are needed before these issues can be answered with confidence. Nevertheless, it is apparent that current developments in reforming agricultural support policies provide an important opportunity for climate-change issues to be better integrated into mainstream agricultural policy.

POLICIES AND INSTRUMENTS FOR GREENHOUSE-GAS REDUCTION FROM AGRICULTURE

While reform in agricultural support policies has important implications in terms of GHG emissions, the adoption of these reforms in many OECD countries remains limited. There is, therefore, also a need for more specific measures to be adopted which target the reduction of GHG emissions more directly, in the short term. An evaluation of the various policy measures that are being used in various OECD countries reveals a wide range of possible measures that can be used or adapted to serve GHG-mitigation objectives. For example, information on best agricultural practices can be augmented to include advice on how to minimize GHG emissions. Regulatory instruments aimed at reducing fertilizer nitrogen (N) surpluses and programmes to encourage the adoption of organic-farming techniques could be strengthened. Table 2.1 presents a summary of different categories of policy measures. This is by no means a complete list, but rather an illustration of the different policy tools that exist which can be beneficial in terms of reducing GHG emissions.

The section which follows presents a brief description of some of these measures. A detailed assessment of the relative merits of each of the different types of measure is not attempted in this chapter. The most appropriate solutions are likely to be a mixture of these different types of policy instruments, and the most appropriate mix will vary from country to country.

Economic Instruments and Broader Policy Measures

Reduction or redirection of agricultural support
The reduction or retargeting of support to agricultural producers has been discussed to a large extent in the previous section. It is mentioned again here to emphasize the context in which other policy measures should be considered. In seeking to identify and implement specific measures to reduce net GHG emissions from agriculture, attention must first be given to the policy framework and financial incentives that are influencing the agricultural

Table 2.1. Policy instruments and measures that influence GHG emissions from agriculture.

Economic instruments
   Reduction/reform of agricultural subsidies
   Cross-compliance schemes
   Taxes on fertilizer use
   Subsidies for converting to organic agriculture
   Set-aside payments
   Incentives to forestry

Regulation and legislation
   Regulations on fertilizer application
   Standards on nitrogen levels in the soil
   Limits on burning of straw in open fields
   Restrictions on livestock density
   Requirements for farmers to adopt mineral accounting systems

Information and education
   Information on ways to improve agricultural productivity
   Codes of good agricultural practice

Research and development
   Research and development on ways to reduce enteric fermentation and liquid-manure production and on more efficient ways to apply fertilizers
   Research and development on the uptake and loss of carbon in forests
   Research and development on the production and recovery of energy from waste

Voluntary approaches (VAs)
   VAs with farmers to use energy from methane recovered from manure
   VAs to reduce the use of fertilizers

practices in the first place. If not, there is a danger of only addressing the symptoms of the problem rather than the underlying causes.

Cross-compliance with agricultural support
Cross-compliance schemes aim to tie agricultural support payments to the requirement to adhere to defined agricultural management practices. Cross-compliance schemes are possible mechanisms to achieve GHG emission reductions in the agricultural sector while at the same time targeting other economic, social and environmental objectives. Examples of cross-compliance could include tying eligibility for Area Support Payments (introduced under the 1992 reform of the CAP) to adhering to defined agricultural management practices (possibly based on codes of good agricultural practice).

Problems arise with cross-compliance measures, however, because they attempt to achieve multiple objectives, which may provide conflicting signals to land-users. Consequently, they are often difficult to target and evaluate in terms of their effectiveness. Nevertheless, as a transitory measure between high levels of production-linked assistance and more clearly targeted policies, they offer a potentially important policy instrument, which can help better incorporate climate-change and other environmental objectives.

Taxes or levies on mineral nitrogen fertilizers
Reductions in the use of N fertilizers can be achieved through the use of a levy or N tax. Fertilizer levies and taxes have already been introduced in Austria, Finland, Norway and Sweden (Enquete Commission, 1995). Austria's levy, which was abolished when the country joined the EU, amounted to a 40% increase in the price of N. In 1988, following the introduction of the levy, fertilizer use fell by 11%. However, in Sweden, despite an effective 30% price levy on N being introduced, N demand remained constant – partly because of falling energy prices and rising product prices.

The problem with a tax on fertilizer, however, is that it is an inefficient instrument for mitigating the environmental effects of fertilizer use. These effects can vary across a wide range, depending on the type of crop to which the fertilizer is applied, soil conditions, temperature and rainfall. Also, the damage function is typically non-linear with respect to the rate of fertilizer application, starting at a low level and rising steeply as the rate begins to exceed the needs of the growing crop. In comparison, a fertilizer tax is linear, which means that it risks imposing disproportionate costs on farmers who may in fact be causing very little damage to the environment. This is why many countries are now looking at ways to tax only that portion of the (usually organic) fertilizer applied that is surplus to crop requirements. Such a 'surplus' levy was introduced in the Netherlands in 1996.

Promotion of organic systems of agricultural production
The promotion of lower external input, more extensive and organic or ecological systems of agricultural production could contribute to lowering agricultural emissions of all three principal GHGs. In relation to C emissions, organic systems, for example, tend to have and maintain a higher organic and humus content in their soils. Consequently, these systems can enhance or maintain the C storage potential of the soils. Policy initiatives that promote organic and lower external-input systems of production also offer the opportunity to reduce GHG emissions associated with the production and use of synthetic fertilizers and pesticides (Howes, 1995).

Set-aside payments and land-conversion schemes
Currently, about 25 million hectares in the USA, Canada and the EU have been taken out of production in government set-aside programmes (IPCC, 1996a). In the EU, set-aside programmes are temporary, for 1–5 years, and have been basically targeted to reducing agricultural production. If soils are left uncultivated and left to return to native vegetation, C contents in upper

horizons could eventually reach levels comparable to their precultivation condition (IPCC, 1996a).

Promotion of energy crops and biofuels
The Intergovernmental Panel on Climage Change (IPCC) Second Assessment Report (IPCC, 1996a) identified the use of plant material, or biomass, for energy production as the agricultural option offering the greatest potential to mitigate CO_2 emissions (assuming there are no land-availability constraints). If grown sustainably, some forms of biomass can provide a virtually CO_2-neutral energy source. When used to displace fossil fuels, the large-scale use of energy crops, notably wood fuel, could contribute to reducing CO_2 emissions.

Maximum stocking rates
Setting maximum restrictions on the number of cattle or stocking densities is a measure that has been promoted in some countries, in particular in countries where intensive livestock-production systems are causing serious pollution problems. The limitation of stocking numbers by this type of policy instrument could also have potentially beneficial impacts in terms of reducing CH_4 emissions. Problems arise with this measure, however, because it is difficult to adapt to specific agroecological conditions.

Education and Information

Best-practice guidelines
Developing codes of good agricultural practice is a means of disseminating practical advice to land-users on agricultural practices that help mitigate against adverse environmental impacts. For example, certain management practices and activities – such as retaining crop residues, using reduced tillage techniques and restoring agricultural land – can contribute to C sequestration. Significant decreases in CH_4 emissions from agriculture can be achieved through improved nutrition of ruminant animals and better management of paddy-rice fields. Additional CH_4 decreases are possible by altered treatment and management of animal wastes and by reduction of biomass burning. Nitrous oxide emissions from agriculture can be reduced through careful attention to frequency, timing and appropriate placement of fertilizer applications.

Research and development
The promotion of research to develop technologies and practices to mitigate against GHG emissions is an enormous topic and no attempt to summarize some of the major research areas is made in this chapter. Suffice it to say that there is a clear need for improved information on the

relationship between modern farming practices and emissions of the principal GHGs.

Voluntary Approaches

With increased awareness of environmental problems in the past decade and the need for sustainable agriculture, there have been new developments in several OECD countries with government encouragement of voluntary organizations. These organizations have often originated in single-issue campaigns but have widened out to tackle a number of environmental-quality issues in an integrated or targeted way.

Voluntary agreements are seen to be most effective in dealing with issues that are locally important, where the cost of remedial action is low and where individual behaviour or outcomes can be readily observed (OECD, 1995b). The most well developed of these approaches is the National Landcare Programme (NLP) in Australia, which involves programmes at all levels of governance. A major theme of the Landcare movement is to shift the focus from individual land-users to groups and the involvement of the people whose daily decisions shape the land in developing the strategies for more sustaining practice. It is a 'bottom-up' approach, aimed at encouraging land-user groups to develop their own land-management strategies in response to their particular environmental and economic conditions. Similar schemes are under development in Canada, New Zealand and the Netherlands.

CONCLUSIONS

Agricultural policy reform, aimed at reducing overall levels of support to agriculture and redirecting remaining support to better meet wider socio-economic and environmental objectives is under way in several OECD countries. While the main driving forces of these reforms have been budgetary and trade-related concerns and, to a lesser extent, other environmental issues, current developments in agricultural policy have important implications in terms of their potential to reduce GHG emissions. In general, a shift towards less intensive farm-management practices, arising as a result of either reduced levels of production-linked support or increased emphasis on sustainability objectives, is likely to have beneficial effects in terms of net national reductions in GHG emissions. The magnitude of these potential impacts is, however, highly uncertain, and not all the impacts will be positive. Further analysis is needed in order to identify with confidence the likely impacts of these policy changes. There is also a need for the questions to be looked at from a global scale. While net national reductions in GHG emissions may result from a move to less intensive farming systems in many OECD countries, this could possibly be offset by increased production and increased emissions in other countries.

Furthermore, while current agricultural policy reforms appear to offer some positive benefits in terms of reducing GHG emissions from agriculture, the adoption of these reforms remains slow and limited in many OECD countries, including the EU. There is, therefore, also a need to identify other policy instruments that more directly target the reduction of GHG emissions in the short term. A brief examination of existing policy instruments in OECD countries reveals a wide range of approaches, ranging from regulatory approaches specifying, for example, limits on fertilizer application to voluntary approaches encouraging the adoption of best-practice guidelines. The appropriate mix of these instruments will depend on political and socio-economic factors and also on location-specific characteristics of agroecosystems. The central theme of this chapter is that current changes in the direction of agricultural policy in most of the industrialized world present a valuable opportunity for climate-change policy objectives to be better recognized and targeted. It is important that this opportunity is taken.

NOTES

1. The views presented in this paper are those of the authors and do not necessarily represent the views of the OECD or of its member countries, or of World-Wide Fund for Nature (WWF) International.
2. The total PSE is an indicator of the value of the monetary transfers to agricultural producers resulting from agricultural polices in a given year. Both transfers from consumers of agricultural products (through domestic market prices) and transfers from taxpayers (through budgetary or tax expenditures) are included (OECD, 1996).

Forestry Options for Mitigating Predicted Climate Change

William M. Ciesla

INTRODUCTION

The relationship between forests and predicted global climate change due to increased levels of greenhouse gases (GHGs) in the Earth's atmosphere is complex. Forests, especially human activities associated with forests, contribute to increases in the level of GHGs in the Earth's atmosphere. Net emissions from deforestation and associated land-use change in the tropics are currently estimated at 1.6 billion tonnes of carbon per year or approximately 23% of total annual carbon emissions (IPCC, 1994). Forests can also be affected by climate change. Changes in growth and yield, distribution of forests, composition of forested communities and the incidence and intensity of wildfire episodes and insect and disease outbreaks are among the predicted effects of global climate change (Ciesla, 1995). Because of their ability to absorb carbon dioxide (CO_2) from the atmosphere during photosynthesis and store carbon in woody tissue, trees and forests offer an opportunity to reduce the net annual rate of increase of GHGs in the Earth's atmosphere and thus mitigate the predicted effects of global climate change.

Forest options to mitigate climate change can be classified into three broad strategies; reducing rates of GHG emissions from forests, maintaining their ability to store carbon and expanding their carbon-storage capacity. Options under each of these strategies are reviewed. According to current estimates, through actions compatible with traditional forest management, there is a potential to reduce net CO_2 emissions from forests to the point where forests are no longer a net carbon source and to sequester 11–15% of total carbon emissions due to burning of fossil fuels over the next 50 years. The greatest opportunities to sequester carbon in forests are in the tropics.

FOREST-SECTOR OPTIONS FOR CLIMATE CHANGE

Adaptation Versus Mitigation

There are two overall approaches, or strategies, for responding to predicted climate change: adaptation and mitigation. Adaptation is concerned with responses to the effects of climate change. It refers to any adjustment that can be undertaken to ameliorate the expected or actual adverse effects of climate change. Silvicultural practices, such as species and site matching, regular thinnings and timely harvesting of trees, and protection of forests from wildfire, pest or disease activity, are adaptive strategies. Mitigation, on the other hand, attempts to address the causes of climate change. Forest-sector approaches for mitigating effects of climate change can be classified into three broad areas of: reducing sources of GHGs; maintaining existing sinks of GHGs; and expanding sinks of GHGs. Both adaptive and mitigation strategies should be considered in an integrated approach when designing forest-sector responses to climate change.

Forests as Carbon Sinks

Trees and forests are temporary carbon sinks. This is often overlooked, perhaps because of the perception of forests as long-lived. Forests are dynamic systems, however, and are subject to change. Often, changes in forests are gradual: a mature tree in a mixed forest dies and is replaced by other vegetation, including young seedlings. In other cases such as when a wildfire occurs, thousands of hectares of forest can be destroyed within a matter of hours. Regardless of whether individual trees die or entire forests are destroyed or harvested by humans, some of the carbon stored in woody tissue is once again released into the atmosphere. Forest-sector policies and strategies should, therefore, be orientated toward prolonging the carbon-storage capacity of trees and forests for as long as possible, as opposed to attempting to store carbon in forests in perpetuity.

Characteristics of Forest-sector Mitigation Measures

Actions taken to mitigate the predicted effects of climate change should be sustainable, economically viable, adaptable, technologically simple and socially acceptable. Forest-sector actions to mitigate effects of climate change should be sustainable. Sustainable resource management is broadly defined as 'providing for the needs of the current generation as well as future generations'. Sustainable forestry has recently received much attention on a global scale. The reasons for this extend over and above the issue of forests and climate change and include concerns about the conservation of biodiversity; water and soil resources; ecosystem productivity; wildlife habitat and populations; forest contributions to global ecological cycles; and increases in the flow of forest benefits to human society (Rotherham, 1996).

In 1987, the World Commission on Environment and Development, commonly referred to as the Brundtland Commission, recognized the need for a broad approach to sustainability in all sectors, including forestry. The Commission defined 'sustainable development' as a process of change in which the exploitation of resources, the direction of investments, the orientation of technology and institutional change are all in harmony and enhance both current and future potential to meet human needs and aspirations (FAO, 1993a).

There are many definitions and concepts associated with sustainable forest management. In its simplest form, sustainable forest management is concerned with providing an even flow of timber and other wood products. However, this concept focuses only on wood production and does not address the wider issues of the ecological and social functions of forests (FAO, 1993a). In a broader sense, sustainable forest management aims to ensure that all of the values derived from forests (economic, ecological, social) meet present-day needs while at the same time ensuring their continued availability and contribution to long-term development needs (FAO, 1994).

At present, a number of intergovernmental processes are addressing issues concerning sustainable forestry. These are in reaction to growing public and international concerns about increased demand for forest products, forest management, the power of technology, loss of nature and diminishing forests. In addition to the report of the Brundtland Commission already mentioned, they include the United Nations Conference on Environment and Development (UNCED) and Agenda 21; Helsinki and Montreal Process criteria and indicators for sustainable forestry; the work of the International Tropical Timber Organization (ITTO); and the international conventions on climate change and biodiversity. Most recently, an open-ended, *ad hoc* Intergovernmental Panel on Forests (IPF) has been organized under the UN Commission on Sustainable Development (CSD). The work of this panel relates to all the themes of sustainable forestry and provides a focus for current international action in this area.

Economic factors must be taken into account in the design, implementation and evaluation of projects and public-policy issues related to climate change, including the analysis of costs and benefits. Although costs and benefits cannot all be measured in monetary terms, various techniques exist which offer a useful framework for organizing information about alternative actions for addressing climate change. The family of techniques for examining economic environmental policies and decisions include traditional project-level cost–benefit analysis, cost-effectiveness analysis, multicriteria analysis and decision analysis. Cost–benefit analysis attempts to compare all costs and benefits expressed in terms of a common monetary unit. Cost-effectiveness analysis seeks to find the lowest-cost option to achieve a specified objective. Multicriteria analysis is designed to address problems where some benefits or costs are measured in non-monetary units. Decision analysis focuses specifically on making decisions under uncertainty.

In the forestry sector, costs for conserving and sequestering carbon in biomass and soil are estimated to range widely but can be competitive with other mitigation options. Factors affecting costs include opportunity costs of land, initial costs of planting and establishment, costs of nurseries, cost of annual maintenance, monitoring and protection and transaction costs. Direct and indirect benefits will vary with national circumstances and could offset costs.

The net amount of carbon per unit area conserved or sequestered in living biomass under a particular forest-management practice and present climatic conditions is relatively well understood. The most important uncertainties associated with estimating a global value include: the amount of land suitable and available for afforestation or regeneration; the rate at which tropical deforestation can actually be reduced; long-term use of these lands; and continued sustainability of practices for particular locations, given the possibility of changes due to predicted climate change (Watson *et al.*, 1996).

Forest-sector measures should be adaptable and have sufficient flexibility to adapt to changing economic, political, social, ecological and climatic conditions. It will be particularly important to consider the possible effects of climate change on the success or failure of mitigation actions. These include effects of increased levels of CO_2 on plant growth (fertilization effects), changes in soil moisture, length of growing season, natural ranges of trees and forest communities and incidence of wildfire, insect and disease outbreaks (Apps *et al.*, 1993; Ciesla, 1995).

Other desirable characteristics of forest-sector climate-change mitigation actions include technological simplicity and social acceptability. These actions should be implementable under a variety of conditions with a minimum of specialized equipment, training or complicated procedures. In addition, they should have immediate and clear benefits, especially to local residents.

REDUCING SOURCES OF GREENHOUSE GASES

Reducing Deforestation

Felling and burning of forests to make land available for agriculture or livestock grazing is the major forest-sector contributor to increases in the levels of atmospheric GHGs. Accelerated deforestation in the tropics has become a major environmental concern in recent years, not only from the standpoint of its contribution to climate change but also because of potential loss of biodiversity. According to the Food and Agriculture Organization (FAO) (FAO, 1993b), the annual rate of deforestation in the tropics during the decade of the 1980s was 15.4 million hectares. This is roughly equivalent to the total land area of Nepal, Nicaragua or Greece, and it has resulted in a reduction of the area covered by tropical forests from 1,910 million hectares at the end of 1980 to 1,756 million hectares at the end of 1990. Tropical deforestation has increased in comparison with the previous decade, when the estimated annual rate was 11.3 million hectares (Lanly, 1982).

Deforestation has typically been considered as a forest-sector problem when in reality it is a multisectoral issue. Reducing the current rates of deforestation requires actions that reduce pressures to convert forest lands to other uses and that protect remaining areas of forest so that they can be managed sustainably. Most deforestation is the result of the expansion of agriculture, which is in direct response to expanding human populations. Efforts to reduce rates of deforestation must, therefore, be accompanied by efforts to increase productivity and sustainability of existing agricultural lands, so that production keeps pace with increasing demands. Implementation of such cropping systems is best determined through the development of strategic, multisectoral land- and resource-management plans at the country level.

Reducing Area and Frequency of Forests and Woodlands Consumed by Biomass Burning

Biomass burning is associated with forest clearing, burning of savannah vegetation to stimulate regeneration of grasses for livestock, burning of fuel wood and charcoal and consumption of agricultural residues. The area of savannah vegetation burned each year is estimated at 750 million hectares. Approximately half of this area is in Africa and is often based on traditions which are thousands of years old. Prescribed fires often escape and burn areas that are not intended to be burned, thus becoming wildfires. The number and area of wildfires can be reduced through the implementation of integrated fire-management programmes. While the primary benefit of these programmes is the protection of forests and other wild land resources, they will also result in reduced carbon emissions.

Many developing countries have no wildfire-management capacity. Consequently, fires often consume extensive areas of forest and savannah woodland. When planning forest-sector development projects in such areas, it is critical that a wildfire-protection component be included, in order to protect investments made in resource development. An important factor to consider when designing wildfire-protection programmes is the natural role of fire in the ecosystems to be protected. In many semiarid ecosystems, fire plays a key role in plant succession and biomass regulation. Fire exclusion resulting from wildfire-protection programmes can result in the establishment of more vegetative biomass than sites are capable of carrying. This can lead to increased incidence of insect and disease activity and wildfires of a greater intensity.

Increasing Efficiency of Burning Fuel Wood and Other Biofuels

Fuel wood and other biofuels, including charcoal, crop residues and animal dung, are used in many parts of the world for cooking, heating and processing of raw materials at the household level. Biofuels are currently the fourth most important source of energy worldwide, with fuel-wood and charcoal consumption accounting for 10% of overall world energy consumption. Use of household biofuels is estimated to contribute between 2 and 7% of annual emissions from GHGs from human sources.

Household biofuels are often converted into energy using methods that are inefficient and produce a low energy output. They also produce a high GHG output per unit of energy produced. The use of more efficient production systems provides an opportunity to increase energy output and reduce the per unit output of GHGs. They provide the added benefit of reducing pressure on existing biofuel resources, a critical consideration in semiarid regions, where rapidly expanding populations are putting increased pressure on existing biofuel resources, leading to deforestation and desertification.

Introduction of more efficient cooking stoves and industrial processes could reduce fuel-wood requirements by 25–70% at a low investment cost, also resulting in reduced GHG emissions. In addition, the use of biomass of better quality, in terms of size, moisture content and heating value, can contribute to increased efficiencies and reduced GHG output.

Increased Use of Wood and Other Biofuels in Place of Fossil Fuels

The substitution of biomass in place of fossil fuels as a modern energy source has the potential to change the climate-change implications of rising energy consumption dramatically, especially in tropical countries. Opportunities exist to consume efficiently agricultural and forest residues that would otherwise go to waste. There are also opportunities to develop biomass crops, including trees, specifically for energy production. Biofuels can not only help close the CO_2 cycle and reduce GHG emissions but, when established on lands that are currently fallow, they can also expand carbon reservoirs.

Increasing Efficiency of Timber-harvesting Practices

Inefficient timber-harvesting operations result in excessive soil disturbance, logging residues and damage to residual trees. This causes increased GHG emissions and reduces the capacity of the residual forest to sequester carbon. Timber harvesting is a damaging process, no matter how carefully planned and implemented; however, there are practices which can be implemented to reduce the disruption of forest processes. These include (Dykstra and Heinrich, 1992): development of forest-management and harvesting plans; reduction of logging damage by prefelling of vines where they bind trees together; directional felling; reduction of yarding damage by restricting bulldozers to skid trails and maximizing log-winching distances; increased utilization of felled trees – according to a study conducted in several tropical countries, less than 50% of the mainstem of harvested trees is currently utilized, compared with 78% in industrial countries (Dykstra, 1994); and postharvest practices, such as removal of stream crossings, proper slash disposal and treatments to promote vegetative growth in logged-over areas.

An example of the benefits of improved timber-harvesting techniques comes from a study recently completed in dipterocarp forests in Malaysia and is shown in Fig. 3.1. Through the use of controlled harvesting techniques, it has been demonstrated that, if damage to the residual stand can be reduced from 40% to 20%, the additional amount of carbon remaining in the residual forest after 10 years could be over 65 t ha^{-1} (Pinard, 1994).

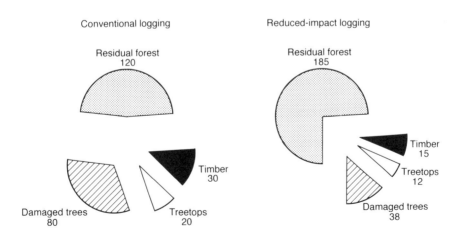

Fig. 3.1. A comparison of residual carbon storage (tonnes of carbon per hectare) between conventional and reduced-impact logging in Malaysia (from Pinard, 1994).

MAINTAINING EXISTING GREENHOUSE-GAS SINKS

Management and Conservation of Natural Forests

Improved management of natural forests can increase their productivity, and hence their potential to sequester carbon, through accelerated growth, maintenance of optimum stocking levels and protection from fire, insects and disease or invasive weeds. There are many opportunities, worldwide, to improve the management of natural forests. In the tropics, for example, an estimated 137 million hectares of logged-over forests could benefit from enrichment planting, because present selective logging practices have reduced long-term productivity (Grainger, 1990). Development and expansion of non-wood forest products would provide increased incentives to maintain and protect forests. This could have the added desirable effect of increased carbon-storage capacity. Examples of opportunities to develop non-wood forest products include production of latex, nuts, resin, mushrooms, wild meat and plants of medicinal value.

Establishment of reserved or protected forests, which are excluded from timber harvesting, will also help maintain existing natural forests as carbon sinks, provided that they are given adequate protection from encroachment, wildfire and insect and disease outbreaks. Reserved or protected forests still provide many opportunities for the management of non-timber resources, including habitat for wildlife, protection of rare and endangered plants or animals, *in situ* conservation of genetic resources, development of outdoor recreation opportunities, protection of unique scenic or aesthetic values and soil and water resources and gathering of non-wood products, such as wild fruits and mushrooms. These forests will probably provide a limited increase in carbon absorption potential, however, because they often contain extensive areas of mature forests, where carbon absorption is roughly equal to carbon release.

Long-term Uses of Forests and Forest Products

From the perspective of carbon storage, the most desirable uses of forests and forest products are those which extend rotation ages and production of goods that are durable and long-lasting. This will allow for carbon to be stored in woody tissue for as long as possible. Obviously, local and national needs for goods and services derived from forests will prevail over global concerns for carbon sequestration. Therefore, from the socio-economic standpoint, forests should be put to whatever uses are required to support local and national needs, provided they are sustainable. If forests provide for the needs of people, there are built-in incentives to manage them on a long-term, sustainable basis and their chances of long-term survival are increased. Problems arise, however, in contexts such as global warming, where effects external to the forest economy at local or national level become important.

EXPANDING GREENHOUSE-GAS SINKS

Forest Planting

Because trees and forests have the capacity to absorb carbon from the atmosphere, afforestation has been suggested as a cost-effective approach to mitigating predicted effects of global climate change. Establishment of plantations of fast-growing trees in the tropics is a particularly attractive option, because they can fix from 12 to 70 tonnes of carbon per hectare over rotation ages ranging from 7 to 20 years. Afforestation programmes have been initiated in the tropics by public utilities based in industrial countries to offset carbon emissions. The Framework Convention on Climate Change provides for a process known as 'joint implementation' to reduce GHG emissions. Tree planting and other forestry activities have been proposed and carried out on a pilot basis under this option and are discussed in Chapters 20 and 21 in this volume.

An overriding constraint in afforestation or reforestation is to match tree species and provenances to the sites on which they are to be planted. Species should also be selected which meet the objectives of the afforestation programme and are acceptable to the local people. If these conditions are met and with proper management and protection, they should be assured of high survival rates and good growth and will meet the objectives for which they are planted, including carbon fixation.

Two key points which must be addressed in the planning and execution of large-scale afforestation projects are the land area required for planting to absorb anthropogenic GHGs and its availability. Several studies have been conducted to address the question of land area available for afforestation. Sedjo and Solomon (1989) report that the current annual increase in atmospheric carbon could be sequestered for about 30 years in approximately 465 million hectares of planted forests. This would require an increase of more than 10% in the current area of forests on the Earth's surface and an increase of more than four times the present plantation area in the world. This estimate is based on the assumption of an average annual growth rate of 15 m^3 ha^{-1} $year^{-1}$. An annual growth rate of 5 m^3 ha^{-1} $year^{-1}$ is more realistic for plantations in boreal and temperate zones and for many tropical areas. Consequently, this estimate of 10% required increase in forest area would have to be substantially revised upwards if the annual growth rate assured was lower.

Other calculations of land-area requirements are presented by Grainger (1990) for several afforestation scenarios. These suggest a number of findings:

planting 60 million hectares per year for 10 years would establish sufficient forests to absorb 2.9 billion tonnes of carbon, which is roughly equal to the present net increment of CO_2 from all sources; planting 20 million hectares per year beginning in 1990 would achieve a carbon absorption equal to the current net annual contribution by the year 2020; continuing the present rate of forest planting for the next 40 years will offset less than 10% of the current net CO_2 increase over the whole period; and afforestation of 2–5 million hectares per year could offset CO_2 emissions from present loss of tropical forests by the year 2020.

These estimates of the impact of various scenarios of global afforestation compare with the current rate of forest planting in the tropics of approximately 1.8 million hectares per year (FAO, 1993b). Several estimates of land available for afforestation in the tropics have been made. Grainger (1990) estimates that there may be 621 million hectares of land 'technically' available, but this estimate does not take into account socio-economic considerations. Of these 621 million hectares, 418 million hectares would be in dry, montane regions and 203 million hectares would be forest fallow in humid areas. According to Houghton (1990), however, up to 865 million hectares of land are available in the tropics for afforestation. Of this total, there may be about 500 million hectares of abandoned lands that previously supported forests in Latin America (100 million hectares), Asia (100 million hectares) and Africa (300 million hectares). The additional area would be available only if increases in agricultural productivity on other lands allowed these marginal lands to be removed from production.

In the mid- and northern latitudes, rates of land-use change have slowed in most areas over the past century. There are still, however, ample opportunities for reafforestation or new afforestation projects. In some temperate-zone countries, the area of forest land has actually increased, as marginal agricultural lands which have been abandoned have reverted to natural forests or have been reforested. Within the past 15 years in Europe, for example, approximately 600,000 ha $year^{-1}$ have left agricultural production, of which about 40% were transferred to forest and other wooded land (FAO, 1992).

In the USA, there are an estimated 46.8 million hectares of crop and pasture lands which may be physically capable of growing trees and may be physically better suited to this purpose. There are also opportunities for planting fast-growing trees for wood-energy production on 14–28 million hectares and establishment of windbreak plantings on 1.37 million hectares. If planted, such areas would provide an additional carbon-storage capacity of from 0.066 to 0.210 Gt carbon $year^{-1}$ (Sampson and Hamilton, 1992).

Other constraints to large-scale afforestation include its social, economic and ecological impact, as well as the human-capital constraints of implementing such a large-scale land-use change. Limited institutional capacity, lack of research on appropriate species to plant, unrealistic government incentives and increased pressure on available land can constrain afforestation initiatives. The interests and needs of local people living in areas proposed for afforestation are of paramount importance. When goals and objectives of forestry projects do not coincide with those of local resource users, conflicts will result. Farmers can easily perceive government-sponsored afforestation or reforestation efforts as an encroachment on customary land-use rights and a challenge to their welfare. Reactions of this nature have often led to active opposition and even sabotage through setting of intentional fires (Trexler et al., 1992).

Costs of afforestation programmes are highly variable, depending on the nature of the terrain to be planted, labour costs and the tree species to be planted. A global estimate indicates that tree-planting costs can range from a low of $US200 to a high of $US2,000 ha^{-1}. Projected socio-economic benefits derived from plantations may not always justify planting costs (Bernthal, 1990). It would be difficult, if not impossible, for countries receiving loans from development banks for afforestation projects to repay the loans. This calls into question the economic viability of large-scale afforestation programmes. Ecological constraints include the potential for introducing low levels of genetic variability, which can characterize large areas of single-species plantations. This can reduce their resistance to site- or climate-related stress or attack by insects and disease. Forest planting can also strain existing water resources in areas that are already experiencing overdrafts and escalating demands.

Agroforestry and Urban Tree Planting

The magnitude of the carbon-storage contribution from agroforestry planting will depend on its scale and the ultimate uses made of the wood. Given suitable economic, social and environmental conditions, as alluded to above and in other contributions in this book, farmers worldwide have readily adopted agroforestry systems. In countries where public forest lands or government-sponsored tree-planting efforts are limited, agroforestry planting can represent a significant contribution to tree planting and carbon sequestration. There are opportunities for increased agroforestry planting in the tropics and in temperate zones. Large-scale agroforestry schemes, such as the 'Four Around' schemes in China, were reportedly carried out over 6.5 million hectares of agricultural land during the 1980s. Projects of this magnitude, if successful, would sequester large amounts of carbon, as well as providing other benefits such as protection of soil from wind and water erosion. Agroforestry systems are being considered in Western Europe as an alternative to the intensive-production systems which are currently in place and produce large crop surpluses. In the USA, there are large areas that would benefit from increased wind-break and shelter-belt planting.

Urban tree planting would provide a more limited carbon-storage benefit than rural planting, because, by their nature, they would not be very extensive. However, they have the potential to contribute other benefits in the climate-change context that are much more significant. Urban trees have a significant and quantifiable effect on local climates. They are known to break up urban heat islands by providing shade. Trees can also have a significant beneficial effect on the cost of winter heating and summer cooling of buildings. Depending on its location, the energy-conservation efforts of a single urban tree can prevent the release of 15 times more atmospheric carbon than would be locked up directly in its biomass. The shade provided by strategically placed trees per house can reduce summer home air-conditioning needs by 30–50% in warm climates. Trees planted as windbreaks around buildings in temperate and boreal regions can reduce winter heating-energy use by 4–22% (Sampson *et al.*, 1992).

CONCLUSIONS AND SUMMARY

Regardless of climate-change considerations, protection of forests should be an integral part of all forest management. In the future, however, it will be increasingly important to consider the potential effects of climate change on the incidence of wildfire and outbreaks of insects, disease or other damaging agents in the development of strategic forest-sector development plans and in the execution of projects designed to achieve those plans. It will also become more critical to provide for protection of forests from illegal human encroachment.

Programmes designed to protect the health of forests should include a monitoring component, decision criteria for the management of fire, pests and disease, based on ecological, economic and social criteria, and the availability of environmentally friendly tactics (biological, chemical, cultural, mechanical and regulatory) which can be used in integrated fire- and pest-management systems. These would create conditions unfavourable for the development of damaging agents and provide for effective responses to protect investments made in forest-sector development.

There are many uncertainties associated with the global climate-change issue. In addition, in many countries, the amount of land available and suitable for agriculture or forestry is limited. Any forest-sector responses to adapt to or mitigate the potential effects of climate change should therefore represent sound policy independent of predicted global warming and produce net benefits separate from those which may ultimately arise in the climate-change context. The position of FAO, with respect to afforestation for CO_2 absorption is to encourage tree planting in areas where forest cover is the appropriate vegetation. This should be defined by land-use plans and forest strategies and not by theoretical targets (FAO, 1990). Instead of focusing narrowly on afforestation or reforestation, the adoption of an approach should be considered which includes the management and protection of existing areas of natural forest to provide for long-term sustained-yield productivity of a wide range of commodity and non-commodity resources, including carbon absorption. Such action, coupled with the development and implementation of realistic criteria and indicators of sustainability at regional and national levels, should provide for both immediate and long-term needs of human society and contribute toward increased carbon storage by trees and forests.

Analysis conducted under the Intergovernmental Panel on Climate Change (IPCC) (IPCC, 1996a) assessment reported in Brown (1996) indicates that, through implementation of forest-management options that are compatible with traditional objectives of forestry (such as improved forest management, slowing deforestation and agroforestry), there is a potential to reduce net CO_2 emissions from forests to the point where forests are no longer a significant source of carbon and to sequester an amount of carbon equivalent to 11–15% of total fossil-fuel emissions over the next 50 years (60–87 billion tonnes). Most of the opportunities to increase the carbon-sequestration potential of forests are in the tropics, where there is a potential to sequester or convert 45–72 billion tonnes of carbon between 1995 and 2050.

A Critical Review of the Scientific Basis of Projected Global Warming

Antonio Zecca and Roberto S. Brusa

INTRODUCTION

The assessment of projections for the future global warming is a prerequisite for the study and design of mitigation strategies. Constraints on decision-making in the presence of uncertainties are directly related to the magnitude of the future global warming and to the time evolution of this phenomenon. In spite of a mean global surface-temperature rise of 0.6°C in the last 150 years, a minority in the scientific community still questions the attribution of this trend to human activities. A larger group, although believing in a possible anthropogenic influence on the climate, has stressed the insufficient statistical significance of available data to discard fully the possibility that the observed temperature increase has been produced by a natural fluctuation in the Earth's climate. A majority of the researchers in the field (Houghton *et al.*, 1996) assume that, in spite of the fact that the statistical indicators do not support a definitive claim of anthropogenic warming with complete confidence, the evidence concerning ongoing changes is more than enough to support or justify measures to counteract this trend and to mitigate its consequences.

The growth of knowledge in the last few years, in the areas of the detection of the greenhouse warming and reliable projections of the warming to be expected in the next decade, is therefore of major significance. The goal of this chapter is to summarize the present scientific knowledge on global warming and to contribute to more reliable projections.

DETECTION OF THE GREENHOUSE EFFECT AND GLOBAL WARMING

The most recent Intergovernmental Panel on Climate Change (IPCC) (IPCC-1995) report (Houghton *et al.*, 1996) gives the most comprehensive review on the evidence for global warming to date. The conclusion concerning 'detection and attribution' of the past warming to human activity has been outlined in rather baroque language, in the *Summary for Policymakers – Working Group I*: 'The balance of evidence suggests a discernible human influence on global climate.' In other words, the existence of an anthropogenic contribution to the past (and future) warming is demonstrated with a 95% confidence limit. The report of the scientific working group states explicitly that much of the progress since the 1990 report has been made by taking into account the role of the sulphur dioxide (SO_2) aerosol in the Earth's energy balance, as discussed below.

The global warming projections in the IPCC-1995 report predict a lower level of mean warming than in the first IPCC report (Houghton *et al.*, 1990). Referring to the 'low', 'best' and 'high' estimates, as defined in the first report, the projections are 1°C, 2°C and 3.5°C, respectively, at the target year 2100. This reduction of the predicted warming is claimed to be due to the lower emission scenarios used in the 1995 report and to the inclusion of the SO_2 action in the climate models. The second statement contrasts with previous results in this area (Zecca and Brusa, 1990, 1991; Wigley and Raper, 1992). The uncertainties concerning temperature and other parameters remain large, both on the low and on the high side of the error bars. It is important to stress that all the projections made in IPCC-1995 implicitly assume parallel growth of carbon dioxide (CO_2) and SO_2 emissions. In other words, no measure intended to reduce SO_2, because of it resulting in acid rain, has been included in the scenarios; the resulting projection would imply an atmospheric SO_2 doubling sometime between 2040 and 2100. Such a scenario would obviously have consequences for both forestry and agriculture, especially in the industrialized, temperate countries. The later sections of this chapter show that actions intended to reduce the anthropogenic SO_2 atmospheric burden would induce warming, a factor which has not been accounted for in IPCC-1995. In the extreme case of complete abatement of the SO_2 emissions, the numbers given by IPCC-1995 at the year 2100 should be increased by about 0.8°C, giving 1.8°C, 2.8°C and 4.3°C for the low, best and high estimates, respectively.

INTRODUCING SULPHUR DIOXIDE INTO THE GLOBAL ENERGY BUDGET

The impact of SO_2 on the global climate has been recognized for some time in theory, but was largely overlooked until a few years ago. This oversight was possibly caused by the belief that the effect was insignificant, due to the low atmospheric concentration of SO_2. An important step in rectifying the oversight was made by Charlson and colleagues (1987), who studied the possible climatic action of SO_2 from marine natural sources. Their conclusion was that biogenic SO_2 was probably acting as a cooling agent on the global temperature, with a global average value close to −1°C. A further step was to consider the effect of anthropogenic SO_2 emissions (Zecca and Brusa, 1990, 1991), which were known to be of the same order of magnitude as the natural flux. In subsequent years, a number of papers considered the effect of SO_2 (see, for instance, Charlson *et al.*, 1992; Wigley and Raper, 1992; Charlson and Wigley, 1994). The SO_2 effect has only recently been inserted as a parameter in the global circulation models (GCMs), which project global climate parameters (Taylor and Penner, 1994; Mitchell *et al.*, 1995; Santer *et al.*, 1996). These models use a detailed mapping of anthropogenic sulphur emissions over the planet, as one input parameter, in producing GCM projections. The projections show, as expected, areas of cooling over the industrialized regions of Asia, Europe and North America.

Sulphur dioxide has a secondary impact on the climatic system through the formation of aerosol particles, which scatter back the incoming shortwave radiation and act as cloud-condensation nuclei, thus increasing the cloudiness. Both actions result in an increase of the Earth's albedo, the 'reflectivity' of the Earth. Sulphur dioxide is also a greenhouse gas, but its warming potential is small enough for the net effect to be cooling.

Figure 4.1 shows a plot of the annual mean global average temperatures in the years 1860–1995. This time series is marked by the very well-known but unresolved anomaly in the years 1940–1970, where the temperature was slightly decreasing. This anomaly can be explained only in terms of the competition between greenhouse warming and SO_2 cooling. In Fig. 4.2, the evolution of the anthropogenic SO_2 emissions is plotted over the period 1860–1980 (Moeller, 1984). A discontinuity is visible in this record, from the 1940s. This jump marks the onset of a rapid increase in fossil-fuel use in transport and other sectors. It has been demonstrated (Zecca and Brusa, 1990, 1991) that anthropogenic SO_2 emissions are responsible for counteracting

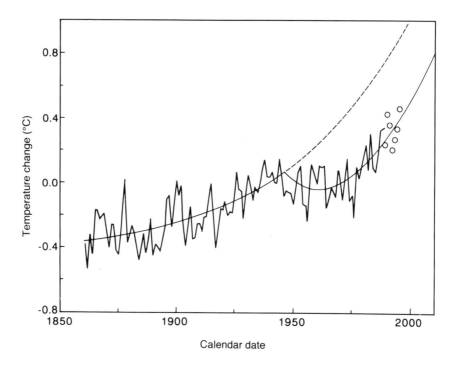

Fig. 4.1. Global average temperature anomalies, 1860–1990. The mean value for the years 1940–1970 is assumed as temperature zero. The broken line is measured data (Jones and Raper, 1990). The dashed line is calculated greenhouse forcing. The continuous line is greenhouse warming, minus an SO_2 aerosol cooling in 1945. Open circles are measured data in the years 1989–1995.

greenhouse warming, possibly by as much as 0.5°C in 1990. Figure 4.3 shows the partitioning of the observed global average temperature increase among the different anthropogenic gases according to the first IPCC report (Houghton *et al.*, 1990) and the partitioning as proposed by Zecca and Brusa (1990). In the latter, the SO_2 cooling contribution is apparent.

It is important to stress that SO_2 is produced simultaneously with CO_2 when fossil fuels are combusted. It is also essential to note the large difference in the atmospheric residence times of CO_2 and SO_2 – 100 years and 7 days, respectively. This implies that any climatic effect, if averaged over periods of the order of 1 year, depends mainly on the emission rate for SO_2 and on the realized concentration for CO_2. The consequence of this balance, on a time-scale shorter than 100 years, is that any discontinuity in the burning of fossil fuels will show up as a momentary (time-scale of the order of years) slowing

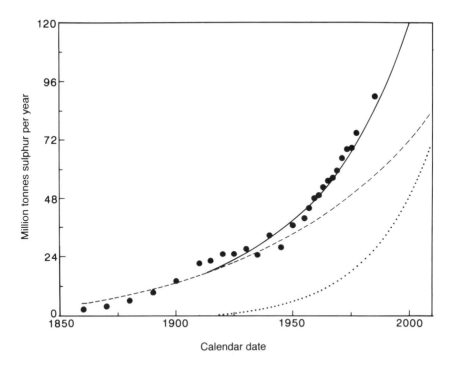

Fig. 4.2. Global anthropogenic sulphur emission, 1860–1983 (from Moeller, 1984). Dotted line, from oil combustion. Dashed line, total minus oil. Continuous line, sum of the two fitted lines. A discontinuity is present in the emission starting from the 1940s.

down of the temperature rise, such as the observed anomaly in the years 1940–1970. In Fig. 4.1, the trough in the curve shows how the model developed by Zecca and Brusa (1991) fits the measured data, giving for the first time an explanation of the 1940–1970 minimum. The circles in Fig. 4.1 are the observed average temperatures for the years 1989–1995.

There are today a minority of scientists who still maintain that there is no evidence of the human impact on the warming realized in the last 150 years. Nevertheless, the study of the climatic action of SO_2 leaves no doubt about its anthropogenic contribution. In fact, the IPCC-1995 report states that there is a low probability of the 1860–1990 temperature increase being completely natural, since the probability that natural fluctuations mimic simultaneously the greenhouse warming and the aerosol cooling, as we expect them from anthropogenic emissions, is negligible.

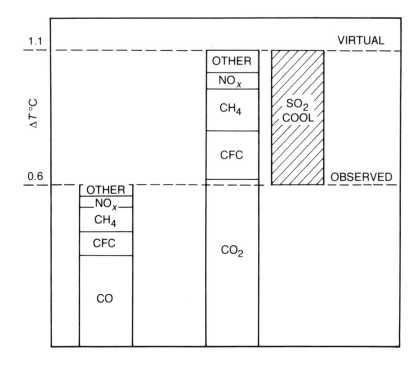

Fig. 4.3. Proposed partitioning of 1860–1990 observed temperature increase by anthropogenic greenhouse gas (from Zecca and Brusa, 1990, 1991). The left box is according to Houghton *et al.* (1990); the observed increase (= 0.6°C) is attributed to greenhouse warming only. Right: the observed increase is the result of a 1.1°C greenhouse warming minus a 0.5°C SO_2 albedo cooling. NO_x, nitrogen oxides; CH_4, methane; CFC, chlorofluorocarbon; CO, carbon monoxide.

WARMING PROJECTIONS: BEYOND THE ANNUAL AVERAGES

The analysis of the global climate contained in IPCC-1995 deals with average climatic properties and gives only a hint about future trends of climatic extremes (Nicholls *et al.*, 1996). These future trends are more important for the impact on agriculture than year average values. Studies in the field of the prediction of weather extremes are progressing, but, while awaiting the results of present research on climate extremes, the IPCC projections can be interpreted, using simple, fundamental considerations.

As discussed above, the cooling effect of SO_2 is mediated through an increase of the planetary albedo. Such cooling is therefore active during daytime only (Zecca and Brusa, 1992), when sulphate and aerosols reflect the incoming radiation. During night-time, the cooling mechanism is almost

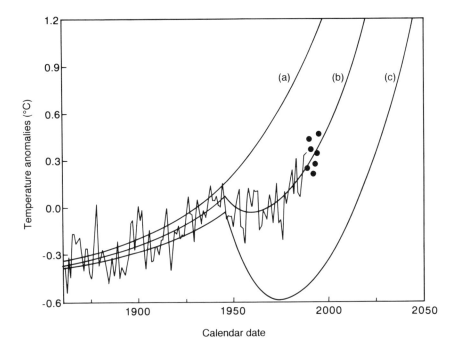

Fig. 4.4. Behaviour of minimum daily (global annual average): curve (a) and maximum daily: curve (c) temperature (Zecca and Brusa, 1992). Curve (b) is the same as in Fig. 4.1. Under certain meteorological conditions, sudden temporary switching from curve (c) to curve (a) will be possible during daytime.

non-effective. As a consequence, the global temperature rise due to greenhouse gases is expected to be faster during night-time than during the day, as illustrated in Fig. 4.4 (Zecca and Brusa, 1992) and confirmed by Karl et al. (1993). This could partially offset the greenhouse consequences for both agriculture and forestry. However, the geographical distribution of anthropogenic SO_2 emissions introduces further complications (Kiehl and Briegleb, 1993). Emissions are mostly concentrated over the eastern part of North America, Central Europe and South-East Asia. In these regions, aerosols can completely cancel the greenhouse warming or even produce a cooling trend. This means that over the remaining continental regions (mostly developing countries) the warming will be higher (by a factor of 1.3, on average) than the values given in the above IPCC projections. As stated above, any action intended to reduce the SO_2 anthropogenic emissions would cause an immediate global warming, above the IPCC average levels, because of its relationship to acid-rain impacts. The global average curve would shift from curve (b) in Fig. 4.4 towards curve (a).

In addition, over Europe, North America and South-East Asia, the day-to-day variability of the sulphate cooling will be very high; a form of temporary switching from curve (c) to curve (a) in Fig. 4.4 will take place in particular weather conditions. In such situations, there would be summer days in which the temperature extremes would approach those pertaining to the anthropogenic SO_2-free regions (i.e. again higher than the average values projected by IPCC-1995 by a factor of 1.3). It is evident that, from the point of view of crop productivity and agricultural risk, extreme values are much more important than annual average values (Semenov and Evans, 1996). Therefore, this result, regarding the relationship between SO_2 and heat stress due to temperature extremes, should be carefully considered when dealing with both climate-change consequences and mitigation strategies.

WARMING PROJECTIONS: FURTHER CORRECTIONS

Here we shall mention two mechanisms related to forestry which have not been considered in the IPCC projections and which could introduce upward corrections in those forecasts. Recent studies of the effects of forests on global climate (Bonan *et al.*, 1992; Kirschbaum *et al.*, 1996) have examined boreal forests in particular, but results are shown to be qualitatively valid also for all forest types. The conclusions indicate that the location of forests is the outcome of coupled dynamic interactions, in which the geographical distribution of boreal forests or altitude of montane forests affects the climate and vice versa. The important consequence of this result is that the migration of forests toward polar latitudes or toward higher altitudes in response to global warming may produce further warming. This is a positive-feedback loop with a relatively long time constant which has not been fully considered to date.

A further feedback involves forested areas: it is observed that the maximum rate at which carbon can be stored in forests is much lower than the rate at which carbon can be released to the atmosphere. Forest changes in response to climate changes will lead to a carbon release to the atmosphere, which can be compensated by forest regrowth only after a delay, which can be of the order of decades. This time-lag allows the occurrence of a further positive-feedback loop, which could possibly be effectively counteracted with appropriate forest-management policy, as discussed in various chapters in this volume. However, the feedbacks associated with natural adaptation of forests to climate parameters do not provide opportunities for policy intervention.

Finally, recent research regarding the impact of climate changes on Alpine flora has been carried out by Grabherr and colleagues (1994). This study highlights the shift of plant species and communities as they have been forced to higher altitudes by the warming realized on the Alps during the last 70–90 years. They state that even moderate warming induces species migration and that this migration is being observed in the Alps. The velocity of such migration for the observed species is two to ten times lower than the speed of the temperature change, leading the authors to fear for 'disastrous extinctions in these environments'. In order to avoid such impacts, some breathing-space must be provided for these ecosystems to avert their extinction. This breathing-space can be provided for mitigation of the current greenhouse-gas emissions. As a first step, such impacts of climatic changes should be added to the potential costs of anthropogenic greenhouse global warming.

5 The Economic Costs of Climate Change and Implications for Land-use Change

Samuel Fankhauser

INTRODUCTION

While many unknowns remain, a better picture is gradually starting to emerge of the impacts of climate change. There is now wide agreement that the impacts will be diverse, with winners as well as losers. Some effects may be positive, others may at least be easy to adapt to. For the majority of people, however, the consequences of climate change will probably be negative. For some regions, they could be disastrous. Working Groups II and III of the Second Assessment of the Intergovernmental Panel on Climate Change (IPCC) have extensively reviewed the potential impacts of climate change (Pearce *et al.*, 1996; Watson *et al.*, 1996). The best-studied regions remain the developed countries, in particular the USA. Studies usually deal with only a subset of damage and are often restricted to a description of impacts in physical terms. By far the best-studied aspects of climate-change impacts are agricultural impacts (for example, Rosenzweig and Parry, 1994; Adams *et al.*, 1995) and the costs of sea-level rise (for example, Fankhauser, 1995b; Yohe *et al.*, 1996). Attempts at a comprehensive monetary quantification of all impacts are relatively rare and usually restricted to the USA (Nordhaus, 1991; Cline, 1992a; Titus, 1992). Preliminary global assessments are provided by Fankhauser (1995a) and Tol (1995).

This chapter reviews the available economic assessments of climate-change impacts, drawing on the work of IPCC Working Group III (Pearce *et al.*, 1996). It is structured as follows. The next section starts with a brief summary of expected impacts in qualitative terms. The next two sections then summarize the main findings of IPCC Working Group III, dealing with the results of equilibrium analysis and with dynamic aspects. The main shortcomings of the current models and analyses of how economic assessments might change in the light of recent findings are then presented. The chapter concludes with some remarks on the implication of damage estimates for land-use change policy.

CLIMATE-CHANGE IMPACTS

Global warming will have a variety of effects on both human and natural systems. The impacts working group of IPCC (Watson *et al.*, 1996) foresees a shift in the current agricultural production pattern away from current production areas to more northern latitudes. Together with changes in soil-water availability, the increased occurrence of climatic extremes and crop diseases, this may lead to an overall reduction in agricultural yields and could result in serious regional or year-to-year food shortages. The forestry sector may have to adjust to altered growing conditions and a change of species mix. Fisheries will face a similar challenge. The IPCC also highlights the possibility that the increased stress on unmanaged ecosystems may lead to the extinction of species unable or too slow to adapt.

The rise in sea-levels connected with a warmer climate will threaten low-lying coastal areas. Sea-level rise will particularly affect densely populated coastlines and small island states. Health experts expect a rise in climate-related diseases, such as heat stroke, and an increased incidence of vector-borne diseases, such as malaria. Adverse effects like these may trigger a stream of climate refugees away from the worst-affected regions.

Others have warned about the consequences of increased water shortages. Climate-dependent economic activities, such as construction, transport and tourism, will also be affected, with improved conditions for some activities and deterioration for others. The need to adapt could affect energy consumption, with higher demand for air cooling in summer and lower heating needs in winter.

A large unknown is the effect of climate change on extreme weather events, such as droughts, floods and storms. Many experts predict an increased incidence of such events, which would entail greater vulnerability for areas such as coastal zones. An increase in extreme events would also have repercussions on the insurance industry (Downing *et al.*, 1996).

Damage Assessment I: Equilibrium Analysis

The scientific research on global warming impacts has focused predominantly on the so-called $2 \times CO_2$ scenario, that is an atmospheric CO_2 concentration of twice the preindustrial level. Most of the figures reported in this section are based on the $2 \times CO_2$ scenario. Climate-change impacts related to this warming can be classified as either market-related (effects that will be reflected in the national accounts) or non-market-related (impacts affecting 'intangibles', such as ecosystems or human amenities). Table 5.1 categorizes the expected impacts from global warming. It also assesses how carefully they have been estimated in the literature so far.

Climate-change impacts can be expressed either in physical units or in a common unit of measurement, such as money. Monetary estimates of both market and non-market damage are ideally expressed in the form of willingness to pay (WTP) or willingness to accept compensation (WTA), as described in Box 5.1. Unfortunately, WTP/WTA estimates are not always available for the assessment of global-warming impacts and approximations have had to be used. In addition, some forms of damage that could not so far be estimated have been ignored altogether in the aggregated damage estimates (Pearce et al., 1996).

Based on an extensive survey of the literature, IPCC Working Group III expects the following aggregate damage for $2 \times CO_2$:

World impact:	1.5–2.0% of world gross national product (GNP)
Developed-country impact:	1–1.5% of national GNP
Developing-country impact:	2–9% of national GNP

A wide range of sectoral and ecosystem impacts contribute to this total damage. Table 5.2 shows the relative importance of different damage categories, using figures for the USA. Note that estimates include both adaptation costs and residual damage. Examples of the former include the costs of coastal protection, the costs of migration, and the change in energy demand due to alterations in space heating and cooling requirements. Examples of residual damage include agricultural impacts and the loss of dry- and wetlands. The underlying adaptation assumptions, however, are not explicitly stated for most impact categories.

A caveat is necessary. The figures in Table 5.2 provide best-guess estimates, as found in the literature. The range does not reflect a confidence interval. As noted above, estimates are neither accurate nor complete, and a considerable range of error can be expected. Figures on developing countries, in particular, are clearly less reliable than those for developed regions.

The IPCC (Pearce et al., 1996) predicts considerable regional differences, with potentially higher impacts for some countries, such as small island states. Table 5.3 shows some of the estimates underlying these conclusions in more detail, highlighting the substantial differences between regions. This shows that some developing countries could face extremely high damage. Although developing countries will face less warming than industrialized countries of the northern latitudes, developing countries tend to be more vulnerable to climate change. Their economies are more dependent on

Table 5.1. Overview of climate-change impacts (from Pearce et al., 1996).

	Market impacts				Non-market impacts		
Damage	Primary economic-sector damage	Other economic-sector damage	Property loss	Damage from extreme events	Ecosystem damage	Human impacts	Damage from extreme events
Fully estimated based on willingness to pay	Agriculture		Dryland loss Coastal protection		Wetland loss		
Fully estimated, using approximations	Forestry	Water supply		Hurricane damage	Forest loss		Hurricane damage
Partially estimated	Fisheries*	Energy demand Leisure activity	Urban infrastructure	Damage from droughts	Species loss	Human life Air pollution Water pollution Migration	Damage from droughts†
Not estimated		Insurance Construction Transport Energy supply		Non-tropical storms River floods Hot/cold spells Other catastrophes	Other ecosystem loss	Morbidity Physical comfort Political stability Human hardship	Non-tropical storms River floods Hot/cold spells Other catastrophes

* Often included in wetland loss.
† Primarily agricultural damage.

Box 5.1. Can climate change impacts be valued?

Calculations such as those in Tables 5.2 and 5.3 are controversial. How can impacts like the loss of ecosystems or an increased likelihood of premature death be expressed in monetary terms? The notion is not as exotic as it may seem. Monetization of economic well-being is standard in applications of economics in various fields.

Since welfare cannot be measured directly, like temperature or weight, an indirect measure has to be used. The most common measures are based on the concepts of willingness to pay (WTP) and willingness to accept compensation (WTA). Willingness to pay measures the amount of income a person is willing to forego in exchange for an improved state of the world. For example, it measures what fraction of their income people would be willing to spend to preserve an endangered species or a unique ecosystem. Correspondingly, WTA estimates the compensation that would be required for people to accept its loss. In this way, WTP/WTA can serve as a monetary measure of people's preferences.

There are several techniques to obtain empirical estimates of people's WTP/WTA. In many cases, the easiest way is to elicit estimates directly in surveys or interviews (the contingent valuation method). Alternatively, estimates can be deduced from observed differences in the price of market goods with different environmental characteristics, e.g. from property prices in noisy and quiet neighbourhoods (the hedonic approach). In the same way, the wage differential between high-risk and low-risk jobs is often used to estimate people's WTA for safety. The money spent to visit a park gives an indication of people's valuation of the visited sight (travel cost method). Similarly, money spent on defensive measures (e.g. on noise reduction) can sometimes serve as an indication of WTP by consumers.

The empirical estimation of WTP and WTA is difficult and inexact. The analysis of climate change, with its variety of different and usually uncertain impacts, is particularly complex. Estimates of WTP/WTA are therefore not always available, and proxies, such as the return on input factors (e.g. on capital or land), and other indicators are frequently used to approximate the welfare impacts of climate change (see Table 5.1). Moreover, where estimates are available, they indicate the WTP/WTA of the current generation. However, climate-change impacts will only occur gradually, with significant effects not expected for another 30–50 years. By that time, the demographic and socio-economic structure will have changed, and the welfare impacts of global warming will have changed with them.

Available damage figures can therefore only indicate the likely order of magnitude of the true welfare costs. Even so, they provide useful insights into the relative importance of different climate-change impacts on human welfare.

Table 5.2. Economic damage from $2 \times CO_2$: present US economy (base year 1990; billion $US) (from Nordhaus, 1991; Cline, 1992a; Titus, 1992; Fankhauser, 1995a; Tol, 1995).

	Cline (2.5°C)	Fankhauser (2.5°C)*	Nordhaus (3°C)*	Titus (4°C)	Tol (2.5°C)†
Agriculture	17.5	8.4	1.1	1.2	10.0
Forest loss	3.3	0.7	Small	43.6	–
Species loss	4.0 +	8.4	§	–	5.0
Sea-level rise	7.0	9.0	12.2	5.7	8.5
Electricity	11.2	7.9	1.1	5.6	–
Non-electric heating	–1.3	–	–	–	–
Human amenities	+	–	§	–	12.0
Human morbidity	+	–	§	–	–
Human mortality	5.8	11.4		9.4	37.4
Migration	0.5	0.6	§	–	1.0
Hurricanes	0.8	0.2	§	–	0.3
Construction	±	–	§	–	–
Leisure activities	1.7	–	§	–	–
Water-supply					
Availability	7.0	15.6	§	11.4	–
Pollution	–	–		32.6	–
Urban infrastructure	0.1	–	§	–	–
Air pollution		7.3	§		–
Tropospheric ozone	3.5			27.2	
Other	±			–	
Mobile air-conditioning	–	–		2.5	–
Total	61.1 ±	69.5	55.5	139.2	74.2
% of GDP	1.1	1.3	1.0	2.5	(1.5)†

* Transformed to 1990 base.
† USA and Canada, base year 1988.
‡ Identified, but not estimated.
§ Not assessed categories, estimated at 0.75% of GDP.
GDP, gross domestic product.
+ or ± denote identified but not quantified impacts.

climate-sensitive sectors, in particular agriculture. They have less technical, institutional and financial capacity for adapting to changing conditions. In addition, they tend to be more exposed to extreme weather events, such as tropical cyclones. The combination of these effects could result in particularly severe damage. However, it should be noted that, to avoid long-term predictions, figures were derived by imposing $2 \times CO_2$ on to a society with today's structure. Vulnerability is likely to change as regions develop and population grows.

Table 5.3. Monetary 2 × CO_2 damage in different world regions and share of tangible damage (annual damage) (from Pearce et al., 1996, based on Fankhauser, 1995a; Tol, 1995).

	Fankhauser		Tol	
	bn $US	% GDP*	bn $US	% GDP*
European Union	63.6	1.4		
USA	61.0	1.3		
Other OECD	55.9	1.4		
OECD America			74.2	1.5
OECD Europe			56.5	1.3
OECD Pacific			59.0	2.8
Total OECD	180.5	1.3	189.5	1.6
E. Europe/former USSR	18.2†	0.7†	−7.9	−0.3
Centrally planned Asia	16.7‡	4.7‡	18.0	5.2
South and South-East Asia			53.5	6.6
Africa			30.3	8.7
Latin America			31.0	4.3
Middle East			1.3	4.1
Total non-OECD	89.1	1.6	126.2	2.7
World	269.6	1.4	315.7	1.9

* The GDP base may differ between the studies.
† Former Soviet Union only.
‡ China only.
GDP, gross domestic product; OECD, Organization for Economic Cooperation and Development.

Damage Assessment II: Dynamic Analysis

The analysis so far has been confined to comparative statics. All figures in Tables 5.2 and 5.3 are estimates of the impact of one specific change of the climate $(2 \times CO_2)$ on the current economy. This is clearly insufficient. Not only shall we, for the larger part of the future, be confronted with climate change substantially different from $2 \times CO_2$, but socio-economic vulnerability to climate change will also shift as a consequence of economic development.

What would be relevant to know from a policy point of view are marginal figures or, in other words, estimates of the extra damage done by one extra tonne of carbon emitted. Unfortunately, the requirements for marginal damage calculations go far beyond the information available from $2 \times CO_2$ studies. Greenhouse gases are stock pollutants. That is, a tonne of gas emitted will affect climate over several decades, as long as fractions of the gas remain in the atmosphere. Calculating marginal costs therefore requires the comparison of two present-value terms: the discounted sum of future damage associated with a certain emission scenario is compared with the sum of damage in an alternative scenario with marginally different emissions in the base period.[1]

The current generation of models deals with this challenge in a rather *ad hoc* manner, using very simplistic representations of the complex dynamic processes involved. In previous studies, damage costs were typically specified as a polynomial (usually linear to cubic) function of global mean temperature, calibrated around the $2 \times CO_2$ estimates. Damage is usually fully reversible and assumed to grow with GNP. Only recently, studies have started to emerge that explicitly incorporate regionally diversified temperatures and sea-levels and model individual damage categories, such as agriculture, separately, or at least distinguish between damage related to absolute temperature level and that related to the rate of change (for example, Dowlatabadi and Morgan, 1993; Hope *et al.*, 1993; Tol, 1996). Table 5.4 provides estimates of marginal damage obtained from polynomial damage models. Figures range from about $US5 to $US125 per tonne of carbon, with most estimates at the lower end of this range. The wide range reflects variations in model assumptions, as well as the high sensitivity of figures to the choice of the discount rate. Calculations based on a 'high' or descriptive discount rate (e.g. Nordhaus, 1994) tend to be in the order of about $US5–15 per tonne of carbon emitted now. Assuming a 'low' or prescriptive discount rate (Cline, 1992a) yields estimates that are about an order of magnitude higher.[2]

Shortcomings and Extensions

Comprehensive damage assessments have been fiercely criticized by many authors (for example, Grubb, 1993; Ekins, 1995). While not all criticism was based on sound analysis, the damage estimates of Tables 5.2–5.4 do have a number of shortcomings. The most important points of contention are as follows.

Valuation

Probably the main objection concerns the monetary valuation of climate impacts. While some authors have fundamentally questioned the applicability of economic valuation techniques as such, the main problem for those involved in developing the approach appears to be the low quality of many available estimates. As Table 5.1 shows, estimates are frequently based on approximations and extrapolations instead of original valuation work. Better estimates are particularly needed for such damage aspects as ecosystems loss, health, mortality and morbidity. Further research is also needed with respect to the damage costs to developing countries.

Table 5.4. The marginal social costs of CO_2 emissions (current value 1990 $US per tonne carbon) (from Pearce et al., 1996).

Study	Type	1990–2000	2001–2020	2011–2020	2021–2030
Nordhaus (1991)	MC			7.3 (0.3–65.9)	
Ayres and Walter (1991)	MC			30–35	
Nordhaus (1994)	CBA				
	DICE				
	Certainty/best guess	5.3	6.8	8.6	10.0
	Uncertainty/expected value	12.0	18.0	26.5	n/a
Cline (1992b)	CBA	5.8–124	7.6–154	9.8–186	11.8–221
Peck and Teisberg (1992, 1993b)	CBA	10–12	12–14	14–18	18–22
Fankhauser (1994)	MC	20.3 (6.2–45.2)	22.8 (7.4–52.9)	25.3 (8.3–58.4)	27.8 (9.2–64.2)
Maddison (1993)	CBA/MC	5.9–6.1	8.1–8.4	11.1–11.5	14.7–15.2

DICE, Dynamic Integrated Climate Economy model; MC, marginal social-cost study; n/a, not available; CBA, shadow value in cost–benefit study. Figures in parentheses denote 90% confidence intervals.

Catastrophic events

The estimates of Tables 5.2 and 5.3 concentrate on the most probable damage scenario, i.e. they merely provide a best-guess assessment of what damage is most likely to be. Given the complexity of the climatic system and the unprecedented stress imposed on it, this focus may be too narrow. Other, more disastrous scenarios cannot be excluded with certainty. Rather than with only one point, we are confronted with an entire damage probability distribution. Unfortunately, only a little is known about the shape of this distribution and, in particular, about the probability of an extremely adverse outcome (the upper tail of the damage distribution). Several catastrophic scenarios have been portrayed in the IPCC report so far.

- The melting of the polar ice caps (for example, a possible disintegration of the West Antarctic ice sheet), which could eventually lead to a rise in sea-level of several metres.
- Changes in ocean circulation patterns (for example, a shut-down of the ocean conveyor belt), which could – somewhat ironically – lead to significant cooling in Western Europe.
- The runaway greenhouse effect: initial warming may be amplified through feedback effects, such as the liberation of methane from previously frozen sediments into the atmosphere.
- Abrupt, non-linear changes in climate patterns. Evidence from ice cores points at the possibility of a highly unstable climate, with temperature changes of several degrees within only a few years.

In addition to such worst-case impacts, there may also be surprises, events that are impossible to predict beforehand and for which no probability of occurrence therefore exists. In considering neither the entire damage distribution nor the possibility of surprises, existing estimates are clearly incomplete. This shortcoming clearly has to be highlighted when using damage estimates for policy analysis.

New Findings

New results and recent methodological advances have led to some revisions of earlier findings, although so far the broad picture has remained unchanged. Important recent developments include an increased emphasis on adaptation and on climate variability and extreme events. The importance of non-climate change-related stress factors and of integrated climate-change assessment is also increasingly stressed (Fankhauser and Tol, 1996). As a consequence of these and other scientific developments, Fankhauser and Tol (1996) identify three broad directions in damage assessment.

1. Increasing regional and sectoral differences. Recent findings stress the regional diversity of impacts. The notion that a warmer world will know winners as well as losers now features far more prominently than in the first generation of assessments.

2. Lower market impacts in developed countries. Reassessments of market-related impacts in developed countries have, in many cases, led to a reduction in expected damage compared with earlier estimates. Adjustments in estimates have occurred for a variety of reasons. One of the most important factors is the better incorporation of adaptation into impact models. Whether this trend to decreasing market impacts can be extended from industrialized countries to other regions, where the adaptive capacity is often lower, is therefore not clear.

3. Increasing importance of non-market impacts. While estimates of market impacts are often corrected downwards, new results on non-market impacts suggest that these effects may initially have been underestimated. Improvements in this area have not so much occurred with respect to the accuracy of figures – it remains low – than with respect to their comprehensiveness. Some non-market impacts that were neglected in earlier analysis for lack of data can now be quantified, in particular some of the indirect health effects, such as the climate-change-induced change in the incidence of malaria.

CONCLUSIONS

What are the implications of the above results for land-use change policy? The primary application is perhaps the use of marginal-damage figures (see Table 5.4) in project appraisal. Such figures are needed to measure the benefits associated with greenhouse-gas-mitigation projects. However, care needs to be taken when applying the estimates of Table 5.4 in this way, for at least two reasons.

First, and most importantly, available estimates are still preliminary. They are based on rather simplistic models, as noted above, and are subject to large uncertainties. Analysts are therefore well advised to back up their results with extensive sensitivity analysis. Even so, and despite the relatively wide range of results, making use of the available information may provide useful insights. For example, projects that pass a cost–benefit test based on a relatively conservative shadow price in the order of $US5–20 per tonne of carbon are likely to be worth undertaking whatever the true social costs of carbon emissions. Research reported in this volume and elsewhere suggests that a large number of land-use projects could fall within this category.

Second, it should be noted that the shadow price of carbon emissions depends on the chosen policy objective. By choosing a shadow price based on actual damage costs, analysts implicitly opt for a policy objective based on economic efficiency. Other objectives, derived, for example, from such notions as sustainable development or the precautionary principle, are also conceivable. The shadow price of carbon implied by these objectives will be different from the figures reported here and will represent the costs of the carbon constraint imposed on the economy. Anderson and Williams (1993), for example, chose as their objective the development of a carbon-free energy technology early in the next century. The result is a shadow price of carbon rising rather sharply at the rate of discount so as to bridge the difference in costs between carbon-based technologies and the cheapest back-stop.

On a related point, social-cost estimates can also shed some light on the question of carbon discounting. In the context of reforestation projects, where carbon benefits occur relatively late in time, it is sometimes argued that carbon ought to be discounted in the same way and at the same rate as other costs are. Not doing so would overemphasize costs and discriminate against projects with early carbon benefits. The counter argument is that discounting is a financial concept that cannot be applied to physical units such as tonnes of carbon. Both arguments seem correct. However, the dilemma is easily resolved by applying shadow prices. Once carbon savings are valued at their shadow price, the resulting benefit stream is expressed in monetary units and can therefore be discounted in the usual way. Costs and benefits are treated symmetrically without having to discount (physical) units of carbon. It should be noted, though, that because social costs vary over time (see Table 5.4) the outcome of this procedure will be different from discounting carbon directly (which essentially implies a constant shadow price equal to one). Because social costs rise over time, the discount rate effectively applied to (physical) carbon directly will be lower than the rate of return on capital.

As the above considerations show, the assessment of the carbon benefits of greenhouse-gas-mitigation projects requires a careful analysis of marginal climate-change damage over time. This chapter has attempted to provide an overview on the state of the art in this field.

NOTES

1 In estimates based on optimal control models, the marginal costs are calculated as the shadow price of carbon, i.e. the carbon tax necessary to keep emissions on the socially optimal trajectory (see, for example, Peck and Teisberg, 1993b; Nordhaus, 1994).

2 A descriptive approach assumes a pure rate of time preference (utility discount rate) consistent with historical savings data (usually about 3%), while a prescriptive approach sets the rate of time preference equal to zero for reasons of intergenerational equity. For a detailed discussion, see Nordhaus (1994), Arrow et al. (1996), and Price in Chapter 6 of this volume.

6 Analysis of Time Profiles of Climate Change

Colin Price

INTRODUCTION

Relationships between carbon fluxes and the discounted value of greenhouse damage are complex. They encompass profiles of carbon fixing by forests and decay of wood products; uptake by oceanic and terrestrial sinks; concentrations of carbon dioxide (CO_2) and temperature; and temperature and discounted damage. In attempting to simplify the contributing processes, economists have seriously underestimated potential costs and overstated the effectiveness of mitigation strategies. Discounting climate-change effects at commercial rates is particularly questionable, as will be shown in detail in this chapter.

While scientists and economists share the objective of assessing the importance of climate change, their approaches to the problem have been different. Scientists, as is their wont, have pursued an ever more realistic representation of complex systems. On the other hand, economists, eager to deliver results with policy indications, have simplified the science that is already known, so that it can be incorporated in tractable cost–benefit models. In this endeavour, they have been unwittingly assisted by ambiguous statements of scientists and incorrect interpretations appearing in the scientific literature.

This chapter is particularly concerned with how the benefits of mitigation strategies might develop through time. However, it does not focus on development of greenhouse-gas emissions and sinks under various policy scenarios, but rather on the profile of outcomes associated with given change in greenhouse-gas levels. This is the relevant profile in evaluating individual projects and land-use programmes. Timing is particularly important because of the effect of discounting – the process by which future costs and benefits are converted to a present equivalent according to a negative exponential relationship.

The chapter is itself not immune from the perils of compromise between theoretical accuracy and practical applicability, and it suffers from compression of complex arguments. At best, it seeks to indicate where economists must refine their models, if they are to help rather than hinder a balanced response to climate change.

CARBON FIXING: A FOREST PROFILE

For 150 years, foresters have struggled with the problem that their products, including non-market benefits, are mostly derived after a long period. The benefits of fixing carbon, achieved by young crops and continuing through life, at last offer forestry a benefit profile comparable with agricultural and industrial investment (Price, 1990).

But the profile of carbon fluxes, shown in Fig. 6.1, is by no means simple, for a number of reasons. Firstly, biomass increment increases to a

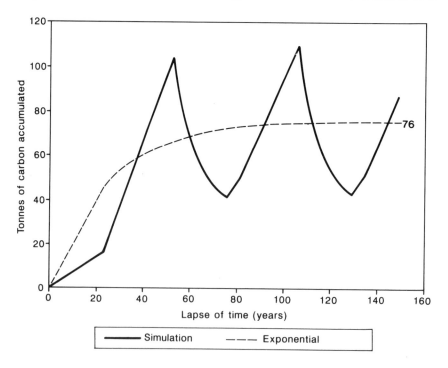

Fig. 6.1. Net carbon-fixing profile for unthinned Sitka spruce yield class 12 over two rotations.

peak, and then declines again during crop life. Secondly, successive removals of biomass result in carbon losses, as products decay or combust. Thirdly, carbon accumulates in or is lost from the soil as the crop develops. (The shape and direction of these processes depend on the soil type, and are not included in Fig. 6.1.) Fourthly, carbon fluxes may arise indirectly. These factors are now elaborated.

1. Attempts have been made to simplify the uptake/loss picture. The dashed line in Fig. 6.1 shows Pearce's (1991a) approximation: an equilibrium level, about which actual levels fluctuate, is approached according to a negative exponential function. If concern lies only with the final amount of carbon held in forests, the approximation may suffice. However, the timing of accumulation is important in discounting calculations. Pearce's formulation gives carbon fixation as maximal at the beginning of the crop's life, when actual fixation is minimum, and in rapid decline at a time when actual fixation is accelerating. Using a 6% discount rate, benefits are overestimated by a factor of 3.

2. Rates of volatilization differ between wood products. Even if each product decays exponentially, the overall profile of decay is not exponential: as Fig. 6.2 shows, early decay is dominated by fast-decaying products, later

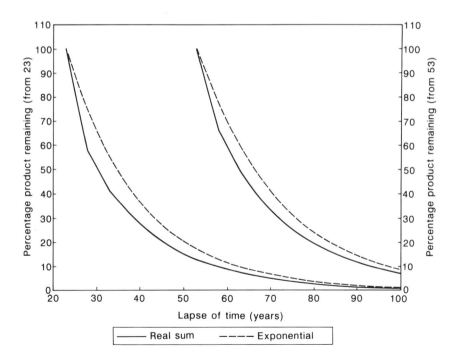

Fig. 6.2. Decay profiles for biomass for felling at 23 and 53 years.

decay by slow-decaying ones. A general figure (as in Dewar and Cannell, 1992) substantially understates early losses. Moreover, the mix of products from early harvest is less durable (especially pulpwood). Again, net early retention is exaggerated by standardized attrition profiles.

At a 6% discount rate, however, decay has little significance, compared with initial fixing. Thus, a single 53-year unthinned rotation, followed by release of all fixed carbon, has a discounted carbon-flux value only 12% less than a rotation whose products never decay. This procedure substantially exaggerates the long-term effectiveness of short-term plantation programmes.

3. Accumulation of soil carbon is expected following afforestation of mineral soils. But on peat soils short-term net fixing through afforestation may turn to long-term loss of carbon through oxidation of peat (Cannell *et al.*, 1993).
4. When afforestation causes loss of hydroelectric generation, replacement fossil energy used sometimes exceeds forest energy fixed (Barrow *et al.*, 1986); the resultant CO_2 balance may also be negative. On the other hand, inelastic demand for countryside recreation (Price *et al.*, 1986) means that locating forests close to urban centres reduces total recreational use of fossil fuels.

Afforestation improves the carbon balance only slowly. Faster results are gained by reduced-impact logging (RIL), a technique designed to retain forest biomass during selective logging (Pinard *et al.*, 1995). Power-generating companies have shown interest in RIL as a means of making rapid compensation for CO_2 emissions. However, if forest regrowth would eventually occur, the long-term difference between RIL and conventional logging (CL) may become insignificant. Figure 6.3 gives a notional comparison, showing profiles of carbon: fixed by afforestation over a rotation of 78 years; lost in CL (75% of mature stock), but recovered over 78 years; and lost in RIL (25% of mature stock), but also eventually recovered. The discount rate used determines whether the immediate benefit of replacing CL by RIL or the long-term gain of afforestation is the more valuable (Fig. 6.4). Perhaps surprisingly, the 'environmentally benign' RIL becomes less attractive at discount rates that give significant weight to the distant future.

OCEANIC UPTAKE OF CARBON DIOXIDE

Atmospheric CO_2 is taken up by the ocean, but data discrepancies show that another carbon sink exists, now believed to be terrestrial ecosystems: net uptake is attributed to recolonization of land, degraded, for example, by deforestation, and to fertilization by elevated CO_2 and nitrogen. The greater the atmospheric concentration, the faster the uptake by these sinks. Thus, an autonomous system is in place which mitigates the effects of emissions. But,

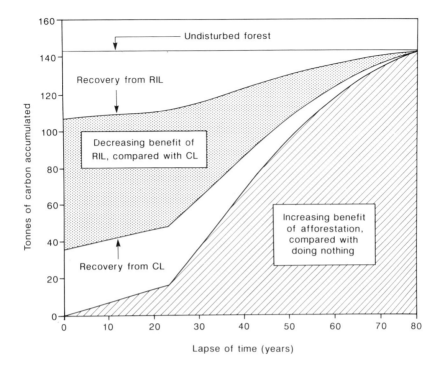

Fig. 6.3. Benefits of accumulation programmes with alternative forest-management regimes.

obversely, when programmes are implemented to remove CO_2 or to reduce emissions, the benefits of these mitigating programmes are themselves mitigated by reduced uptake by autonomous sinks.

Great confusion surrounds the long-term uptake profile. Many economists apparently believe that a substantial percentage of emissions is swiftly removed, or simply dematerializes: 'nearly half the additional CO_2 released by mankind is removed within a year' (Read, 1990); 'only about half of the emissions remain in the atmosphere beyond a decade or so' (Cline, 1992a); 'close to half [of annual CO_2 emissions] disappears somewhere' (Schelling, 1992). These quotations display a strange lack of curiosity about the processes involved. But the result is that, in economic models, the significance of CO_2 emissions – and, conversely, fixing – appears to be halved almost immediately. This mistake seems to have two origins.

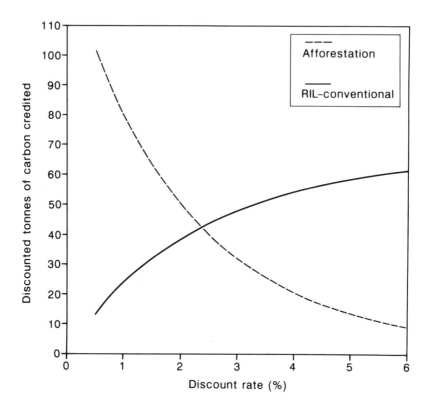

Fig. 6.4. Effect of discount rate on choice of forest-management programme.

The Emissions/Uptake Ratio: Average is Not the Same as Marginal

Statements from scientists encourage a belief in swift disappearance: 'increases in the atmospheric CO_2 concentration can account for about half of the CO_2 emissions' (Siegenthaler and Sarmiento, 1993); '46% ± 7% of the CO_2 emitted remains in the atmosphere' (Cannell, 1995). But rate of uptake by sinks results largely from historical factors (excess of atmospheric CO_2 over preindustrial levels; area of forest land recovering after exploitation). It happens that at present this historically determined rate is about half the present rate of emissions. But uptake at this rate would occur irrespective of further emissions. There is no implication that half of any *extra* tonne of carbon emitted would rapidly disappear into sinks: the consequent processes have a much slower time-scale.

Econometric Misanalysis

There have been serious errors in economists' analysis of CO_2 emissions and atmospheric concentrations. Uptake of CO_2 has been represented as a simple negative exponential relationship. This is inadequate, for three reasons. Firstly, the ocean sink for CO_2 has a greater capacity (five to six times) than the atmosphere to absorb extra emissions, but the sink is not infinite, and about 15% of any extra emissions is expected to remain in the atmosphere. In contrast, exponential uptake makes long-term atmospheric residence asymptotic to zero.

The second reason why the negative exponential is inadequate for describing CO_2 uptake is that initial uptake is dominated by a relatively rapid process, involving the surface ocean; later, a slower process, involving the deep ocean, takes over. Figure 6.5 shows two representations of the consequent atmospheric residence profile, together with two simple exponentials. 'Parallel' uptake uses the model of Maier-Reimer and Hasselmann (1987), in which emissions to the atmosphere are partitioned into five differently sized 'boxes', each losing carbon at an independent exponential rate. As the

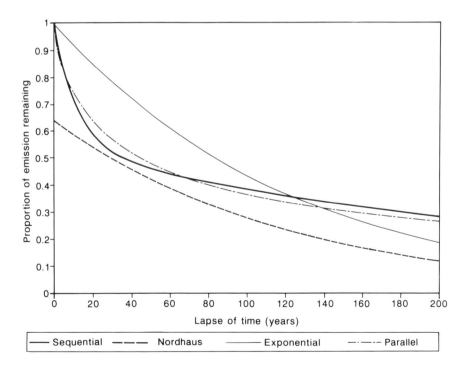

Fig. 6.5. Various representations of oceanic uptake of CO_2.

atmosphere is well mixed, this cannot be a causally accurate model. 'Sequential' is a more thermodynamically coherent, non-exponential process, in which successive transfer between three sinks tends to a long-term equilibrium.

The final complication arises because the terrestrial sink has its own characteristics and parameters.

Much economic analysis of climate-change profiles has implicitly followed the model of CO_2 uptake given by Nordhaus (1991, 1992, 1993b, 1994). This model ignores terrestrial uptake altogether; takes the capacity of the ocean sink to be infinite; and derives its single parameter from historical fluxes of CO_2, which are heavily influenced by the rapid initial uptake phase. Denied a terrestrial sink and rapid initial ocean uptake, the model 'explains' faster-than-expected falls in atmospheric concentrations as 'disappearance' (for detailed discussion, see Price, 1995). Given the widespread reliance placed on Nordhaus's analysis, it is worrying that this contravention of matter and energy conservation has gone unremarked so long.

Nordhaus's formulation underpredicts atmospheric CO_2 at all future times, but the discrepancy is particularly serious in the short and very long term. Figure 6.6 shows the profile of damage projected under Nordhaus's model and the 'sequential' uptake model. Damage is taken to be proportional to warming, an assumption reviewed later in the chapter. The following

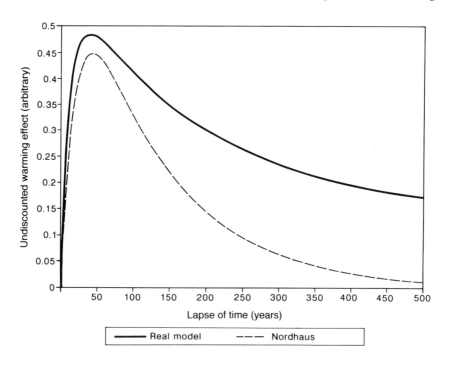

Fig. 6.6. Relative global-warming undiscounted-damage profiles.

section examines the delay in reaching maximum damage. As would be expected, under moderate discount rates Nordhaus's formulation underestimates damage costs less seriously than under high or low rates: by 24% at 10%; 18% at 4%; 34% at 0.5%; 67% at 0.1%.

THE TERRESTRIAL SINK

While oceanic and terrestrial sinks both counteract change of atmospheric carbon, they are otherwise very different in character. Oceanic sinks are essentially passive, responding to thermodynamic dictates of equalized partial pressure. Terrestrial ecosystem uptake is active, utilizing sunlight energy to work against entropy. This gives theoretical capacity to remove, over a relatively short time, all CO_2 accretions in the atmosphere, whereas passive physical uptake only removes a portion of accretions, over a long time span. On the other hand, ocean sinks have much greater total capacity. Only on a geological time-scale could biological systems remove the CO_2 expected from burning fossil fuels and kilning limestone, by the processes that originally created these resources from a CO_2-rich atmosphere.

Current terrestrial uptake rates of CO_2 are probably of similar magnitude to oceanic uptake, but this effect has so far been little quantified in long-term models. The elements due to recovery of degraded ecosystems and the effect of nitrogen fertilization are not responsive to increased CO_2 levels, and so provide no additional sink for incremental carbon fluxes. Moreover, while increased atmospheric CO_2 seems to increase *rates* of terrestrial carbon uptake, there is less certainty that in the long term there will be a higher *equilibrium* level of sequestered terrestrial carbon. This uncertainty results partly from the effect of temperature changes associated with higher atmospheric CO_2 concentrations. On the whole, terrestrial uptake seems likely to play a small role in long-term mitigation, and its relative economic significance will be low at low discount rates.

Given these factors, it would be unwise (when economic modellers do recognize the terrestrial sink) to model a 'composite sink': the two uptake processes should be modelled separately. However, the uptake is interactive: if terrestrial systems take up more CO_2, atmospheric concentrations are reduced and the oceanic system will take up less.

OTHER GREENHOUSE GASES

Carbon dioxide is not the only biologically generated greenhouse gas. Methane, with a more powerful radiative forcing effect, molecule for molecule, than CO_2, is an important product of rice production, livestock husbandry and wetland soil processes. The importance of other gases is

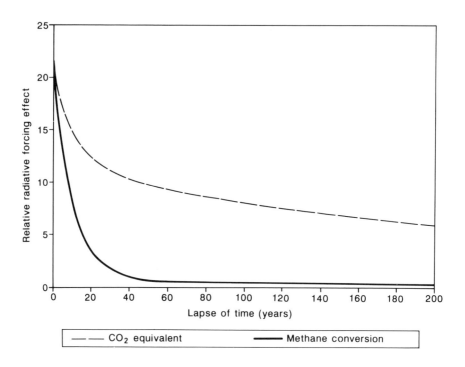

Fig. 6.7. Profiles of methane radiative forcing.

usually recorded in the scientific literature (see, for example, Houghton *et al.*, 1990) as their warming potential relative to CO_2, over a specified period. Unfortunately, from these aggregated data, the relative *discounted* effects cannot be determined (Reilly, 1992). Nordhaus (1991) states that 'this complication is of second-order importance'. Figure 6.7 suggests otherwise. It shows warming-potential profiles for a quantum of CO_2 whose immediate warming effect is equivalent to a unit of methane, and for a unit of methane rapidly oxidizing to CO_2 (water vapour is ignored here). Clearly, the long-term impact of the CO_2 equivalent is much greater. At a 10% discount rate, methane has 14 times the discounted effect of CO_2, molecule for molecule; at a 1% rate, five times; and at a 0.1% rate, two times.

BALANCING UPTAKE AND EMISSIONS: COULD FORESTS CONTRIBUTE TO A STABLE ATMOSPHERE?

Because CO_2 uptake by oceanic and terrestrial sinks is caused by historical atmospheric accretions, it is technically possible to stabilize atmospheric CO_2

by simply reducing emissions, although the indicated reduction in industrial emissions would be severe: 'The long-lived gases would require immediate reductions in emissions from human activities of over 60 per cent to stabilize their concentrations at today's levels' (Houghton *et al.*, 1990).

Allowable emissions could be enhanced by programmes of forest fixing. This has prompted a belief that indefinitely continuing emissions are compatible with a halt to climatic-change effects. The weaknesses of this view are as follows.

- Climate change would persist, even with stable atmospheric CO_2, as a result of thermal inertia (see next section).
- Oceanic uptake is driven by excess concentration in the atmosphere compared with the oceans. As uptake brings oceanic concentration closer to the stabilized atmospheric one, uptake will decline. The surplus of emissions over fixing must decline also, as in Fig. 6.8. Uptake by terrestrial ecosystems allows a further, balancing emission in the medium term, but this sink is likely to fill relatively rapidly.
- Uptake by plantations slows oceanic uptake further. Plantation uptake mitigates *emissions* tonne for tonne, but in the very long term plantations need to take up 6 tonnes for every 1 tonne reduction in atmospheric *content*.

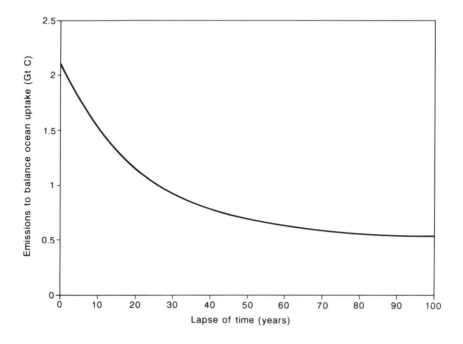

Fig. 6.8. Declining carbon uptake at stabilized concentration.

- There are physical limits to the land available for afforestation (Nilsson and Schopfhauser, 1995). Once forests reach dynamic equilibrium, there is no further offset of emissions.

The idea of balance between emissions and uptake is also misunderstood by Cline (1992a). He proposes that effects of current emissions need be valued only to AD 2275, because by then uptake will have risen to equal emissions. Despite the overall balance, however, the *incremental* effect of present emissions will still be a higher concentration of CO_2 at all future times.

THE PROFILE OF WARMING AND SEA-LEVEL RISE

The heat capacity of the surface ocean (down to around 1 km) imposes some thermal inertia on any global warming, resulting in the profile shown in Fig. 6.6. However, over hundreds of years slower processes will warm the deep ocean and melt some Antarctic ice. Thus, even with stabilized atmospheric CO_2 there would be firstly decades-long change of temperature and sea-level, followed by centuries-long further adjustment (Wigley, 1995), with rather stable atmospheric temperature but continuing sea-level rise. There may, moreover, be positive feedbacks on temperature: reduced ice-sheet extent reducing reflected radiation; reduced ice sheets and permafrost soils releasing trapped methane hydrates; warmer deep oceans boiling off dissolved CO_2. Thus, neither temperature nor sea-level adjustment can be represented as a simple negative exponential (contrast Nordhaus, 1992).

In conventional discounting terms, even the smaller thermal inertia of the surface ocean halves the significance of costs. But long-term effects have no importance at all. Losing the entire gross world product, discounted for 500 years at a 6% rate, is equivalent to losing just a few pounds now.

DAMAGE PROFILES

While global warming has some benefits, the general expectation is of negative net effects. The profile of damage is less agreed. Temperature increase is expected to be less than proportional to concentration of CO_2 but damage more than proportional to temperature increase (Cline, 1992a), with the overall effect shown in Fig. 6.9. The variation in incremental damage means that the future impact of any given present flux depends on the overall future CO_2 level.

Damage costs are of two kinds: those associated with a given state of climate (loss of agricultural production with sea-level rise, ongoing cost of

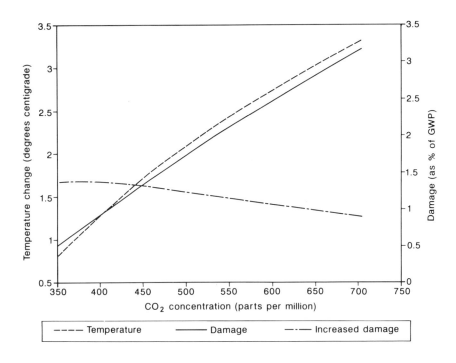

Fig. 6.9. Speculative profiles of global-warming damage with increased CO_2. GWP, global-warming potential.

maintaining sea-defence works) and those associated with adjusting to change (relocation of settlements, adoption of new production technologies). Figure 6.10 shows the relative effect of the two types of cost under regimes where climate-induced damage is increasing at 1% year^{-1}. State costs, particularly, are serious at low discount rates, adjustment costs less so. State costs are *mitigated*, but adjustment costs are *reversed* by subsequent uptake.

Most past damage estimates have been expressed as a proportion of gross world product: for example, the opportunity cost of lost agricultural land would increase with agricultural productivity. It is worth observing, however, that national economies grow at very different rates, and the sum of several positive exponential growth processes is not itself a positive exponential.

THE RATIONALE OF DISCOUNTING

While the magnitude of climate-change effects depends on physical systems and social adaptation, their importance in economic analysis depends on how they are discounted. Mitigation strategies are often seen as 'buying

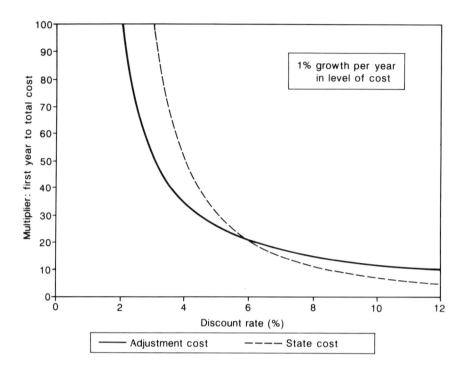

Fig. 6.10. Relative effect of costs of state and costs of adjustment (equalized at 6%).

time' – holding off the worst effects of change until a time of lesser significance. Discounting affects not only the relative importance of carbon fluxes at *different* points in time, but also the absolute price of a flux at a *given* point in time. This twofold influence makes discounting particularly important in valuing long-term mitigation strategies (Price and Willis, 1993). It has been ignored in studies that vary the discount rate without also adjusting the carbon price (Pearce, 1991a; van Kooten *et al.*, 1993). Obversely, carbon-flux values are often discounted at an unadjusted rate, even when the discount rate used to derive a carbon price was deflated by the rate of increase of gross world product.

At customary discount rates, effects in 100 years have virtually no importance, and unabated greenhouse-gas emissions seem justified (for example, it does not matter that high-carbon fuels are rescheduled for combustion next century). This has prompted re-examination of the justification for discounting (Broome, 1992; Cline, 1993; Price, 1993, 1996). The main arguments for discounting climate-change effects, and counter arguments, follow.

1. Revenues obtained early (as from extra industrial production, or carbon taxes) may be reinvested at interest to compensate for future costs, and so are more valuable than later revenues. Obversely, present costs represent an early debit against the growing compensation fund. Doubts have, however, been expressed about whether such a fund would operate in practice (d'Arge *et al.*, 1982). Significantly, carbon taxes have been welcomed (Cline, 1992a) for their capacity to replace present-day taxes on wealth, income and expenditure, which are seen to distort economic efficiency. (Having sold off public capital accumulated in past generations, in lieu of current taxation, many market-orientated governments now want to cash in the fund nominally reserved for future generations' compensation claims.)

2. Inherent preference for early consumption should be respected by democratic governments. But preference of individuals for consumption early rather than late within their own lives is irrelevant to choice between consumption in future generations' rather than the current generation's lives. Moreover, time preference, if interpreted as 'preference for now over past or future', rather than 'preference for early rather than late', provides no justification for discounting within any time frame.

3. Future generations will be richer, so monetary and opportunity costs of climate change will have diminished marginal utility. This factor acts in a contrary direction to the increased significance of climatic-change costs, calculated as a percentage of gross world product. Note, however, that it is consumption per head that determines the marginal utility discount. Even if consumption grows faster than population, discounted damage costs may increase, as Fig. 6.11 shows. Moreover, discounting based on the mean characteristics of two (or all) nations understates the importance of nations with slower-growing consumption per head: the 'average' negative exponential in Fig. 6.11 completely fails to identify this effect, which soon increases aggregate importance of climate-change costs with lapse of time. The caution against using world aggregate statistics should be plain.

However, these refinements are minor, compared with the extraordinary shift of ground whereby discounting justified by diminishing marginal utility is applied to the total basket of consumption and experience encompassed in a human life. Thus, loss of the entire present population of the world, 400 years in the future, has the same significance as the loss of one life now, at a 6% discount rate.

4. Technological change brings prospects of carbon-free fuels; it facilitates adaptation to new conditions, making them less burdensome. In this context, 'buying time' is entirely sensible. (It is, however, important not to double-count by inserting specific technological mitigation and then discounting the mitigated costs.) But optimistic technological assumptions

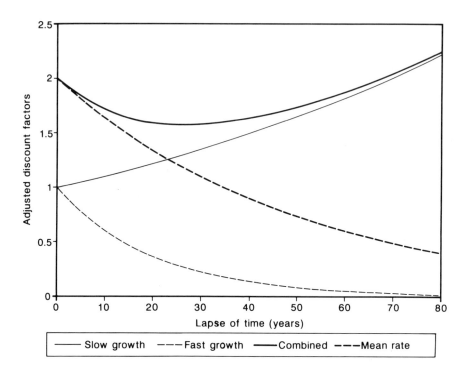

Fig. 6.11. The illegitimacy of discounting based on aggregated growth parameters. Elasticity of marginal utility of consumption = –2; growth of gross national products (GNPs) = 6% and 2%; growth of rates of populations = 0% and 1.5%.

have not always been correct. In 1955, von Neumann (1955) predicted that 'within a few decades' energy would be free. Climate change is among numerous unforeseen limitations that have obstructed realization of that expectation.

CONCLUSIONS

Time profiles are important in every process relating land use to the values associated with climate change. Economists often represent these profiles by combined negative exponential functions. However, illustrative material in this chapter shows that functions should not be aggregated or averaged, either for the cumulative effect of sequential processes, such as the uptake by sinks, or in the addition of contemporary processes, such as the decay of wood products. Some errors of simplification are important at high rates (the profile of forest fixing), others, at low rates (oceanic uptake). The alternative of constructing more detailed simulation models is more expensive in time and computer space; but some of the simplifications in economic analysis of climate change, such as oceanic uptake and standard discounting, have been dangerously misleading.

In the long term, the options for mitigation are limited. Terrestrial sinks, natural or created, have restricted capacity. Oceanic sinks leave a permanent residue of emissions in the atmosphere. We may be able to accelerate rates of uptake, but the consequent equilibrium position is not at our command. Most discussed mitigating strategies therefore offer only a short-term solution or a reduction of impact. In contrast, potential emissions are limited only by the reserves of carbon in fossil fuels and limestone. Releasing all these would return the atmosphere to its state prior to the emergence of carbon-fixing life-forces – an atmosphere lethal to present-day life forms. Unless we believe that discounting makes it acceptable to ignore future problems, we have two viable mitigating strategies: to replicate the geological-time-scale processes which fixed this carbon, or else to desist from releasing the carbon now locked up.

7 Public Policies and Incentives to Accelerate Irreversible Green Investments

Cesare Dosi and Michele Moretto[1]

INTRODUCTION

Uncertainty and irreversibility are two crucial issues when dealing with global environmental risks, such as anthropogenic climate changes induced by greenhouse-gas (GHG) emissions. As far as uncertainty is concerned, apart from the general agreement that continuous inflow of GHGs is likely to result in an increase in atmospheric temperature, there is no consensus about how significant global warming will be, how rapidly it will occur, what its local, regional or global impacts will be, or how the costs (or benefits) from environmental changes will be distributed across the world.

Apart from the lack of scientific knowledge, uncertainty arises from the intrinsic nature of the phenomena under consideration. In fact, on the one hand, the dynamics of climate change is either stochastic or chaotic and therefore intrinsically unpredictable (Chichilnisky *et al.*, 1996). On the other hand, the risks under consideration are not exogenous, but are affected by our actions or inactions; since the relationships between human activities and atmospheric warming, as well as the ecosystem's and the socio-economic response to climate change, are either far from being fully understood or intrinsically unpredictable, it is difficult not only to predict if and when climate change will occur but also to evaluate its ultimate impacts on future human welfare.

In this maze of uncertainty, since expected costs following possible rises in the sea-level, desertification, species extinction and other detrimental impacts largely exceed the potential benefits from climatic change (such as development of new activities in today's cold regions), curtailing GHG immediately would undoubtedly be the best and most rational choice, at least from the global point of view, if pollution abatement were relatively inexpensive. However, abatement certainly requires changes in lifestyle and economic activities, which would involve socio-economic costs, at least in the short and medium term.

Since GHG abatement costs, although difficult to assess, are almost certain, while the benefits are still uncertain, it would be prudent to wait for some resolution of the uncertainties surrounding anthropogenic climatic change and related impacts, before undertaking irreversible and expensive abatement programmes. However, waiting means that we continue to accumulate GHG, with consequences that could be irreversible once they have occurred, at least on a relevant time-scale. In other words, the irreversible costs of pollution abatement and the potentially irreversible environmental impacts of business-as-usual practices imply a challenging trade-off between preserving and reversing options, i.e. between the value of postponing and the value of undertaking pollution abatement immediately.

Although there are authors (see, for example, Manne and Richels, 1992; Kolstad, 1993) who have noted this trade-off, in general, when considering the risk of irreversible ecosystem response to pollutant inflow, the value of controlling pollution immediately, rather than the value of postponing it, has been more frequently emphasized, in both environmental economic literature and public debate. The rationale behind this is quite obvious: 'if we learn that the ecosystem is not at a threshold of irreversible damage, we can always resume pollution later; but if we do not control now and observe irreversible changes in the ecosystem, we cannot undo them by controlling later' (Fischer and Hanemann, 1990, p. 400).

In this chapter, on the other hand, instead of focusing on the social value of the control-pollution-now option, we concentrate on the value – in particular, the private value – of postponing abatement measures involving sunk costs ('irreversible green investments'). This does not imply that we disregard or undervalue the social benefit of curtailing pollution immediately when facing the risk of irreversible environmental changes. Rather, our aim is to draw attention to an issue that has been somewhat overlooked in environmental-economics literature, namely, the need for an appropriate regulatory framework when policy-makers – who wish to accelerate abandonment of polluting technologies, so as to avoid possible long-run irreversible damage – are faced with private agents who – on the other hand, believe that postponing green investments is the way to preserve flexibility.

The following stylized analysis is carried out with reference to a single firm, facing the decision whether or not to switch from a polluting technology to a 'green' one, through, for example, adoption of an energy-saving technology or adoption of agricultural and forestry practices able to increase carbon fixation. The technological switch is assumed to be irreversible in that it requires a sunk capital cost.

To clarify the nature and the implications of the private (option) value to delay an irreversible technological change, we consider a quite optimistic, though not too unrealistic, scenario, by assuming that, besides the social benefits arising from curtailing emissions, the technological change under consideration is also expected to provide private benefits.[2] However, the actual increase in the firm's future profits is uncertain; formally, we assume that the green investment's additional cash flow per unit of time is driven by a diffusion process.

To keep the analysis as simple as possible, while the private switching timing is assumed to be stochastic, because of the volatility of the investment's private benefits, the pollutant accumulation process is assumed to be known and deterministic.[3]

We shall first examine the optimal private timing of technological change without public intervention, when the investment timing simply reflects the innovation's private benefits, without taking into account the technological switch's social benefits. We then introduce public intervention, which is based on the need to internalize within private decisions the social cost deriving from the firm's continued use of a polluting technology. More specifically, two alternative 'policy scenarios', reflecting different public-authority perceptions of this social cost, will be considered.

The first scenario is built around the assumption that, although the policy-makers recognize the existence of social damage deriving from accumulation of pollutants, they do not feel that such accumulation may reach a 'critical' level, for example a level at which irreversible damage and a dramatic fall in social utility might occur. In this case, the objective of public intervention consists of speeding up technological change, not so much as to prevent a specific, predetermined pollution threshold from being reached, but simply to reduce the social damage deriving in any case from pollutant emissions.

The second scenario, on the other hand, is built around the assumption that the planners are particularly concerned to ensure that a critical pollution threshold is not exceeded. In other words, it is assumed that the planners consider the presence of pollution to be to some extent socially tolerable, and hence take no corrective action, as long as the stock of pollutants is not expected to reach a critical level at which they feel that serious, irreversible damage could occur, with a notable loss of social utility.

In both cases, the purpose of public intervention is clearly to make the private timing of technological change compatible with a social-welfare function that would also take account of environmental externalities. However, whereas in the former case the regulators will in any case act to internalize the social cost of pollution, in the latter case they will attempt to speed up technological change only if they feel that the spontaneous timing of technological change is incompatible with the objective of preventing a critical pollution threshold level from being exceeded.

In a later section, we shall consider these two 'policy scenarios'. The aim is to identify the general properties of environmental policy tools, in line with the two above-mentioned public attitudes and perceptions of the social costs of pollutant accumulation. The analysis will be carried out only with reference to non-coercive regulatory measures, namely public subsidies aimed at reducing private switching costs. However, with appropriate arrangements, the results can be extended to other policy instruments as well, such as taxes or penalties for non-compliance, with 'command-and-control' regulatory measures.

THE OPTIMAL PRIVATE TIMING OF GREEN INVESTMENTS

Consider a firm[4] using a constant-returns-to-scale technology – involving pollutant emissions accumulating over time according to a deterministic process – which has a monopoly right to switch to a 'zero-emission' technology. Throughout the chapter we assume that the firm's rate of emissions per unit of time cannot be significantly reduced without adopting the green technology, at a sunk capital cost K. Consequently, should the technological switch take place, pollution would not increase and would eventually decline through self-purification processes.

Besides improving environmental quality, we assume that the technological switch is expected to increase the firm's cash flow. We assume that the firm's additional cash flow per unit of time (π_t) is described by an arithmetic Brownian motion:[5]

$$d\pi_t = \alpha dt + \sigma dB_t, \quad \text{with } \pi_0, \alpha, \sigma > 0 \tag{1}$$

where dB_t is the increment of a standardized Wiener process, satisfying the condition of zero mean, $E(dB_t) = 0$ and variance $E(dB_t^2) = dt$.[6] We can consider eqn (1) as the infinitesimal random increment of benefits $d\pi_t$ over the infinitesimal time dt.[7]

Using an exogenously specified discount rate ρ, the investment's expected present value is defined by:

$$V(\pi_T) = E_T \left\{ \int_T^\infty \pi_t e^{-\rho(t-T)} dt \right\}$$

where T is the green technology's starting time. By eqn (1), starting from the initial value π_T, the random position of the process π_t at time t has normal distribution, with mean $\pi_T + \alpha_t$. Then a simple integration gives the following (Harrison, 1985, p. 44):

$$V(\pi_T) = \frac{\alpha}{\rho^2} + \frac{\pi_T}{\rho} \tag{2}$$

Let us now evaluate the option value to adopt the green technology, F. Since the return from introducing the green technology at time T is simply $V(\pi_T) - K$, the option value can be evaluated through maximization of the following expected present value:

$$F(\pi) = \max E_0[(V(\pi_T) - K)e^{-\rho T}/\pi_0 = \pi]$$

However, as the investment opportunity yields no cash flow up to time T, when the green investment is undertaken, the only return from keeping this opportunity alive is the option value's capital appreciation ('capital gain'). By imposing the arbitrage condition (Bellman equation) that the capital gain has to be equal to the natural return, over an infinitesimal period dt, when it is optimal not to invest, we get:

$$\rho F(\pi)dt = E[dF(\pi)]$$

Expanding dF and applying Itô's lemma yields the following differential equation (Dixit, 1993):

$$\frac{1}{2}\sigma^2 F''(\pi) + \alpha F(\pi) - \rho = 0 \qquad (3)$$

Imposing that, if π goes to $-\infty$, the opportunity value to invest must go to zero, the general solution of eqn (3) becomes:

$$F(\pi) = Ae^{\beta_1 \pi}, \qquad \text{for } \pi \in (-\infty, \pi^*], \qquad (4)$$

where A is a constant to be determined, and β_1 is the positive root of the following quadratic expression:

$$\Phi(\beta) \equiv \frac{1}{2}\sigma^2\beta^2 + \alpha\beta - \rho = 0$$

Since eqn (4) represents the option value to switch optimally in the future to the green technology, the constant A must be positive and the solution is valid over the range of π for which it is optimal to keep the option alive $(-\infty, \pi^*)$. Furthermore, both the constant A and π^* are obtained from consideration of optimal investment. First, when switching, the firm must be indifferent between keeping the option alive (i.e. being stuck with the polluting technology) and exercising it (i.e. adopting the green technology at the sunk cost K); this is the matching value condition:

$$F(\pi^*) = V(\pi^*) - K \qquad (5)$$

Second, at π^*, any possible arbitrary exercise of the option at a different point should be ruled out; this is given by the smooth-pasting condition:

$$F'(\pi^*) = V'(\pi^*) \qquad (6)$$

By substituting eqn (4) in eqn (5) – eqn (6), we get:

$$\pi^* = \left(\frac{1}{\beta_1} - \frac{\alpha}{\rho}\right) + \rho K \qquad (7)$$

The optimal private trigger value, π^*, indicates the green investment's innovation benefits (per unit of time) for which the firm will find it profitable to abandon the polluting technology. In other words, the firm will abandon the polluting technology and adopt the green one the first time π, randomly fluctuating, hits the upper threshold level π^*.

Notice that π^* is greater than the usual investment's rental price, ρK. In fact, the term $\left(\frac{1}{\beta_1} - \frac{\alpha}{\rho}\right) > 0$ (the option value factor) indicates the difference the firm requires before making an irreversible investment providing uncertain benefits. Moreover, as $\frac{\partial \beta_1}{\partial \sigma} < 0$, an increase in uncertainty over future realizations of π leads to an increase in the wedge between π^* and the rental price ρK, with a further increase in waiting time before adoption of the green technology.

Finally, as far as the timing of technological change is concerned, as the innovation's benefits are driven by the stochastic process of eqn (1), the switching time $T = \inf(t > 0 | \pi_t = \pi^*)$ at which π will first reach the upper trigger level π^* also becomes a stochastic variable, holding the following probability density function (Cox and Miller, 1965, p. 221):

$$f(T | \pi_0, \pi^*) = \frac{\pi^* - \pi_0}{\sigma\sqrt{2\pi T^3}} \exp\left[-\frac{(\pi^* - \pi_0 - \alpha T)^2}{2\sigma^2 T}\right] \text{ for } T > 0 \qquad (8)$$

with $E(T) = \frac{\pi^* - \pi_0}{\alpha}$ and $V(T) = \frac{(\pi^* - \pi_0)\sigma^2}{2\alpha^3}$ as the first and second moments, respectively.

THE RATIONALE FOR PUBLIC INTERVENTION AND INNOVATION GRANTS

So far, the analysis has been carried out assuming the absence of public intervention. If and when the green technology is adopted thus depends exclusively on the private sunk-capital cost and the firm's expectations regarding the possibility of obtaining adequate returns from the investment in the market-place. In other words, without public intervention, private investment decisions, in particular the timing of environmental innovation, will simply reflect the innovation's private benefits, without taking into account its social benefits (avoidance of pollutant emissions). The rationale for public intervention is thus based on the need to internalize within private decisions the social benefits deriving from abandonment of a polluting technology (or, taking a different perspective, internalizing the social cost deriving from extended use of a polluting technology).

As anticipated, the following analysis will be carried out with reference to two alternative 'policy scenarios', which in some ways may be considered representative of two opposing, though rather common, attitudes in environmental policy-making. The first situation (case 1) is that of a public decision-maker who, though recognizing the existence of social damage deriving from accumulation of pollutants, does not however feel that such accumulation might reach a 'critical' level, that is a level entailing irreversible consequences. In this case public intervention is aimed at speeding up technological change, not so much as to prevent a given level of pollutant accumulation from being reached, but rather to reduce the social damage from pollution which is thought to occur disregardless of whether a critical level is reached. In the second situation (case 2), a public decision-maker is more concerned with the damage that may derive from exceeding a specific pollution threshold, rather than reducing the social damage deriving from pollution.

Case 1: The Social Optimal Switching Rule and the 'Pigouvian Subsidy'

Recalling that current technology involves pollutant emissions while the 'green' one – a 'zero emission' technology – does not involve social damage, we define:

$$D(\pi) = E\left\{\int_0^T se^{-\rho t}dt \mid \pi_0 = \pi\right\} = \frac{s}{\rho}\left[1 - E\left(e^{-\rho T} \mid \pi_0 = \pi\right)\right] \tag{9}$$

where s represents the (say, annual) social damage, calculated with reference to a constant (annual) rate of use of the polluting technology.[8] This damage will disappear only at time T, that is, when the technological change takes place. For the sake of simplicity, in eqn (9) we assume that the social-damage function is additive, so that the total damage is equal to the sum of the annual damage attributable to the emissions generated under the current technology.

Under the above hypothesis, the social value of the environmental-innovation option, i.e. the value of waiting before adopting the green technology, becomes:

$$\hat{F}(\pi) = F(\pi) - D(\pi) \tag{10}$$

Then, by imposing an arbitrage condition and transforming both components of eqn (10) into differential equations similar to eqn (3) with boundary conditions:

$$\hat{F}(-\infty) = -\frac{s}{\rho} \tag{11}$$

$$\hat{F}(\hat{\pi}^*) = V(\hat{\pi}^*) - K \tag{12}$$

$$\hat{F}'(\hat{\pi}^*) = V'(\hat{\pi}^*) \tag{13}$$

and, going through the same steps as in the previous section, we derive the optimal social trigger value $\hat{\pi}^*$. That is:[9]

$$\hat{\pi}^* = \left(\frac{1}{\beta_1} - \frac{\alpha}{\rho}\right) + \rho\left(K - \frac{s}{\rho}\right) = \pi^* - s \tag{14}$$

As long as $K - \frac{s}{\rho} > 0$, the green technology should be adopted earlier than the firm would spontaneously do, i.e. $\hat{\pi}^* < \pi^*$. In order to ensure that the private timing of technological change is compatible with the social goal of improving environmental quality, the regulator should intervene in such a way as to fill the gap between the private and the social trigger value. By the linearity of the process in eqn (1), such a difference is simply s, which then represents the 'Pigouvian subsidy' (per unit of time) determined with reference to the operating benefits which would induce the firm to abandon the polluting technology as soon as π hits $\hat{\pi}^*$.[10] Notice that, instead of continuously providing subsidy s, the regulator could offer a *una tantum* green investment grant, equal to $\frac{s}{\rho}$.

Numerical solutions

Some numerical solutions will help illustrate the above results. Let us normalize the capital cost $K = 1$, and set $\pi_0 = 0$, the annual payout $\alpha = 0.05$, the discount rate $\rho = 0.1$, and $s = 0.0075$.[11] Table 7.1 shows the option-value factor and the expected timing of the investment for both the private and social optimum.

Table 7.1. Simulated private and social trigger values.

σ	$\frac{1}{\beta_1} - \frac{\alpha}{\rho}$	π^*	$E(T)$	$\hat{\pi}^*$	$\frac{s}{\rho}$	$E(\hat{T})$
0	0	0.1000	2	0.0925	0.075	1.850
0.1	0.0855	0.1855	3.710	0.1780	0.075	3.560
0.2	0.2624	0.3624	7.247	0.3549	0.075	7.098
0.3	0.4659	0.5659	11.318	0.5584	0.075	11.168

Case 2: Innovation Grants Under a (Preselected) Pollutant-accumulation Target

Unlike case 1, let us now suppose that the public authorities have identified a 'critical' pollution threshold, $\bar{P}$, and wish to keep the actual pollutant stock below the level at which either a temporary significant reduction in social utility or an irreversible breakdown in the ecosystem's assimilative capacity is expected to occur.[12] Assuming, again, a deterministic pollution-accumulation process, the policy-maker is also able to identify the date ($\bar{T}$) beyond which, without technological change, the stock of pollutants would exceed $\bar{P}$.[13]

The need for public intervention arises every time the date at which the firm would find it appropriate to abandon the polluting technology goes beyond $\bar{T}$. This comparison is easy (at least, as long as there is no asymmetry of information) if the investment's private benefits are not affected by uncertainty; in this case, in fact, it is possible to identify with certainty the optimal private switching time and hence compare it with $\bar{T}$ (see case 2a).

The situation is different when private benefits are uncertain: in this case, since the private switching time becomes a stochastic variable, comparison with the policy-maker's desired time becomes meaningless and the regulator has to identify a policy rule referring to T's probability distribution (see case 2b).

Case 2a: the innovation grant under certainty
As a point of reference, let us first consider the case where the firm does not face uncertainty about the green investment's private benefits ($\sigma = 0$). Without uncertainty, eqn (1) can be easily integrated and the optimal time of innovation would be:

$$T_c = \frac{1}{\alpha}[\pi_c^* - \pi_0] = \frac{1}{\alpha}[\rho K - \pi_0]$$

Therefore, the required capital-cost subsidy is simply given by:

$$S_c = K - K_c = \frac{\alpha}{\rho}(T_c - \bar{T}) \tag{15}$$

That is, the subsidy should cover the difference between the actual capital cost (K) and the cost which the firm would be willing to invest at time $\bar{T}(K_c)$. Obviously, if $T_c \leq \bar{T}$, $S_c = 0$.

Case 2b: the innovation grant under uncertainty
Let us now go back to the uncertainty case. As we have seen, uncertainty about the technological switch's private benefits makes the private switching time a stochastic variable. Consequently, the regulator has to identify a policy rule referring to the probability distribution of eqn (7). To keep the analysis as simple as possible, we assume that the regulator adopts the following simple criterion:

$$E(T) - \bar{T} = m, \text{ with } m \leq 0$$

where m is a predefined constant which takes on the sense of a 'safeguard interval'. In other words, we assume that, when defining his/her lines of intervention, the regulator aims at the objective of ensuring that, at least in terms of expected value, the technological switch will take place m periods before the date at which the level of pollution $\bar{P}$ is reached.[14] By eqns (7) and (8), the following proposition holds.

Proposition 1: under the criterion $E(T) - \bar{T} = m$, the optimal capital subsidy is:

$$S_u = K - K_u = K - K_c + \frac{\alpha}{\rho}m + \frac{1}{\rho}\left(\frac{1}{\beta_1} - \frac{\alpha}{\rho}\right) > S_c \qquad (16)$$

In other words, if the firm faces uncertainty over the irreversible innovation's benefits, the public subsidy required to bridge the gap between the (expected) private switching time and $(\bar{T} - m)$ should be such as to compensate it for giving up the option to delay the investment to wait for new information. As, depending on m, the subsidy moves the density function $f(T|\bar{T},m)$ to the left, the probability rate of failing to induce the firm to innovate before $\bar{T}$ reduces. Obviously, if $E(T) - \bar{T} < m$, the regulator considers the current failure rate socially satisfactory and no subsidy will be granted.

Finally, as an increase in uncertainty increases the option to delay the investment, $\frac{\partial S_u}{\partial \sigma} > 0$.

Numerical solutions
As well as the values of the parameters already adopted above, let us suppose $\bar{T} = 2$ (years) and $m = 0$. In Table 7.2, we show, for different values of σ, the private trigger value (π^*), the expected private time of the technological switch $E(T)$ and the required subsidy to meet the policy objective $E(T) - 2 = 0$.

Given the values hypothesized for the various parameters, if there were no uncertainty on private benefits ($\sigma = 0$), the firm would adopt the new technology in two periods; in this case, since $T = \bar{T}$, $S_c = 0$. If, on the other hand, private benefits were uncertain ($\sigma > 0$), the expected private timing would be incompatible with the above-mentioned policy goal and a subsidy would have to be granted. In particular, since $E(T)$ increases with increased uncertainty, the amount of the subsidy S_u required to bridge the gap

Table 7.2. Simulated private trigger values and subsidies inducing technological switching before $\bar{T} = 2$.

σ	π*	E(T)	S_u
0	1.000	2	0 (S_c)
0.1	0.1855	3.710	0.855
0.2	0.3624	7.247	2.623
0.3	0.5659	11.318	4.659

between $E(T)$ and $\overline{T}$ is positively correlated with σ. Moreover, it is worthwhile to point out the non-linearity characterizing the relationship between uncertainty and the increase of σ, i.e. S_u increases at a higher rate with respect to σ.

CONCLUSIONS

It is well known that, if its rentability is uncertain, making an irreversible investment has an opportunity cost, represented by giving up the option of waiting for new information about its profitability. In other words, there is a premium on being stuck with, say, a polluting technology, in that this preserves future flexibility and does not preclude exploitation of additional information about, say, a green technology's private benefits.

It follows that, even when green technologies are theoretically profitable, uncertainty about private benefits is likely to induce potential adopters to delay environmental innovation. As a consequence, agents may go through a transition period, entailing continuing pollutant accumulation, which public authorities may consider socially undesirable, especially when there is a risk of irreversible environmental impacts.

If public authorities wish to induce private agents to reduce the transition period from polluting to green technologies, the private premium on being stuck with the current technology should not be ignored, especially when firms face irreversible investment decisions. In other words, public incentives should be designed so as to either lower the private value of postponing technological change, such as by supporting production under the green technology or, equivalently, by penalizing production under the polluting one, or directly finance the green investment, in such a way as to compensate firms for giving up their option to wait for new information about the investment's private rentability.

Needless to say, non-coercive policy instruments, such as environmental-innovation grants, do not guarantee that the private timing of technological change will actually be compatible with the 'socially desirable' one. Nevertheless, appropriate public incentives designed to account for the investment irreversibility and firms' uncertainty about private benefits may contribute to reducing the probability that technological change will occur after the policy-maker's desired date.

NOTES

1. We wish to thank Davide Pettenella for useful suggestions.
2. According to an (increasing) number of successful experiences, environmental innovation not only may not conflict with, but is often complementary to, the

private objective of improving productivity and enhancing industrial competitiveness; this occurs because the initial investment required to abandon a polluting technology can be more than offset by lower production costs (material or energy savings, better utilization of by-products, etc.) or higher revenues, due to the widespread demand for a better environment, which, in affluent societies, consumers express not merely in the ballot-box but also through an increasing willingness to pay for 'clean' products (Dosi and Moretto, 1996a).

3   For the purpose of the following analysis, assuming that even the pollutant-accumulation process is stochastic would complicate the model, but would add few insights.

4   The following analysis about the optimal private timing of technological change can be extended to a competitive industry consisting of a large number of small (infinitesimal) identical firms, each employing a standardized constant-returns-to-scale polluting technology (Dosi and Moretto, 1997).

5   For an introductory treatment of Brownian motions, see Dixit (1993) and, for a more in-depth development, Cox and Miller (1965) and Harrison (1985).

6   Equation (1) implies that π may be negative, that is, at some point in time, the polluting technology might be more profitable than the green one (e.g. because of an exogenous shock involving an increase in the relative cost of the variable inputs used under the green technology).

7   Although different processes (e.g. a geometric Brownian motion) are harder to solve analytically, they yield similar results (see, for example, Olsen and Stensland, 1992). In fact, only the stated dB_t's properties are crucial for the following analysis.

8   Yet, as, in general, emissions depend on the rate of use of polluting technology and hence, indirectly, on the operating profits of the technology itself, in eqn (9) we have hypothesized a profit-inelastic rate of use.

9   Formally, the term $E_0(e^{-\rho T}|\pi_0 = \pi)$ in the social damage represents the Laplace transformation of the first passage time to a single barrier for an arithmetic Brownian motion. It yields:

$$e^{\beta_1(\pi - \hat{\pi}^*)} = E_0 \int_0^\infty e^{-PT} f_0(T) dT$$

where the density, $f_0(T)$, of the first time at which the benefit process π hits $\hat{\pi}^*$, conditional on information at time zero, is given by eqn (8). For more on the optimal-control problem of Brownian motion with terms like $E_0(e^{-\rho T}|\pi_0 = \pi)$, (see Harrison (1985) and Moretto (1995).

10   The fact that the 'Pigouvian subsidy' is not influenced by the infinitesimal variance σ^2 is due only to the linearity of π. A different process, e.g. a geometric Brownian motion, would yield a variance-dependent cash-flow regulation. See Dosi and Moretto (1996b).

11   Such values, which only serve to illustrate the above theoretical results, have been borrowed from Dosi and Moretto (1996c).

12   $\bar{P}$ could be a pollutant stock at which either a temporary significant reduction in social utility or an irreversible breakdown of the ecosystem's assimilative capacity is expected to occur.

13 Following an approach quite commonly adopted in environmental economic literature, we could assume the following pollutant-accumulation process:

$$\dot{P} = L - \eta(P)$$

where L stands for the pollutant flow per unit of time and $\eta(P)$ the assimilative function. Once $\dot{P}$ has been identified, by integrating the above accumulation function, public authorities may identify $\bar{T}$.

14 This policy target might be replaced by alternative criteria. For example, the regulator might determine the grant by referring to both the first and second moments of T's distribution or by fixing a lower boundary for the probability rate of failing the objective of inducing the technological change before $\bar{T}$. This would complicate the model, adding few insights.

Human Adaptation in Ameliorating the Impact of Climate Change on Global Timber Markets

8

Brent Sohngen, Roger A. Sedjo,
Robert Mendelsohn and Kenneth S. Lyon[1]

INTRODUCTION

In this chapter, we discuss the role of humans in adapting to large-scale ecological adjustments implied by global climate change. Three important elements are recognized and addressed: spatial issues, dynamic issues and the global scope. Spatial issues are captured with existing steady-state climate and ecological models. A dynamic climate and ecosystem adjustment pathway is then proposed. Human adaptation to these changes is captured with a dynamic economic model of global timber markets. This model will allow us to measure the value of the ecosystem change and to assess carbon fluxes during critical transition periods.

To date, most climate-change research attempts to estimate the likely changes in atmospheric conditions that will result from an increase in the level of carbon dioxide (CO_2) in the atmosphere. This research has focused on both the scientific evidence for climate change and the models that can be used to predict the environmental impacts by Houghton *et al.* (1990). While many have suggested that climate change will irreparably damage the globe, only recently have researchers begun to consider the economic consequences of climate change (Nordhaus, 1991; Cline, 1992a).

Estimating the economic impacts of climate change brings up several important issues. First, although climate change will affect everybody in some way, global circulation models (GCMs) and ecological models predict that climate change will have a different impact on each region. Economists thus must develop tools to capture the important spatial components of climate change. Second, there is a strong dynamic component to climate change.

© CAB INTERNATIONAL 1997. *Climate-change Mitigation and European Land-use Policies* (eds W.N. Adger, D. Pettenella and M. Whitby)

The IPCC's (Houghton *et al.*, 1990) most likely case suggests that climate change will occur slowly over time, as CO_2 levels slowly increase in the atmosphere. It is entirely possible that discrete changes in climate could occur, but these too exhibit temporal components. Third, climate change will have an impact on the entire globe. To understand what effect it will have on global-level and country-level markets, the geographical scope of enquiry must include the entire Earth.

In this study, we address these three important elements in assessing the economic impact of climate change on forests: the spatial detail, the dynamic components and the global scope. By addressing these issues specifically, we show how ecosystems and humans adjust and adapt during periods of large-scale change. Human adjustment and adaptation are likely to minimize the global damage from climate change, as suggested in a study of US timber markets by Sohngen and Mendelsohn (1996).

The biophysical changes are described by steady-state models that predict equilibrium responses of climate and ecosystems to doubling the CO_2 in the atmosphere. Steady-state analysis begins with GCMs, which predict large-scale adjustments in climate patterns across the globe. Global circulation models provide input into steady-state ecosystem-change models, which predict adjustments in the distribution and net primary productivity of ecosystems. Although these models are steady-state, we propose a dynamic adjustment pathway for the climate and ecosystem change.

A global timber-market model must then be used to provide for human adaptation and to value these changes. The economic model is based on the earlier work of Sedjo and Lyon (1990) and Sohngen and Mendelsohn (1996). These two earlier models were limited by their geographical scope from capturing impacts in all regions of the globe. Here, we incorporate information from every continent, so that we may value all potential changes. Although we shall not present the empirical results of our work at this point, in discussing the model we shall suggest ways that human adaptation will affect the transition during climate change.

STEADY-STATE MODELS

Steady-state models provide the basis for the biophysical changes we seek to measure and value. They predict the equilibrium adjustment of climate and ecosystems to instantaneous changes in the level of CO_2 in the atmosphere. We focus on two types of ecosystem models: (i) those that predict changes in the distribution of ecosystems (biogeographical distribution models); and (ii) those that predict changes in the productivity of those ecosystems (biogeochemical cycle models).

Global Circulation Models

The current distribution of climate variables in 0.5° × 0.5° grid cells across the globe serves as our baseline climate. Global circulation models are used to predict steady-state values for climatic variables for the same 0.5° × 0.5° grid cells across the globe contingent on a doubling of CO_2 in the atmosphere. The gridded nature of these data allows us to capture variation in climate change across the globe.

Biogeographical Distribution Models

Biogeographical distribution models predict the distribution of ecosystem types based on these climate data. These models use mechanistic rules to classify spatial vegetation patterns according to climatic variables (Emanuel et al., 1985; Monserud and Leemans, 1992; Nielson et al., 1992; Prentice et al., 1992; Cramer and Leemans, 1993; Tchebakova et al., 1993; Nielson, 1995; Woodward et al., 1995). Although they each depend on different algorithms and contain different levels of detail, these models are all based on the idea first presented by Holdridge (1947), that climate regulates the types of ecosystems that are able to exist in an area.

These models are useful because they allow us to compare the current distribution of ecosystem types with potential future distributions based on climate change. For many of the world's timber-production forests, the regional distribution of timber types is related directly to the range of ecosystem types. Thus, as climate changes and the boundaries of ecosystem types adjust, the distribution of timber types will change as well.

Biogeochemical Cycle Models

Biogeochemical cycle models predict net primary productivity (NPP) for the ecosystems described above (Melillo et al., 1993). Net primary productivity is the net amount of carbon available for plant growth from photosynthesis in any given period. Of all the carbon fixed by photosynthesis, NPP represents that part which is not used by respiration, but which can be used by plants for growth. Changes in timber growth are inferred from a baseline and doubled CO_2 prediction of NPP.

DYNAMIC MODELS

Unfortunately, the steady state considered above is likely to occur many years in the future, because of the slow changes in climatic variables (Houghton *et al.*, 1990) and lags in the response of ecosystems to these slowly evolving climatic conditions. Rather than assuming that ecosystems change instantly, we implement both a dynamic climate adjustment and a dynamic ecosystem adjustment.

Dynamic Climate Model

The first step is to define the 'dynamic' climatic response to a gradual doubling of CO_2 in the atmosphere. This means proposing a time path for temperature and precipitation change. Given uncertainty involved with predicting both the future path of CO_2 emissions and the future climate change associated with it, we shall consider several alternatives. Our central case is based on Houghton *et al.* (1990), which projects that uncontrolled carbon emissions will result in a linear increase in temperature from now through 2060, the time by which CO_2 (or its equivalent in all gases) will have doubled. Climate (i.e. temperature and precipitation) is assumed to change linearly over 70 years, as additional carbon is added to the atmosphere. Exogenous controls are assumed to limit the continued growth of CO_2, so that climates will stabilize in 2060.

Alternative cases will consider both slower and quicker climate adjustments, including possible non-linearities. Earlier analysis of the impact in the USA (Sohngen, 1995; Sohngen and Mendelsohn, 1996) implies that the economic impacts are sensitive to the particular path chosen. This is particularly true where the ecological changes suggest substantial transient losses of currently valuable timber land. We thus shall pay special attention to how these alternative pathways alter the regional distribution of economic impacts.

Dynamic Model of Ecosystem Change

The dynamic model of ecosystem change begins by assuming that ecosystems adjust proportionally to changes in climate variables. As these variables increase linearly over a given number of years, the ecological adjustment occurs linearly over that time period as well. Dynamic ecosystem change is described by area and growth change, where dynamic area change is derived directly from biogeographical distribution models and dynamic growth changes are described by the biogeochemistry models.

Dynamic area change
Introducing a dynamic shift into the redistribution of timber types is difficult for two reasons. First, there is considerable debate over what will happen to

the old type that is displaced. For example, ecologists question whether the old type will die off or whether it will be replaced competitively by other species. Second, if left alone, ecosystems and trees do not migrate quickly. They rely on mechanisms such as wind or fire to disperse pollen and seed, so that migration rates at the frontier of ecosystem boundaries may be very slow (Overpeck et al., 1991).

We thus consider two methods of stock removal, one of which is dieback. Some modellers have suggested that timber species will die back as the conditions under which they grow change (Nielson et al., 1992). These modellers imply that conditions will become too different from those to which plants are currently accustomed. This, in turn, will stress the plants and cause the standing stock to die.

The second method of stock removal occurs at the regeneration stage, generally through the competitive displacement of ecosystem types. Here, the only way an ecosystem type is removed from the land is through difficulties in regenerating the old type. New types may either be planted or reproduce naturally (although it may take a while for natural migration to occur). No additional mortality is assumed to occur strictly from climate change.

The biogeographical models will provide information on which types die back and which ones are replaced competitively. If leaf area index (LAI) decreases at the same time as a piece of land adjusts from one ecosystem type to another, it implies that dieback occurs. If, on the other hand, LAI increases at the same time as land adjusts from one ecosystem type to another, we assume that the old type is competitively displaced by newer ones that are able to regenerate on the land.

Dealing with tree migration involves both seed-dispersal rates and land-management activities. While natural rates of migration are relatively slow, in many places, humans will react quickly by replanting the most healthy and profitable timber types. Land is thus classified into several different management types. Plantations are managed for timber purposes and are regenerated, immediately after either dieback or harvest, with species that are appropriate to the new climate. Low-intensity managed lands are held for multiple management reasons, and they will be restocked either naturally or by humans. Inaccessible land will follow a completely natural adjustment process.

Ecosystem models in general do not consider human influences. By allowing for different management classes, we account for the impacts of both natural and human adaptation to climate change. On plantation land, managers will replant the right species and will suppress the growth of competing trees or shrubs. On low-intensity managed land, natural adaptation will take longer, and there will be lags before regeneration occurs. This will limit the ability and speed of species to compete effectively when they are invading a new area.

Dynamic yield change
Like the redistribution of ecosystem types, we assume that NPP changes in direct proportion to climate change. Net primary productivity can be linked to timber models through the yield function. Although there is not an exact correlation, we assume that growth of timber is proportional to NPP. In the steady state, the climate-adjusted yearly growth is:

$$\dot{V}_i(a,\bar{t}) = \bar{\alpha}\dot{V}_i(a) \tag{1}$$

where α is the steady-state change in NPP for ecosystem type i. Dynamically, this shift occurs proportionally to a changing climate, so that $\alpha(t)$ is:

$$\alpha(t) = 1 + \gamma t \tag{2}$$

where γ is the yearly change in NPP. After the adjustment period, $\alpha(t)$ is constant.

It is important to recognize that improved growing conditions can only enhance future growth, not past growth. We must follow trees during the period of transition and attribute ecological effects to each year's growth. The existing stock of trees is not affected, only growth that occurs after the instantaneous moment. When there is a continuous and slow shift, as with climate change, the same principle is applied. Mathematically, this can be expressed as

$$V_i(a_i,t) = \hat{V}(a_{t^*-1}) + \int_{t^*}^{T}\left\{\alpha_{i,(t)}\hat{V}_i(a_{(t)})\right\}dt \tag{3}$$

where $\hat{V}_i(a_t)$ is the base yield function for a particular region, $V_i(a_p,t)$ is the climate-adjusted yield function and t^* is the age of the timber at the time of the shock. This adjustment accounts for the age of the trees when the climate shock begins, as well as the amount of time the trees have had to grow at the new, changed rates before they are harvested. We implement the linear change this way because the NPP at any particular point in time is a function of both the past growth rates and the current period's growth rate.

Dynamic Economic Model

The large-scale ecosystem adjustments predicted by the ecological models suggest that timber markets will experience changes that are beyond the scope of historical experience. As such, it is important that the economic model captures basic aspects of human activity, particularly how humans adapt and adjust to the changing conditions. To do this, we rely on dynamic optimization processes that attempt to maximize the net present value of net consumer surplus in timber markets. Elements from the work of Sedjo and

Lyon's global Timber Supply Model (1990) and the more recent work of Sohngen and Mendelsohn (1996) will form the basis for our economic model.

Dynamic models such as this rely on the rational-expectations approach to determine harvesting and replanting activities over time. In this study, rational expectations suggest that the consumers and producers in timber markets are forward-looking. By forward-looking we mean that consumers and producers formulate price expectations about future market activity, which are correct, on average. Decisions made today must be consistent with those expectations.

The model is a non-linear, dynamic programming problem, where the objective is to maximize the net present value of net consumer surplus in timber markets over time:

$$\underset{H_{i,a,t},\ b_{i,t}}{Max} \sum_0^T \rho^t \left\{ W\left(\sum_i \sum_a Q_{i,a,t}\right) - \sum_i c_i \sum_a H_{i,a,t} \right. \\ \left. - \sum_i b_{i,t} \sum_a H_{i,a,t} - \sum_{i \in l} R_{i,t} \sum_a X_{i,a,t} \right\} \quad (4)$$

where $W(\cdot)$ is total consumer surplus; $Q_{i,a,t}$ is the total quantity harvested out of land class i, age class a in period t; $H_{i,a,t}$ is the number of hectares harvested; c_i is the harvesting cost; $b_{i,t}$ is the replanting cost; $R_{i,t}$ is a rental value for certain classes (l) of timberland; and $X_{i,a,t}$ is the number of acres in each land and age class at time t. Land rent is the annual capital cost associated with holding land in timber rather than allowing it to flow to other managed uses, such as agriculture or housing developments. Land rental values are determined as the annualized value of bare land in the particular timber type. They will vary over time, depending on the price and yield of each timber type. Although we do not formally introduce a competitive land market in this model, including land rent is important for measuring welfare, because the stock of land in each timber type adjusts during climate change.

In this model, regeneration costs are chosen endogenously; they increase or decrease over time as a function of future timber prices. Timber yield at the time of harvest will be related directly to regeneration expenditures made when the land was planted (as in Sedjo and Lyon, 1990). In addition to the multiple species, three land classes are incorporated: plantation land, low-intensity land and inaccessible land. We determine the initial age and acreage distribution for these three types, based on site indices, proximity to markets and other accessibility criteria. We shall account for differences between the land through access, harvest and transportation costs, yield functions and returns to regeneration expenditures.

The maximization above is subject to several constraints:

$$X_{i,a+1,t+1} = X_{i,a,t} - H_{i,a,t} \quad \forall\ i,a,t \tag{5}$$

$$X_{i,1,t+1} = \sum_{a} H_{i,a,t} \quad \forall\ i,a,t \tag{6}$$

$$Z_{i,1,t+1} = b_{i,t} \quad \forall\ i,t \tag{7}$$

$$Z_{i,a+1,t+1} = b_{i,a,t} \quad \forall\ i,a,t \tag{8}$$

In addition to these constraints, we must have initial and terminal values for the stock of timberland, as well as non-negativity constraints for the stock and control variables. Equation (5) is the equation of motion for stock of any timber type, any age class and any time period. It shows what will happen to the remaining stock after harvests occur. Equation (6) accomplishes the regeneration of timber. Equations (7) and (8) are equations of motion for management intensity, described by $z_{i,a,t}$. Yield at the time of harvest is affected by $z_{i,a,t}$, so that investments made today to assist in regenerating timberland will affect yields when those acres are harvested.

Written in such a way, this is a non-linear programming problem, which can be solved using the maximum principle (Pontryagin *et al.*, 1962), as shown by Sedjo and Lyon (1990). In the forward-looking framework we have chosen to use in solving this problem, all time periods are solved simultaneously. When all periods are solved this way, it should achieve the same transitional price path, as discussed by Lyon (1981), Brazee and Mendelsohn (1990) and Sohngen (1995). Similarly, it should respond to shocks which may occur as a result of such phenomena as climate change.

ISSUES IN INTEGRATING THE MODELS

Several additional elements are important for linking the two types of models in a dynamic framework. The first issue revolves around the link between broad ecosystem types predicted by the ecological models and forest types for the economic model. In general, we have found that a close relationship exists between the two. The most important differences occur for the non-indigenous plantation species that have been introduced across the world.

Although it appears that plantation types grow in conditions that are very similar to their original range, changes in the distribution of ecosystems do not necessarily imply that plantation areas will increase or decrease. While biogeographical models may predict that tropical forests expand, and biogeochemical cycle models may predict that ecosystem productivity increases, plantation establishment and maintenance within the tropical regions will also depend on how conditions change in other regions of the globe. In particular, if productivity increases everywhere and global prices are depressed, it may not be profitable to plant additional acreage in the tropics.

The second issue involves what happens at the northern extremes of the globe. For example, global warming suggests that some land currently in tundra will convert to taiga-type evergreen forests. While this change is important, particularly ecologically, much of this timber may remain inaccessible to markets. The economic model, by placing this area of timberland into inaccessible regions, will allow us to control harvests there by broader economic concerns, such as price.

The third issue is how non-forest land uses are considered in this model. In many regions of the world, sectors like agriculture and forestry are linked closely together, either through agroforestry practices, as in some tropical countries, or through competition for land, as in the USA. Although we shall not build a competitive land supply model in this effort, we recognize that this is one potential extension for future research at the global level. We shall utilize the Olson *et al.* (1983) database to calibrate our baseline economic scenario.

We deal with other land uses during climate change in two ways. First, we borrow from other studies that have assessed the impact of climate change on agricultural areas, such as Darwin *et al.* (1995) and Leemans and Solomon (1993), in order to determine which regions remain productive in agriculture and are therefore likely to remain in agriculture. These areas are masked from our ecological analysis of forest changes. In other areas, where forested ecosystems are intermixed with other land uses, we allow for adjustment of land use as well.

At the moment, biogeographical distribution models predict only the area of land that potentially could contain forests (and grasslands, although we concentrate on forests in this study), but the area of land that actually does contain forests, such as the quantity obtained from inventory statistics, will probably be less than that (particularly in highly populated regions). We assume that the ratio of actual forest land to potential forest land remains constant during climate change. Thus, if the potential area of forested ecosystem types across the world doubles due to climate, the actual area of forested ecosystems must double as well.

CONCLUSION

In this chapter, we describe how ecosystem and economic models can be integrated at the global level to determine the economic impact of climate change on timber markets. Our analysis captures three important elements: spatial issues, dynamic issues and global scope. We pay special attention to providing a framework for joint ecological and economic analysis and to describing the particular models that are available for understanding these impacts.

Although it is beyond the scope of our research to present global empirical results at this time, the work of Sohngen and Mendelsohn (1996) on US timber markets suggests that human adaptation will limit damage in timber markets during climate change. Over a broad range of climatic- and ecosystem-change scenarios, they found that timber markets benefited from climatic change within the USA. Their estimates suggest that benefits range from less than 1% of the value of the market to greater than 10% of the value.

Several important points result from the discussion in this chapter. First, ecosystem modellers predict fairly large adjustments across the globe, ranging from large-scale redistribution of ecosystem types to changes in the underlying productivity of particular ecosystem types. Like climate change, which is predicted to vary from place to place across the globe, the ecological changes imply forest dieback in some regions and forest expansion in others. Any analysis of how humans adapt to the change must incorporate these spatial differences.

Second, the ecological adjustment will entail substantive and complex dynamic processes. This is one of the most interesting areas of this analysis, because it requires a close working relationship between economists and ecologists. The dynamic processes of ecosystems are sure to be influenced by humans and the dynamic processes of markets are likewise going to be influenced by changes in ecosystems. In the future, it will be important to recognize that these two systems are inherently integrated, and to continue utilizing and perfecting models that incorporate aspects of both natural and social sciences.

Third, the economic model is global, and it captures important processes involved with human adaptation and adjustment to climate change. Humans will adjust harvests from region to region to deal efficiently with large-scale dieback events (as shown in Sohngen and Mendelsohn, 1996). Humans will also adapt efficiently by replanting species where they are best able to grow over the next rotation. In a market that typically must look 7–100 years ahead for investments, this type of foresight is entirely possible.

NOTE

1   The authors would like to acknowledge the help of Ron Neilson, who has provided invaluable insight into methods for modelling the dynamic adjustment of ecosystems to climate change. Funding for this project was obtained from the US Department of Energy.

Economic Instruments and the Pasture–Crop–Forest Interface

G. Cornelis van Kooten and Henk Folmer

INTRODUCTION

Government policies to address anthropogenic contributions to climate change often focus on economic incentives that reduce fossil-fuel use, lower rates of deforestation and increase tree planting. In agriculture, research has centred on the use of carbon taxes and subsidies to encourage forest plantations as a method for sequestering carbon (Sedjo et al., 1995).

To date, three areas of policy concern have been neglected. First, policy has focused on the role of land use in preventing or mitigating climate change, but it has ignored land-use changes that occur as a result of climate change itself.

Second, carbon taxes and subsidies are not the only incentives that affect land use. A variety of agricultural programmes have major impacts on terrestrial carbon sequestration or its release to the atmosphere. It is not only the afforestation of agricultural lands that reduces a country's contribution to atmospheric carbon dioxide (CO_2), but also its agricultural practices, such as the extent to which it prevents conversion of wetlands, its ability to maintain grasslands and the incentives that exist for switching land into uses that release or store carbon. For example, Adger et al. (1992) calculate the impact that land-use changes between 1947 and 1980 have had on carbon fluxes in Great Britain, while Brown and Pearce (1994) provide estimates for tropical forest regions. With regard to farm practices, Curtin et al. (1994) and Coxworth et al. (1994) contend that significant carbon uptake by soil will occur by reducing summer fallow, employing conservation tillage and planting permanent pasture or simply returning cultivated land to native grassland, techniques that have the added benefit of reducing soil erosion.

Finally, economic institutions play an important, though often neglected, role in determining land use and total system flexibility (Holling *et al.*, 1993). In particular, it is government as opposed to market failure that prevents landowners from receiving the necessary signals to implement mitigation strategies or to adapt efficiently to climate change.

In this chapter, these issues are addressed by examining the impact of economic policy on land use. We begin by considering the potential of landowners in the primary sectors to adapt to climate change.

IS AGRICULTURE ABLE TO ADAPT TO CLIMATE CHANGE?

If human emissions of greenhouse gases do cause climate change, adaptation will be unavoidable. For example, even the Intergovernmental Panel on Climate Change (IPCC)'s (1994) scenario of reversing deforestation, relying on greater energy efficiency, substitution of lower-carbon-based fuels, controlling carbon monoxide emissions and implementing the Montreal Protocol will only reduce global mean temperatures by 0.3°C from a projected business-as-usual warming of nearly 3°C (or by 0.07°C from the lower business-as-usual scenario of 1°C warming). Projecting the ability of landowners (farmers) to adapt to climate change is the subject of speculation, but it depends on price signals, access to the products of research and development, such as new crop cultivars and pesticides, and freedom to make decisions. These conditions are affected by government agricultural subsidies, barriers to trade, land-use restrictions, patent laws and so on.

Recent studies have combined biological and economic modelling to examine adaptation in agriculture to a double-CO_2-equivalent atmosphere (Tobey *et al.*, 1992; Rosenzweig and Parry, 1993, 1994; Darwin *et al.*, 1995). Tobey *et al.* (1992) use the results from a variety of regional crop-response models to develop yield projection scenarios for a double-CO_2 climate. Yield responses in different regions are linked via a partial-equilibrium trade model of world food markets, known as SWOPSIM and developed by the US Department of Agriculture. Despite assuming yield reductions of up to 50% for the USA, Canada and the European Union (EU)-12 and no CO_2 fertilization effect, the authors conclude:

> even with concurrent productivity losses in the major grain producing regions of the world, global warming may not cause widespread havoc in the agricultural sector, [because] inter-regional adjustments in production and consumption will serve to buffer the severity of climate change impacts.

(Tobey *et al.*, 1992, p. 202)

Rosenzweig and Parry (1993, 1994) rely on estimated potential changes in national grain yields from scientists in 18 countries to develop crop-yield projections under different climate scenarios for 112 sites around the globe.

Projected yields for the enhanced CO_2 atmosphere are calculated for climates from three global circulation models (GCMs) – those of the Goddard Institute for Space Studies (GISS) (1982), the Geophysical Fluid Dynamics Laboratory (GFDL) (1988) and the UK Meteorological Office (UKMO) (1986)[1] – with some scenarios including a CO_2 fertilization effect that increases yield. Wheat yields are projected to decline for the three GCM climates by an average 23.7%, rice by 24.7%, maize by 25.7% and soyabeans by 33.7%. With CO_2 fertilization, average yields of wheat are projected to rise by 0.7%, while those of rice, maize and soyabeans are projected to fall by 3.7%, 19.0% and 4.0%, respectively (Rosenzweig and Parry, 1993, p. 99). These projections do not take into account the potential for technical advances in plant breeding, however, as winter-wheat yields in some regions of Canada and Russia are projected in the model to decline, due to lower flower-bud initiation.[2]

The researchers also linked crop-yield projections for various regions through International Institute for Applied Systems Analysis (IIASA)'s basic linked-system world food-trade model – a global general equilibrium (GE) model. The no-climate-change, or reference, scenario assumes particular rates of population growth and technical change, as well as a specified rate of economic growth (which suggests the model is not truly dynamic). Further, arable-land limits are not adjusted to climate change, water availability for irrigation is ignored, barriers to trade remain in place (at least until 2020) and, surprisingly in light of recent concerns of potential shortfall of food supply in China (Brown, 1995), that country is left out of the model. Without farm-level adaptation and relative to the reference scenario, global output of cereals falls by 10–20%, although with CO_2 fertilization it falls only an average of 3.9% (which is still some 75% higher than production in 1990). With 'extensive' farm-level adaptation and a fertilization effect, the model projects no change in agricultural output compared with the reference scenario.

A major research effort by the US Department of Agriculture's Economic Research Service examined the impacts of climate change on land use and primary production from land (Darwin *et al.*, 1995). The researchers developed the Future Agricultural Resources Model (FARM), which endogenizes crop substitutions, links climate projections of land and water resources, simultaneously estimates the impacts of climate change on crop and livestock production and forestry, and integrates these land-use activities within a global economic model that accounts for all market-based activity. The model consists of a geographical information system and a computable GE model of the global economy. It has eight regions, six land categories and 11 sectors producing 13 commodities (a crops sector, which produces wheat, other grains and non-grains; livestock; forestry; coal, oil and gas; other minerals; fish, meat and milk; other processed foods; textiles, clothing and footwear; other non-metallic manufactures; other manufactures; and services). The climate projections from four GCMs – GISS, GFDL, UKMO and Oregon

State University (OSU) – were employed. In FARM, it is assumed that prices are determined in competitive and international markets and, under one set of scenarios, that landowners are completely able to adapt to climate changes, as opposed to only adjusting on-farm production.

The FARM assumes that only primary production from land is affected by climate change. Results indicate that 'across scenarios, world wheat production increases, while production of non-grains falls. Output of other grains increases or decreases depending on the scenario. Production of livestock and forest products generally increases' (Darwin *et al.*, 1995, p. 23). Output of other food products increases in all scenarios, so that 'climate change's overall impact on world food production is likely to be beneficial' (p. 26). Real global gross domestic product (GDP) is projected to increase slightly.

In most crop-response models where there are no linkages to prices, world cereal supply declines by 19–30% under a double-CO_2 atmosphere, unless CO_2 fertilization is taken into account (see above). Results from FARM (which includes no fertilization effect) indicate that, if farmers are permitted to select the most profitable mix of inputs and crops on existing crop land, some 78–90% of the initial climate-induced reductions in cereal supply will be offset. By allowing for trade and changes in demand, but keeping crop land fixed at prechange levels, more than 97% of the original negative impact is offset. Finally, if farmers can take advantage of new agricultural lands and fully adapt to the climate change, cereal production is actually projected to increase by 0.2–1.2% depending on the GCM that is employed.

Projections are static in FARM, in that they do not attempt to predict future yield increases, economic growth or population, but they also do not include CO_2 fertilization and ignore potential for climate-induced technical change. The model limits substitutability of inputs (e.g. capital for land) and treats imports and domestically produced goods as imperfect substitutes, as is true in the real world. On the other hand, it is optimistic about the potential to expand agricultural production into new areas, determines land classes by climate factors only, assumes land is used only for commodity production and not production of other ecosystem services, and is optimistic about regional water-supply responses.

Despite the differences alluded to above, results from FARM are comparable to those of Rosenzweig and Parry (1993, 1994); the latter projects a decline in global agricultural production, while the former predicts an increase. The difference is attributable primarily to greater flexibility on the part of economic agents in the FARM model. There are no trade barriers in FARM, nor are there restrictions on choice of crops (including planting trees). Farmers are free to adjust their use of inputs and planting dates (within what is feasible), and can expand cultivation into new regions available for crop production because of climate change. Some but not all of these options are modelled by Rosenzweig and Parry (1993, 1994).

CLIMATE-CHANGE IMPACTS ON EUROPE AND CANADA: A COMPARISON

In FARM, there are projections for eight regions, but we focus on two of these – Europe and Canada.[3] These regions are chosen because incentives and institutions in both have resulted in misallocation of land and would not currently permit the types of adjustments assumed in FARM. Europe is projected to experience one of the largest reductions in output of any region, with production of all primary sectors falling, except other grains and forest products, which are projected to increase output by 24.5% and 3.2%, respectively. Since it has a continental climate and is located in the northern latitudes, Canada experiences large absolute changes in land use and the greatest net increase in crop land, mimicking what might happen in Russia.

The results of Darwin *et al.* (1995) serve as an indicator of what might be expected of adaptation in a perfect world. World GDP is projected to increase slightly in this case, with only two regions (out of eight) projected to experience a decline in GDP across all scenarios as a result of climate change – South-East Asia and the EU12. An average reduction in real GDP of 0.8% from 1990 levels ($US2.21 billion and $US45.58 billion, respectively) is projected for each of these two regions. Canadian GDP is projected to increase by an average of 2.2% across the four GCM scenarios, which is the largest increase of any region, although the absolute increase in GDP is relatively small ($US0.13 billion).

Globally, FARM projects an average reduction in land classes 1 and 2 (those with growing seasons of 100 days or less) from 10.17 million hectares to 3.69 million hectares. Land in classes with growing seasons of 101–165 days (principal crops are wheat, other short-season grains and forages) is expected to fall from 33.27 to 22.79 million hectares on average across the four GCMs, while that with a growing season of 166–250 days (so maize can be grown and double-cropping can occur) falls from 117.63 to 42.58 million hectares. Land with growing seasons of 251–300 days, with two or more crops per year (including cotton and rice), increases from 45.07 million hectares to 65.40 million hectares. Finally, land that can be used for crop production all year long (and is used to produce rubber, sugar cane, tropical maize and rice) increases from 16.69 million hectares to 88.36 million hectares. Changes in land use (assuming full adjustment) are reported in Table 9.1 for Europe and Table 9.2 for Canada; these changes take into account both climatic and economic factors.[4]

It is logical to expect that an increase in crop land would also lead to greater incomes, but that is not the case. European wheat production falls by an average 11.6% (from a base of 80.3 million tonnes (Mt)), production of non-grains by 10.6% (from 279.9 Mt) and livestock numbers by 1.5% (from 295.0 million head). Production of other grains rises by an average of 24.5%, but from a lower base of 25.0 Mt, while forestry output increases by 3.2%,

Table 9.1. Projected changes in land use in the EU12 as a result of climate change (from Darwin et al., 1995, and calculation).

Land use	Base (million ha)	With climate change (million ha)	With climate change (% change)
Crops	77.84	83.44	7.2
Pasture	55.07	52.43	-4.8
Forestry	54.41	55.99	2.9
Other	35.50	30.96	-12.8

Table 9.2. Projected changes in land use in Canada as a result of climate change (from Darwin et al., 1995, and calculation).

Land use	Base (million ha)	With climate change (million ha)	With climate change (% change)
Crops	45.96	103.35	+124.9
Pasture	28.20	39.66	+40.6
Forestry	358.00	332.48	-7.1
Other	489.94	446.61	-8.8

from a base of 171.4 Mm^3. Although more crop land is brought into production (at the expense of pasture and other land), the land is used less intensively (e.g. less irrigation and fertilizer), as a result of economic signals from elsewhere in the economy and from other regions of the world and because the land itself is simply less productive, as climate change has an adverse impact on crop production. As a result of these changes in production, commodity prices in the EU12 are projected to increase for all commodities except forest products and other grains. Yet, and this is true only for Europe, revenues accruing to three factors of production – land, labour and capital – are expected to decline, with the owners of land experiencing the largest reductions in incomes.

For Canada, wheat production is projected to rise by an average 130.4% (from a base of 32.1 Mt); production of other grains and non-grains by 273.7% (from 25.0 Mt) and 455.5% (from 13.0 Mt), respectively; livestock numbers by 255.6% (from 23.8 million head); and forestry output by 33.2%, from a base of 155.5 Mm^3. (With the exception of forest products, current Canadian commodity output levels are significantly below those in Europe.) The relative magnitude of these increases in output is a result of significant expansions of crop land and pasture (Table 9.2), and the fact that land itself is more productive, due to an improved climate. Much of the increase in arable-crop production in Canada comes from a northward shift in the western grain belt, at the expense of boreal forest, and a shifting of the highly productive maize belt in the USA into Canada (Arthur and Abizadeh, 1988). The northern limit of the boreal forest shifts some 150–200 km to the north. Canadian outputs in all sectors of the economy, particularly those of the four agricultural sectors, are projected to increase, while prices paid to commodity producers fall. Revenues accruing to land, labour and capital are all expected to rise, while water revenues are expected to increase in some GCM scenarios and fall in others. Landowners are expected to see the greatest increase in revenues.

Importantly, projections assume complete flexibility to adapt all land uses to climate change, which would require changes to institutions. For example, any attempts to implement policies to overcome the projected adverse effects of climate change (e.g. agricultural subsidies, trade restrictions) may benefit farmers, but at the expense of still lower real GDP (i.e. at the expense of other sectors in the economy) and lower world agricultural output. But a policy question remains: should land uses be modified now in an effort to mitigate climate change or should policies be implemented to prepare landowners to adapt to climate change in optimal fashion (van Kooten, 1995)?

CLIMATE CHANGE, LAND USE AND MITIGATION POLICY

Many factors affect a landowner's choice of what to do with land and when to switch from one use to another. In this section, we examine the economic factors that influence such decisions. As in FARM, three land uses are identified – pasture or grasslands (including arid range) that are used for cattle production; arable or cultivated lands that produce principally grains; and the (generally) moister forested areas. The principal characteristic that separates these land uses is not always soil type, but climate. While soil type is important, the soils found in a particular continental region, for example, are affected by factors such as moisture, which affects vegetation and, thereby, the amount of organic matter in the soil. Land use can be linked to climate through plant-available soil moisture or length of growing season, or some combination of these.[5] For ease of exposition, in Fig. 9.1 we use plant-available soil moisture as the climate variable that affects land uses, although soil moisture should be combined with some measure of growing-season length.

As moisture available for plant growth increases, land use changes from desert to range suitable for cattle production. At the intensive margin A, moisture is adequate to permit a shift to crop production.[6] But changes in land use are not symmetric; as available moisture for plant production increases, the switch from range to crop production may occur at a higher moisture level than a switch in the opposite direction (with moisture declining over time). One reason is that cropping requires a higher level of investment

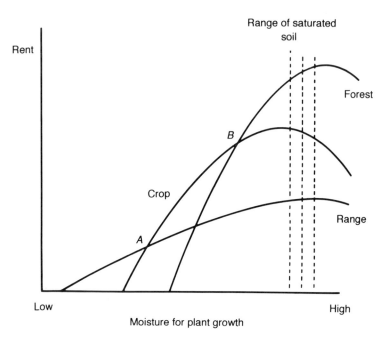

Fig. 9.1. Relationship between rent and available moisture for land use.

in machinery, so that, when moisture falls, it may still be economic to continue with crop production. Likewise, once native range has been converted to crop production, it may not be economically feasible to re-establish the original range ecosystem, in the sense that the costs of doing so exceed the discounted benefits. Thus, while it might be worthwhile to abandon crop production at some future date, it might not be profitable to re-establish the range at the same level of productivity as previously. However, we do not address economic and biological irreversibility here.

The intensive margin B represents the point where moisture has risen to such an extent that arable-crop production is no longer the most profitable activity; rather, land is best used in timber production. Again, point B is unknown and there exists an asymmetry between conversion of arable land to forestry, which requires a significant investment in tree planting, and conversion in the opposite direction, since income available from harvesting trees can offset costs of establishing crops.

The problem of determining when and under what conditions to switch land uses is a difficult one for landowners, but it is also a problem for policy-makers (Crabtree, 1995). Further, there is no reason to presume that adaptation may occur smoothly and may be preferred to mitigation (Kaiser *et al.*, 1993; van Kooten, 1995).

Grassland-crop Production Margin

Consider a mathematical representation of the rent–moisture function for grassland in Fig. 9.1:

$$R_{Grass} = E \sum_{t=0}^{\infty} G(M_t(W_t), P_{Forage,t}, P_{Other,t})(1+r)^{-t} - F_{Grass} \quad (1)$$

Here $G(\cdot)$ represents the annual net returns to grassland after variable costs; M_t refers to available moisture for plant growth at time t, and is itself a function of (unknown but projected) climate conditions at time t, W_t; $P_{Forage,t}$ is the (shadow) price of forage at time t; $P_{Other,t}$ is a vector of the prices of other inputs or outputs; r is the discount rate; F_{Grass} is the fixed cost associated with livestock production; and E is the expectations operator. The rent function for arable cropping can be written as:

$$R_{Crop} = E \sum_{t=0}^{\infty} A(M_t(W_t), P_{Crop,t}, P_{Other,t})(1+r)^{-t} - F_{Crop} \quad (2)$$

where $A(\cdot)$ represents the net returns after variable costs of annual cropping, $P_{Crop,t}$ is the output price of the crop at time t, and F_{Crop} is the fixed cost of cropping. The conversion from grassland to arable cropping at some time occurs when eqn (2) exceeds eqn (1) plus the cost of converting from grass to arable-crop production, $C_{G \to A}$:

$$R_{Crop} - R_{Grass} > C_{G \to A} \quad (3)$$

That is, the expected discounted future benefits from crop production minus the expected discounted future returns from keeping the land in pasture (range) must exceed the costs of converting land from pasture to arable-crop production. Solving for the critical level of available plant moisture gives:

$$M^{*G \to A} = f(P_{Forage}, P_{Crop}, P_{Other}, F_{Crop}, F_{Grass}, r, C_{G \to A}, W) \quad (4a)$$

in the case of conversion from grassland to arable-crop production, and

$$M^{*A \to G} = f(P_{Forage}, P_{Crop}, P_{Other}, F_{Crop}, F_{Grass}, r, C_{A \to G}, W) \quad (4b)$$

in the case of conversion in the opposite direction. In eqns (4a) and (4b), P_{Forage}, P_{Crop} and P_{Other} refer to vectors of expected future prices of the relevant outputs and inputs; r is a vector of expected future interest rates; and W is a vector of expected future climate. As long as conversion costs are symmetric ($C_{G \to A} = C_{A \to G}$), the critical moisture level is the same, whether moisture is improving or worsening due to climate change.

Economic incentives and extant institutions affect $M^{*G \to A}$ and $M^{*A \to G}$ through their impact on all of the variables in eqns (4a) and (4b), except W. For example, government research and marketing institutions, subsidies and publicly funded insurance schemes affect not only the costs of converting pasture to crop land, but also the expected returns to cropping. The only variable that cannot be controlled is future climate.

What public policies are available to make private landowners take into account the externality effects from releasing carbon by changing land use from forage to crop production? Suppose that the only difference between forage and crop production takes place with respect to below-ground biomass. Houghton (1995, p. 48) indicates that soil carbon stored in below-ground biomass in temperate grasslands and pastures averages 189,000 kg ha^{-1}, compared to 128,000 kg ha^{-1} for cultivated land. Assuming that below-ground organic matter decomposes at a rate of 2% per year, converting land from grass to arable-crop production leads to a reduction in the amount of below-ground carbon stored in biomass of 1,220 kg ha^{-1} year^{-1} (= 0.02 × 61,000). The adverse impact of carbon release due to conversion of range to crop production could be internalized in the landowner's decision by imposing a one-time levy (tax) that raises the conversion cost by $\frac{q}{r} P_c$ ha^{-1}, where q is the annual amount of carbon (in tonnes) released upon conversion of the land, r is the discount rate and P_c is the social (shadow) value of carbon in \$US per tonne.[7] If the discount rate is 4%, q = 1.22 t, and P_c = \$US25 t^{-1} (a reasonable value), the landowner would be assessed \$US762.50 ha^{-1} for converting land from pasture to crop land.

Conversion from arable-crop land to pasture would be subsidized according to its climate benefits. In this case, the appropriate strategy would be to provide an annual subsidy based on the amount of below-ground biomass that is sequestered annually (1.22 t ha^{-1}). The annual subsidy would amount to \$US30.50 ha^{-1}, but is likely to have little influence on land-use decisions (Crabtree, 1995; see also Chapter 14, this volume).

Crop-land–Forest Margin

Consider the case where, in order to make landowners aware of the external costs of increasing atmospheric CO_2, the government imposes a levy on carbon released from storage by cutting trees and provides a subsidy for carbon storage. Thus, owners receive an annual subsidy of $P_c \alpha V_T$ for carbon uptake in tree growth, paying a penalty of $P_c \alpha (1 - \beta) V_T$ at harvest for the external damage caused by release of carbon from storage. Here α indicates how much carbon is sequestered per unit of biomass, β is the proportion of carbon that gets stored in wood products after timber harvest, and V_T is the volume of timber at time T (with T equal to rotation age). In addition to carbon subsidies and taxes, forest landowners receive a net income from harvesting trees and incur costs associated with replanting and silviculture during each rotation period. Suppose s is the unchanging real price of the standing timber (stumpage price) and that there are an infinite number of rotations of length T, with rotation length not changing over time.[8] Van

Kooten et al. (1995) show that, under these conditions, the rent function for use of land in forestry is given by:

$$R_{\text{Forest}} = P_c \alpha V_T(M_T)(1+r)^{-T} + rP_c\alpha \sum_{t=0}^{T-1}\Delta V_t(M_t)(1+r)^{-t} + (s - P_c\alpha(1-\beta))V_T(M_T)(1+r)^{-T} \quad (5)$$

In eqn (5), $\Delta V_t = V_{t+1} - V_t$, with V an implicit function of climate W. The rent–moisture function of eqn (5) differs from those for grasslands and arable-crop production, because of the important role that forests play in terms of carbon uptake in above-ground, as opposed to below-ground, biomass.

Now consider intensive margin B. For the private landowner, there is an incentive to convert forest land to agriculture as long as:

$$R_{\text{Crop}} - R_{\text{Forest}} > C_{F \to A} \quad (6)$$

where $C_{F \to A}$ is the cost of converting land from forest to agriculture. Land-conversion benefits the landowner ($C_{F \to A} < 0$) if the returns from cutting and selling existing trees exceed the costs of establishing crop land (e.g. removing stumps). In the case of conversion in the other direction, $C_{A \to F} > 0$, as landowners will generally incur tree-planting costs. The moisture level associated with intensive margin B depends on the direction of land conversion. For conversion from forestry to arable-crop production it is:

$$M^{*F \to A} = f(s, P_c, P_{\text{Crop}}, P_{\text{Other}}, F_{\text{Crop}}, r, C_{F \to A}, W) \quad (7a)$$

while, for conversion in the other direction, it is

$$M^{*A \to F} = f(s, P_c, P_{\text{Crop}}, P_{\text{Other}}, F_{\text{Crop}}, r, C_{A \to F}, W) \quad (7b)$$

It is unlikely that the two values of moisture will be the same – in general, $M^{*A \to F} > M^{*F \to A}$ because $C_{A \to F} > C_{F \to A}$.

What has been ignored is below-ground storage of carbon in the forest ecosystem. According to Houghton (1995, p. 48), below-ground biomass for (boreal) forests exceeds that of arable land by 78,000 kg ha^{-1}, although it is lower for the conversion from temperate forests (since soils in cold environments store more carbon). Again assuming that organic matter decomposes at an annual rate of 2%, converting land from forest to arable-crop production leads to a reduction in the amount of below-ground biomass of 1.56 t of carbon ha^{-1} year^{-1}, and, assuming the same rate of accumulation, there is a gain of a similar magnitude when land is converted from crop production to forestry. On the basis of the below-ground carbon flux, a one-time conversion of land from forest to cultivation should be taxed \$US975 ha^{-1}, assuming $r = 0.04$ and $P_c = $ \$US25 t^{-1}, while a subsidy of \$US39 ha^{-1} year^{-1} needs to be provided when crop land is converted to forestry. This is in addition to the tax/subsidy scheme associated with above-ground biomass.

Assuming 0.2 t of carbon m^{-3} of wood and annual growth of 10–20 m^3 ha^{-1}, the relation in eqn (5) implies that the annual subsidy for above-ground sequestration of carbon would be \$US54.05–\$US104.05 ha^{-1} (if

P_c = $US25 t^{-1}), much lower than the EU's subsidy of about $US800 ha^{-1}, which has been inadequate to bring about major increases in tree planting (see Chapter 14 by Crabtree and Chapter 15 by Lippert and Rittershofer, this volume). To prevent land from being converted from forestry to agriculture, a one-time tax equal to $US1,351.25–$US2,601.25 ha^{-1} should be levied. A tax of this amount, when added to the above tax of $US975 ha^{-1} based on soil carbon, may constitute a significant deterrent to the conversion of forests to arable-crop land. Such taxes are certainly preferred to the regulations that exist in many EU countries to prevent the conversion of forest lands to other uses. These regulations constitute an obstacle to tree planting and reduce the flexibility of landowners to make decisions about the way in which they use their land.

EFFECT OF GOVERNMENT POLICIES AND POTENTIAL FOR ADAPTATION

In Europe, a major obstacle to an appropriate land-use response to global climate change is the Common Agricultural Policy (CAP). The CAP provides distortionary signals to producers of agricultural products, both within and outside the EU. It stimulates production of the 'wrong' products in the EU, while discouraging non-EU landowners from switching to or increasing production of the 'right' products. Under the CAP, farmers are subsidized through support prices and import restrictions to produce grain whose price is determined on international markets. One consequence of such subsidies is that crop production is more intensive (i.e. greater amounts of fertilizer and herbicides are used) and forest and pasture lands have been converted to cultivation (thereby destroying wildlife habitat and releasing carbon into the atmosphere). Location of intensive livestock production is encouraged in regions with easy access to cheap feed (such as non-CAP-protected tapioca from tropical regions) and where land is less suited to grain production – the Netherlands, Belgium and Denmark. Water pollution from intensive livestock operations (from manure spread on fields) and increased CO_2 emissions associated with the transportation of imported livestock feeds and export of grains using subsidies contribute to increasing atmospheric concentrations of greenhouse gases.

Since major reforms in 1992, the EU has gradually replaced price supports with area-based income supports. This is important with respect to climate change, because price signals will no longer be the cause of land-use distortions, although earlier distortions could remain in cases of economic irreversibility. However, income support still leads to distortions associated with non-optimal entry and exit into agriculture (both in the EU and elsewhere) and, because it is area based, with the conversion of marginal lands (shifting the crop rent function upwards in Fig. 9.1).

It is economic incentives that offer a relatively efficient means for reducing agriculture's emissions of greenhouse gases and encouraging farmers to remove CO_2 from the atmosphere through appropriate land-use practices. In reality, however, it will be difficult politically to eliminate subsidies or otherwise change agricultural policies, as evidenced by the EU's reluctance to make even small concessions on agricultural subsidies in the latest General Agreement on Tariffs and Trade (GATT) negotiations (Pearse, 1995). Imposing taxes on farmers, whether these take the form of carbon, energy or land-use conversion taxes, or taxes on fertilizers (to reduce nitrous oxide (N_2O) emissions), seems out of the question. A politically more acceptable strategy might be to target farmers who currently benefit from agricultural programmes by tying benefits to behaviour that results in less atmospheric CO_2. For example, farmers can be encouraged to plant trees on marginal lands, to grow crops for biofuels or to convert cultivated land to forests and grasslands in exchange for CAP support. This approach was adopted in the 1992 revisions to the CAP aimed at reducing agricultural surpluses; farmers are required to set aside land in order to remain eligible for subsidies (Brouwer and van Berkum, 1996). The set-aside programme is likely to be short-lived, particularly as agricultural surpluses disappear. Thus, while set-aside land might be put into pasture, planting trees is not an option because the duration of the set-aside programme is too short, while many countries (for example, Denmark) have regulations that prevent the reversion of forest land to cultivation at a later date. Institutional rigidity, therefore, discourages owners from adopting land uses that are preferred in the context of global change.

As we have already seen, extant EU tree-planting subsidies are generally inadequate to achieve the desired response (see Chapter 14 by Crabtree and Chapter 15 by Lippert and Rittershofer, this volume). The required subsidies for afforestation need to be larger, but these can only be justified on the basis of their potential recreation benefits, about which little is known. Taxes, on the other hand, are likely to be effective in preventing the conversion of forest lands into agriculture and may even encourage landowners to accept tree-planting subsidies, but politicians do not like taxes. Again, these rigidities are detrimental to environmental sustainability.

Canadian agricultural policies are similar to those in Europe. Price supports, output and input subsidies, and subsidies to 'improve' land (by converting wetlands to agriculture or range into arable-crop production) have encouraged greater cultivation and release of carbon from soils. Feed-freight assistance has encouraged livestock operators to locate close to markets, where pollution from livestock is a greater problem, while the subsequent higher farm-gate prices have shifted the extensive and intensive margins in favour of arable cropping. Subsidized crop insurance has had the same impact, because farmers are no longer required to diversify their enterprises to spread risks, which are borne by government. Thus, grain producers no longer maintain pasture for livestock as a risk-diversifying strategy.

The Canadian Wheat Board quota system bases grain deliveries to the marketing system on the area farmers have under cultivation. Since quotas are below average yields, farmers are encouraged to increase their delivery quota by breaking marginal land and, in drier regions, to make greater use of summer fallow in crop rotations to conserve moisture on land that probably should not have been brought into production to begin with (Delcourt and van Kooten, 1995). Not only does fallow lead to more soil erosion, but it also reduces carbon storage in soils. The EU's move towards area-based instead of output subsidies has similar effects.

There are also structural factors that help retain land in arable-crop production and reduce the flexibility of owners to change land uses. To discourage depopulation, for example, Saskatchewan has invested heavily in public infrastructure (such as hospitals and recreation facilities) in the drier regions of that province, where crop production is often unprofitable and climate change is projected to result in yields that are too low to support viable farm enterprises (Delcourt and van Kooten, 1995). Similar structural incentives exist in EU agricultural programmes, such as the special support scheme for less-favoured (i.e. less productive) agricultural areas (Brouwer and van Berkum, 1996). In these areas, investments in public infrastructure are a signal to private investors that the region is viable over the long term. But primary crop production is not sustainable without government support, so any investments in agriculture must be discouraged.

The margin in Canada between agriculture and forestry is found at the northern extent of the great plains. Here the policy concern relates most to land tenure and whether government should focus on mitigation or adaptation. Mitigation strategies imply that planting trees is preferred to converting land to agriculture, while adaptation requires that landowners take advantage of climate change by abandoning crop production in the dry south and expanding cultivation into the boreal forest zone. Most boreal forest land is owned by the provincial government, with large forest companies given rights to harvest trees for periods of 25 years or more in exchange for investments in large pulp mills. If the objective of government is to maximize timber output, while, at the same time, maximizing carbon uptake and hence emission mitigation, replanting, silviculture and sustained-yield forestry are appropriate. If adaptation is a correct policy response, the government needs to devise strategies for privatizing public lands, as public land tenures could be an obstacle to adaptation (van Kooten, 1995).

CONCLUSIONS

The evidence presented in this chapter leads to the conclusion that adaptation to climate change by landowners in forestry and agriculture is not only possible, but can be accomplished without a reduction in global output.

However, as a comparison of results from models by Darwin *et al.* (1995) and by Rosenzweig and Parry (1993, 1994) suggests, any impediments to adaptation can lead to lower global output under climate change. For example, Rosenzweig and Parry (1993, 1994) restrict the extent to which agricultural producers can adapt to climate change (with expansion of agriculture into forested areas not available in their model) and they impose trade barriers, restrictions that are absent in the projections of Darwin and colleagues (1995).

As demonstrated here, policies in most countries have encouraged too much crop production, with subsequent loss of the carbon-sink functions of range and forest ecosystems. But flexibility to adapt land uses to climate change does not exist either. In future, much more attention needs to be focused on incentives, institutions and the role of government if the environmental challenges that climate change poses are to be dealt with appropriately. As demonstrated here, government intervention in agricultural markets and government regulations on land use constitute rigidities not only to adaptation to climate change, but also to implementation of mitigation strategies.

NOTES

1 The dates indicate when the model's projection was made.
2 The authors assume a very low rate for yield increases (less than 1% per year) for their reference scenario (see below), arguing that technological change increased yields by 1.5% annually over the period 1965–1985 and that this rate cannot be sustained. Yield increases actually averaged 2.4% over the period 1961–1992 and there is no reason to suspect that they would be lower in the future, or that technical change could not mitigate some of the adverse effects of climate change.
3 The other regions are the USA, Japan, South-East Asia, Other East Asia, Australia–New Zealand, and the Rest of the World. Data availability determined the choice of regions.
4 If full adjustment is not allowed, all of the negative consequences discussed in the text are made worse, while positive ones are smaller than indicated.
5 A number of measures of climate can be used to determine the margins at which land uses change (Fig. 9.1). For practical purposes, measures depend on the problem at hand. Thus, Darwin *et al.* (1995) employ length of growing season, while others have used growing degree-days (number of consecutive 12 h days when the temperature is above 5°C). The Thornthwaite index uses a combination of temperature and precipitation to separate vegetation classes – rain forest, forest, grassland, steppe and desert. Researchers in western North America employ the Palmer Drought Index, because soil moisture is often the most important constraint on crop production (McGinn *et al.*, 1995).
6 The intensive margin A cannot really be considered a point, because none of the variables that determine the critical moisture at which the switch from grasslands to crop production is profitable is known with certainty. Not only are cattle and

crop prices and prices of inputs uncertain, but climate is both highly variable and uncertain.

7 $\frac{q}{r}$ the familiar bond formula derived by solving

$$\sum_{t=1}^{\infty} q(1+r)^{-t}$$

8 One would expect climate change to affect the rotation age, but only if climate change does indeed occur and affects the rate of growth of trees.

Agricultural Policy Impacts on United Kingdom Carbon Fluxes

10

Susan Armstrong Brown, Mark D.A. Rounsevell, James D. Annan, V. Roger Phillips and Eric Audsley[1]

INTRODUCTION

Under the terms of the United Nations (UN) Framework Convention on Climate Change, signatory countries are committed to reducing their greenhouse-gas emissions to 1990 levels by the year 2000. The measures taken by the UK government to return emissions to the target level are set out in a report entitled *Climate Change: The UK Programme* (UK DoE, 1994). The report describes a package of measures targeted at reducing emissions from the transport, energy, waste-management and industry sectors, covering the major greenhouse gases, carbon dioxide (CO_2), methane (CH_4), nitrous oxide (N_2O) and other trace gases, such as volatile organic compounds, carbon monoxide and halocarbons.

No specific measures have been introduced to reduce greenhouse-gas emissions from the agricultural sector. Nevertheless, agriculture is a significant source of CH_4 and N_2O, as well as acting as a sink for CH_4. Agricultural practices, especially land-use changes, also have a major impact on CO_2 fluxes.

In this chapter, we describe a method to assess the effect of UK agricultural policy on fluxes of carbon in the forms of CH_4 and CO_2. A previous paper used the same approach to evaluate the effects of agricultural policy on UK N_2O fluxes (Armstrong Brown *et al.*, 1996). The effects of current and known forthcoming policies are estimated for the 1990 baseline and projected until the year 2000. The research addresses both the physical and socio-economic aspects of agricultural systems based on an approach that combines economic analysis with the application of models. Consideration is given to policies relating to the arable and livestock sectors, as well as to environmental regulations. The calculations were carried out in 1994, at which time set-aside was 15% (it is now 10% and further changes are anticipated) of the arable area. Changes in world cereal prices have also changed the price levels now expected. This analysis provides the type of information required to make informed decisions about the likely consequences of different policy options.

METHODS

A two-stage approach is applied in the analysis presented here. In the first stage, the impact of policy changes on farm practices is estimated. The second stage involves the assessment of the effects of farm management changes on carbon fluxes. This approach is now outlined for different sectors of UK agriculture.

Modelling the Arable Sector

Figure 10.1 shows the information flows and interactions between the models used in the arable-sector analysis. The most important policy is the Arable Area Payments Scheme. This comprises two parts: (i) cereal-price support; and (ii) set-aside land requirement. As a result of the Common Agricultural Policy (CAP) reform, crop-price support for cereals has been reduced by 30% over the period from 1992 to 1995 (Agro Business Consultants, 1994). The price of oil-seeds has been reduced by 152 European currency units (ECU) t^{-1}, with no further reduction planned. There are compensatory area payments for these crops, with the stipulation that a certain percentage of the total arable area must be set aside each year. The European Union (EU) Agricultural Council sets the base level of set-aside required, and the levels vary from year to year. In 1995 the set-aside requirement was for 15% of the arable area. In 1996 the base level was 18%, and in 1997 it is likely to be reduced, possibly to as little as 5%. We use the 1995 rate of 15% in these calculations.

The effects of the arable regime are assessed, using the two-stage approach. In stage 1 (policy effects on farm practice), the ARABLE model, which calculates optimum cropping areas at different crop prices (Audsley, 1993), is used to assess the impact of the price reduction and the set-aside requirement on the proportions of arable land planted with each of the major crops or set-aside. The England (1984) yield–nitrogen (N) formulae, which calculate the optimum N inputs for different crop prices, are used to estimate changes in N-fertilizer usage. In stage 2 (farm-practice effects on CH_4 and CO_2), an estimate is made of the impacts of N-fertilizer input changes and set-aside on the two greenhouse gases.

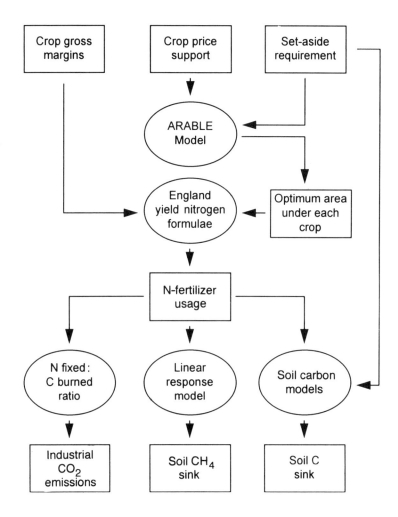

Fig. 10.1. Information flows and models used in analysing the arable sector of England and Wales. N, nitrogen; C, carbon.

Methane from the arable sector
Aerobic soils can act as a CH_4 sink, but it has been observed that the presence of ammonium (NH_4^+), such as from fertilizer, can reduce the sink strength. Given the current lack of available data on CH_4 uptake by arable soils, it is difficult to predict any policy effects. No method for estimating the soil CH_4 sink has yet been developed, even within the large CH_4 emission research programmes in the Organization for Economic Cooperation and Development (OECD) or the WATT Committee in the UK (Williams, 1994). Estimates of the sink reduction caused by fertilizing arable soils are between 30 and 75% (Mosier *et al.*, 1991). It is not clear how a reduction in fertilization

levels will affect CH_4 uptake, because there is some evidence for a permanent inhibitory effect, or at least an extended time-lag, before CH_4 oxidation rates recover. Current estimates of the sink strength of fertilized soils suggest estimates of approximately 1 kg CH_4 ha^{-1} year^{-1} (S.C. Jarvis, personal communication).

Assuming a 50% reduction in the sink strength when a soil is fertilized, an unfertilized soil would have a sink strength of 2 kg CH_4 ha^{-1} year^{-1}. In order to assess the possible impact of changing fertilizer applications on the CH_4 sink, a linear relationship between fertilizer inputs and CH_4 oxidation between 1 and 2 kg ha^{-1} year^{-1} is assumed, and the percentage change in fertilizer applications used to calculate a corresponding change in sink strength. A linear relationship is the simplest assumption in the absence of enough data to construct a true representation of the CH_4 sink response to N levels.

Carbon dioxide from the arable sector
Two widely used soil carbon (C) turnover models, the Century model (Parton *et al.*, 1989) and the Jenkinson model (Jenkinson and Rayner, 1977), are used to assess changes in the soil C sink resulting from fertilizer input changes and set-aside. The Century model is more flexible and able to deal with a higher level of management detail, such as conversions between arable and grassland systems and changing fertilizer levels. The model was developed for North American soils, but has been validated for the UK (Bradley, 1993). The Jenkinson model is used to assess the effect of changing biomass inputs on the long-term C balance.

The effect of changing fertilizer demand on CO_2 release during fertilizer manufacture is estimated, using the relationship of 1 kg N fixed requiring 1.5 kg C burned (Cole *et al.*, 1996). Although not all fertilizer used in the UK is necessarily manufactured in the UK, changes in CO_2 emissions from industry will have the same significance for global change, regardless of their origin.

Modelling the Livestock Sector

Dairy and beef cattle are analysed separately, because different policies apply. Although sheep, pig and poultry farming are also important UK livestock systems, the policies that govern them have remained constant since 1990 and therefore no major changes in greenhouse-gas emissions are expected. Figure 10.2 shows the information flows and interactions between the models used in this analysis.

Dairy sector
The dairy sector is subject to EU-wide milk-production quotas, which regulate production. Any change in cereal price is likely to affect feeding

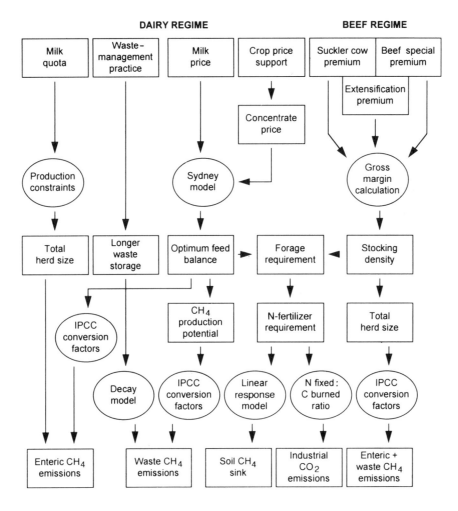

Fig. 10.2. Information flows and models used in analysing the livestock sector of England and Wales. IPCC, Intergovernmental Panel on Climate Change.

strategies, as the economic balance between the cost of feeding concentrates (as opposed to forage) and the improved yields will change. In stage 1, the Sydney model, which calculates optimum feeding patterns for dairy cattle, given the prices of milk and concentrates, is used to estimate a new feeding strategy, assuming concentrate prices will drop by half the grain-price reduction. The probable changes in herd sizes are derived from gross margin calculations and estimated yields, given the limitation imposed by the milk quota.

Stage 2 analyses the likely effect of the new feeding strategy on the two greenhouse gases. The Intergovernmental Panel on Climate Change (IPCC) constants are used to estimate the CH_4 produced by enteric fermentation and from wastes (slurry and manure). Methane released from enteric fermentation is calculated from the percentages of gross feed energy released as CH_4. Gibbs and Leng (1993) suggest appropriate values as 3.5% for forage diets and 6.0% for high-grain diets.

The waste-storage conditions are taken into account, because they influence whether the waste is degraded aerobically or anaerobically. The IPCC CH_4 production potential and conversion factors for waste under different storage conditions were taken (Gibbs and Woodbury, 1993). The requirement for new waste stores to have at least a 4-month capacity is likely to increase storage times, increasing the potential for CH_4 emission from stores. A simple exponential-decay model was constructed to calculate CH_4 release from slurry stores. Finally, the increased concentrate feeding would lower the demand for forage, in turn lowering the fertilizer requirements for forage. The impacts of the reduced fertilizer inputs on the soil CH_4 sink and on CO_2 emissions from fertilizer manufacture are calculated as before.

Beef sector

The beef intervention price has been planned to be reduced by 15% over the 3 years from July 1993, with the beef special premium (BSP) being introduced to compensate for this. The suite of policies affecting the beef sector are likely to promote extensification. The BSP and the suckler-cow premium both have maximum stocking densities. Farmers cannot claim for animals above set limits on livestock units per hectare. Detailed gross margin calculations indicate that it would be more profitable for the farmer to reduce the stocking density than to produce beef without the premiums. The extensification premium further reinforces the trend.

In stage 1 of the analysis, the forage-feeder model is used to estimate the optimum feeding strategy for beef. Gross margin calculations of the total herd size and the forage requirement are carried out as before. In stage 2, the effects on CH_4 emissions from enteric fermentation and wastes are estimated, as well as soil CH_4 sink and industrial CO_2 impacts, using the methods described above.

Environmental Policies in the Agricultural Sector

The likely impacts of four major environmental policies are examined: the ban on straw burning; the Nitrate Sensitive Areas (NSAs) scheme; the Environmentally Sensitive Areas (ESAs) scheme; and the Farm Woodland Premium Scheme (FWPS). Figure 10.3 shows the information flows and interactions between the models used in this analysis. These schemes have been implemented in support of the responsibilities of the Ministry of Agriculture, Fisheries and Food (MAFF) for the integration of environmental

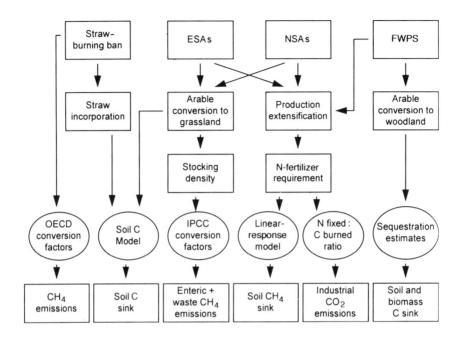

Fig. 10.3. Information flows and models used in analysing the agri-environmental sector of England and Wales.

and agricultural policies in the reformed CAP. All four are analysed, using the models and calculations described above, and are explained in more detail in the section below for clarity.

RESULTS AND DISCUSSION

Arable Policies: Carbon Dioxide and Methane Implications

Output from the ARABLE model and an analysis of fertilizer-usage statistics estimate that, by 2000, N applications to arable areas will reduce by 14% on cropped land from 1990 levels, rising to 27% when land in set-aside is included (Armstrong Brown *et al.*, 1996).

Assuming a linear increase in soil CH_4 sink strength, a 27% reduction in fertilizer inputs will result in an increase in CH_4 sink strength of the order of 0.27 kg CH_4 ha^{-1} year^{-1}. The total crop area in 1990 was 4956 kha (census data), with an assumed sink strength of 4.96 kt CH_4 year^{-1}. Therefore, the overall increase in the CH_4 sink resulting from fertilizer reductions is estimated to be 1.34 kt CH_4 year^{-1}.

The Century model predicts a negligible impact on soil C levels arising from a 14% reduction in fertilizer levels on cropped arable land. It is possible that a slight decrease may occur, but the errors associated with organic-matter responses to N inputs are too large for the effect of such a relatively small reduction to be predicted with confidence.

Estimating the effect of set-aside on soil C sequestration depends on having a realistic estimate of the biomass of a set-aside field, which is highly variable. The Century model is used to estimate a maximum increase rate, using an unfertilized grass cover to simulate the set-aside year. An average increase in C is estimated at 8938 kg ha^{-1} over 50 years, or 179 kg C ha^{-1} year^{-1}. In fact, during the first decade, the sequestration rate is likely to be higher, because initial accumulation is more rapid. To assess lower biomass returns, the Jenkinson model is used to predict C accumulation with inputs of 5 t C ha^{-1} and 4 t C ha^{-1}. The predicted accumulation rates are 106 and 40 kg C ha^{-1} year^{-1}, respectively.

Cereal, oil-seed and protein (COP) crops covered an area of 4,423 kha in 1990 (MAFF, 1994a). Over this area, Century predicts a set-aside C sequestration of 791 kt year^{-1}. The two Jenkinson-model runs predict a total sequestration of 469 kt year^{-1} with a 5 t ha^{-1} biomass input, and 177 kt year^{-1} with a 4 t ha^{-1} biomass input.

These results demonstrate the dependence of estimates of C sequestration in rotational set-aside fields on accurate assessments of the biomass. Until more data are available on the proportions of fields with natural cover and planted cover and an average biomass is estimated for each type, a more accurate assessment of the effect of the scheme will not be possible. At present, it is assumed that an intermediate value would be the most representative figure, and 468.84 kt C year^{-1} has been used. The predicted 139 kt N-fertilizer reduction arising from the arable regime will result in a 208.5 kt year^{-1} reduction in C released as industrial CO_2 during fertilizer manufacture.

Dairy Policies: Carbon Dioxide and Methane Implications

The Sydney model calculates that with current prices (concentrates £142 t^{-1}, £0.222 l^{-1} of milk), the annual optimal concentrate intake is 1.46 t cow^{-1} year^{-1} for a yield of 5,581 l milk. This corresponds closely to the figures of 1.50 t and 5,550 l, respectively, given in Agro Business Consultants (1994). Assuming concentrate prices will drop by half the grain-price reduction, we would expect to see a reduction in the price of concentrates of £20 t^{-1}. According to the model, this would result in an increase in concentrate usage of 1.98 kg day^{-1} to 2.18 t cow^{-1} year^{-1}, with an associated rise in milk yield to 5,883 l.

The Sydney model predicts an average daily intake of 4 kg dry-matter (DM) concentrates and 14.2 kg DM forage, which results in CH_4 production of 64.6 kg cow^{-1} year^{-1}. This figure is significantly below the average value of 87 kg given by Williams (1994), citing Crutzen *et al.* (1986) and Eggleston and Williams (1989). In two more recent publications, Jarvis and Pain (1993, 1994) suggest 49 kg of carbon as CH_4 and 41.1 kg CH_4-C. Our estimate is based on these figures and is 46.2 kg CH_4-C.

For each extra kilogram of concentrates eaten per day, reducing the forage intake by 0.64 kg, 1.6 g extra CH_4 is lost. Over a year, this amounts to 0.59 kg CH_4 cow^{-1}. An extra 1.98 kg concentrates day^{-1} is predicted to be fed. Thus, the increase in CH_4 emissions per cow will be 1.16 kg CH_4 cow^{-1} year^{-1}.

Methane is generated by the waste as it decays. On average, 5.5 kg volatile solids per day is produced by each cow. The current CH_4 production potential for a cow on a typical UK diet is estimated as 0.24 kg CH_4 cow^{-1} day^{-1} (van Amstel, 1993). The proportion of this CH_4 that is actually produced depends on the waste-storage conditions.

The current IPCC estimates for the methane conversion factors (MCFs), for wastes under different conditions are slurry store 10%, solid store 1% and field 1%. A recent ADAS/MAFF survey of animal-waste storage (Nicholson and Brewer, 1994) gives the proportion of wastes in each category. Assuming the cows are housed for half the year, we estimate the fate of waste from UK dairy cows as slurry store 38%, solid store 12% and field 50%. This gives a net MCF of 4.42%. Using all the above constants, we calculate that in 1990 each cow produced 15.2 kg CH_4 year^{-1}.

Using data from Gibbs and Leng (1993), we estimate that the CH_4 production potential of waste will increase by about 7% for each kilogram of concentrates eaten daily. The extra 1.98 kg of concentrates predicted by the Sydney model will therefore result in an increase of 14% in the CH_4-producing potential of the waste.

If the storage and disposal of the waste are unchanged, the CH_4 production would simply increase by 14%. However, farm waste-storage capacity requirements are likely to increase slurry-storage periods. Currently, the average storage capacity on a farm using a slurry system is around 2 months (Nicholson and Brewer, 1994). The *Code of Good Agricultural Practice for the Protection of Water* (MAFF, 1991) stipulates at least 4 months' storage capacity for new constructions. Using a simple exponential-decay model, we predict that the effect of increasing the average storage period from 2 to 3 months would lead to an increase of 17% in the MCF for slurry (from 10% to 11.7%). This would result in an overall MCF of 5.1% (an increase of 15% on the previous figure of 4.42%).

The net increase in CH_4 production from dairy waste is therefore predicted to be 31% or 4.7 kg CH_4 year^{-1} cow^{-1}, taking both the feeding and the waste-storage changes into account. The resulting CH_4 production from a dairy cow in response to changes in feeding and waste storage is shown in Table 10.1. The year 2000 is used to represent post-policy emissions,

Table 10.1. Estimated changes in CH_4 production from different livestock systems.

Source	Methane production per head (kg CH_4 year^{-1})	
	1990	2000
From cow	64.6	65.8
From waste	15.2	19.9
Total	79.8	85.7

assuming no further changes in policy and prices between 1995, when the reforms will have been phased in, and 2000. However, because of fixed quota levels and increased yield per cow, the number of cows should decline. The predicted increase in yield due to the change in feeding is 5.4%, which should lead to an equivalent decrease in the national herd from 3.38 million in 1990 to 3.21 million. Thus, the total CH_4 emissions are predicted to be 270 kt year^{-1} in 1990 and 275 kt year^{-1} in 2000, a net emission increase of 5 kt CH_4 year^{-1}.

The additional 5 kt CH_4 represents a 1.8% increase in emissions from this source. The decrease in herd size (due to increased yield per cow) is almost exactly offset by the increase in CH_4 emissions per cow (if the slurry-storage period was unchanged), giving no net change in emissions from the herd. This remains the case for any change in feeding, and does not depend on the assumed drop in concentrate price of £20 t^{-1}. The increase in emissions can therefore be attributed directly to the assumed increase in the average length of time that slurry is stored.

The soil CH_4 sink may be increased by a reduction in pasture fertilizer levels, as a result of reduced forage demand. The maximum fertilizer change is estimated to be a drop of 27%, resulting in a 0.27 kg ha^{-1} year^{-1} increase in the CH_4 sink, as discussed above. The dairy forage area covers 1.71 million ha, resulting in a total increase in the annual CH_4 sink of 0.46 kt CH_4 from the 1990 level of 1.71 kt. In addition to the above effects, the estimated reduction in fertilizer inputs will result in a reduction in industrial CO_2 released during fertilizer manufacture of 154.5 kt C.

Beef Policies: Carbon Dioxide and Methane Implications

The BSP has been phased in incrementally in the period up to 1996, and is paid where the stocking rate is up to 2.0 livestock units (LU) ha^{-1} (with male cattle from 6 months to 2 years old counting as 0.6 LUs). The suckler-cow premium has been paid from 1995 for each breeding cow in a herd for beef production, with an eventual limit of 2 LU ha^{-1}. There is a further extensification premium if densities are reduced below 1.4 LU ha^{-1}.

Using detailed gross-margin calculations, it is reasonable to assume that all farms with stocking densities between 2 and 3.5 LU ha^{-1} will be reduced to 2 LU ha^{-1}, but densities above 3.5 LU will not be reduced, as these farms are probably almost all intensive beef producers feeding cereal and silage. We also assume that the average reduction in density on farms that apply for the premiums is 0.5 LU ha^{-1}, as most farms will be nearer the lower end of this range. Using the data in the Meat and Livestock Commission (MLC) *Beef Yearbook* (MLC, 1993), we estimate that 13% of cattle are on farms with densities in this range, and so the net reduction in cattle density is equivalent to a 2.6% reduction in cattle numbers.

For the extensification premium, it is assumed that the reduction to 1.4 LU ha^{-1} will be attractive to farms with a density of 1.6 LU ha^{-1} or below. Table 10.2 shows that about 20% of UK cattle are on farms with a stocking density of 1.4–2.0 LU ha^{-1}. Assuming that a third of the farms in the 1.4–2.0 LU ha^{-1} density band lie between 1.4 and 1.6 LU ha^{-1}, 6.5% of cattle are on these farms. If all these farms were to claim the extensification premium, this would amount to a 0.4% reduction in cattle numbers. Taken together, the beef regimes are expected to produce a 3% reduction in cattle densities.

Table 10.2. Percentage of UK farms in various stocking-density bands by region (from MLC, 1993).

	0–1.4 LU ha^{-1}	1.4–2.0 LU ha^{-1}	2.0–3.5 LU ha^{-1}	3.5+ LU ha^{-1}	Percentage of UK cattle (%)
Eastern	47	21	16	16	15
Midlands	50	14	36	0	5
Northern	55	17	20	8	20
Scotland	73	16	9	2	35
South-West	55	32	10	3	12
Wales	73	24	3	0	11
Total	61	19	13	5	100*

*Errors due to rounding.

The beef herd is comprised of beef calves and suckler cows, which, because of their different ages and sizes, need to be assessed separately for CH_4 production. Using the estimates on diet given in Agro Business Consultants (1994) and assuming that beef calves, as for dairy cattle, are similar in this respect, we calculate emission values of 17.4 kg CH_4 per animal per year for cereal beef, 22.1 kg for grass beef and 29.5 kg for silage beef. The wide range of values is caused by variations in the feeding and fattening periods for the different methods. The feeding estimates in the MLC *Beef Yearbook* (MLC, 1993) are also broadly in agreement with the above. The much lower CH_4 output, compared with dairy cows, is reasonable, given that beef cattle are very much smaller for most of their lives. We expect that most beef production is grass beef, and so use 22.1 kg CH_4 year^{-1} as our estimate for emissions from enteric fermentation.

To calculate emissions from wastes, the production of volatile solids (VS) is calculated as a proportion of beef-cattle diet, in the same way as for dairy cows. This amounts to 2.64 kg VS day^{-1}, which agrees well with the OECD figure of 2.7 kg VS day^{-1}. The CH_4 production from beef-cattle waste is calculated as 3.1 kg CH_4 year^{-1}. The total emissions from enteric fermentation and wastes is estimated as 25.2 kg CH_4 beef calf^{-1} year^{-1} in 1990.

For adult suckler cows, the Agro Business Consultants (1994) estimates for the diet of a lowland spring-calving suckler cow were used as the basis for calculating CH_4 output. Cows over 2 years old produce 46.8 kg CH_4 cow^{-1} year^{-1}. The annual CH_4 production from the waste of each of these cows is 8.5 m^3 CH_4 or 6.1 kg CH_4. This gives a total of 52.9 kg CH_4 animal^{-1} year^{-1} in 1990. Using census figures for the age breakdown of the herd, CH_4 emissions from the total beef herd in 1990 were 253 kt CH_4 year^{-1}. The predicted 3% reduction in beef-cattle numbers arising from the three policies would save 7.58 kt CH_4 year^{-1}, reducing the total beef-herd emissions to 245 kt CH_4 year^{-1} in 2000.

Using the assumptions outlined previously, an 11% reduction in fertilizer added to beef pasture is estimated to increase the soil CH_4 sink to 1.11 kg ha^{-1} year^{-1}, an increase of 11%. This would occur on 2.1 Mha of beef pasture, with an original sink of 2.10 kt CH_4 year^{-1}, giving a total CH_4 sink increase of 0.23 kt CH_4 year^{-1}. The estimated 31.5 kt N-fertilizer year^{-1} reduction would also reduce industrial C emissions from the fossil fuel used in fertilizer manufacture by 47.25 kt C year^{-1}.

The United Kingdom Ban on Straw-burning: Carbon Dioxide and Methane Implications

After the 1992 harvest, straw-burning was banned in England and Wales (with some limited exceptions). In 1992, straw-disposal methods were 52% baled, 11% burnt and 37% incorporated (MAFF, 1994b). Averaging the figures for the two post-ban harvests gives 68% baled and 32% incorporated. The total area burnt in 1992 was 330.5 kha, almost all of which was winter wheat. Estimates of the yield of straw vary (Audsley, 1987), but we assume a yield of 5 t ha^{-1} wheat straw, plus 2 t ha^{-1} of stubble, giving 7 t ha^{-1} in total. Using OECD (1991) estimates for the DM content of straw at 83% and C at 48% of DM, this amounts to 2.8 t ha^{-1} C, a total of 930 kt C burnt in 1992. Of this, the OECD guidelines assume that 90% actually combusts (OECD, 1991), giving 830 kt C burnt in 1992.

Amounts of CH_4 released by burning C in straw are not well documented, and estimates vary. In this case, we estimate that 8.3 kt CH_4-C to 11.1 kt CH_4 was emitted in 1992, assuming that 1% of the C in burnt straw is released as CH_4 (OECD, 1991). The straw-burning ban therefore results in annual CH_4 emissions being reduced by this amount.

The Jenkinson model is used to assess the effect of the straw-burning ban on C sequestration. The model predicts that, if the total annual C input to the soil is increased, around 40% of the extra C added will be sequestered over a 10-year period (depending on soil type and cultivations), the remainder being emitted as CO_2. The increase in the soil C bank then drops to roughly half this rate over the subsequent 40 years and gradually approaches a new equilibrium. Baled straw is used mainly as cattle bedding, with some as feed, the remains of which are returned to fields. In the long term, the decomposition rate of this straw is likely to be similar to that added directly to the soil. The C sequestered over the 10 years following the ban is therefore about 330 kt C year^{-1}.

Nitrate Sensitive Areas: Carbon Dioxide and Methane Implications

Data on total livestock numbers in NSAs are not available. There is no evidence for a change in livestock numbers, with excess manure being exported from the NSAs. It is therefore assumed that no change in CH_4 emissions results from livestock changes following the introduction of NSAs. The soil CH_4 sink may, however, be increased as a result of reduced fertilizer inputs. The 26.9% reduction in N-fertilizer applications over the NSAs would increase the soil CH_4 sink from 1 to 1.27 kg ha^{-1} year^{-1}. This 0.27 kg increase, occurring over the 39,462 ha covered by the NSAs from 1994 onwards, would give a total CH_4 sink increase of 10.65 t year^{-1}, or 0.01 kt CH_4 year^{-1}, from the pre-NSA sink of 0.04 kt year^{-1}.

The Century model is used to predict the effect of the NSA fertilizer reductions and conversions from arable to grassland. Under the pilot NSA scheme 1,525 ha were converted to grassland, and 2,600 ha under the new NSAs. Around one-third of the area is in an option where up to 150 kg ha^{-1} year^{-1} N is permitted, the remainder being unfertilized. The Century model indicates that, when arable land is converted to grassland with these levels of fertilizer input, the soil C increases by an average of 32.4 t ha^{-1} after 50 years. This gives an average sink increase of 0.65 kt C year^{-1}. Using the relationship of 1 kg N fixed to 1.5 kg C from fossil fuels, a total fertilizer reduction of 1,499 t N year^{-1} from 1994 onwards will lead to a saving of 2,249 t C year^{-1} in industrial fossil-fuel use.

Environmentally Sensitive Areas: Carbon Dioxide and Methane Implications

Analysing the impact of ESAs on C fluxes is difficult, because each ESA has a different internal structure, with a variety of tiered options. In addition, most monitoring work has been concerned with rural socio-economics, landscape features or ecology, with little information available on fertilizer or livestock practices. The analysis outlined here relies on a monitoring exercise carried out on the first six stage 1 ESAs, which included an assessment of N usage and stocking density (Froud, 1994).

The available data do not indicate any major changes in stocking densities on ESA land (Froud, 1994). Of greater interest to the present study are the total herd sizes. This information is not directly available and must be inferred from the arable-land conversions. A total of 7,419 ha of arable land has been converted under the ESA schemes. The mean post-ESA stocking density on the six stage 1 ESAs is 1.4 LU ha^{-1}. Assuming that all the converted arable land is stocked at this density, the total number of new livestock will be equivalent to 10,387 LU. Using a *pro rata* CH_4 output per LU, based on the rate calculated for grass beef (25.2 kg animal^{-1} year^{-1}), each LU would emit 42 kg CH_4 year^{-1}. Over the whole arable-reversion area, this amounts to 0.44 kt CH_4 year^{-1}.

There may also be an increase in CH_4 oxidation rates in the soil, as a result of reduced fertilizer applications. The 51.2% decrease in N-fertilizer applications from model estimates would increase the soil CH_4 sink from 1 to 1.51 kg ha^{-1} year^{-1}. Occurring over a total ESA area of 271.6 kha, this gives an increase in the CH_4 sink of 0.14 kt CH_4 year^{-1}, from the pre-ESA sink of 0.27 kt year^{-1}.

The Century model is used to calculate the effect on soil C sequestration of arable conversions to permanent grass. A total of 7,419 ha of arable land was converted to permanent grass in the first tranche of ESAs. Given an average accumulation rate of 32,427 kg ha^{-1} over 50 years, a total of 240.57 kt C is sequestered. On an average annual basis, this becomes 4.81 kt C year^{-1}. A 6,816 t N year^{-1} reduction in N-fertilizer will also result in industrial CO_2 emissions being reduced by 10.22 kt C year^{-1}.

Farm Woodland Premium Scheme: Carbon Dioxide and Methane Implications

The FWPS has had an uptake rate of 3,000–5,000 ha year^{-1} since it was introduced in 1992. Crabtree's chapter (Chapter 14) in this volume assesses the economic incentives associated with this scheme and its impact on C fluxes. This section examines the CH_4 as well as the CO_2 implications of the scheme.

Because fertilizer applications would cease under the FWPS, the soil CH_4 sink may recover to its assumed prefertilizer level of 2 kg ha^{-1} year^{-1}. Occurring over 1 year's converted land (5,000 ha), this would be a sink increase of 1 kg ha^{-1} year^{-1}, or 5 t CH_4 (0.005 kt CH_4). Using the same assumptions as previously, the land converted after a decade would have an increased sink of 0.05 kt CH_4 year^{-1}.

The Century model is not used in this case to estimate the effect of converting arable land to forest, because it has not been validated for UK forest soils. Schiffman and Johnson (1991) found that reforestation of old agricultural fields resulted in an average carbon increase of 120 t C ha^{-1} over 50 years. Of this, 14% was in the soil, 10% in the litter, and 76% was in the phytomass, accounting for 91.2 t C ha^{-1}. This value agrees with Wiersum and Ketners' (1989) upper estimate of an accumulation of 90 t ha^{-1} over a 30-year forest life. The total accumulation of 120 t over 50 years gives an average of 2.4 t ha^{-1} year^{-1}, which, on 1 year's converted land (5,000 ha), would equal 12 kt C. The total amount would depend on how long the FWPS was taken up at this rate. For example, if the uptake rate continued at 5,000 ha year^{-1} for a decade, the total C sequestration would be 660 kt.

SUMMARY AND CONCLUSIONS

Table 10.3 summarizes the estimated effects of policies on agricultural CH_4 and CO_2 fluxes, described in the section above. The results are shown as a positive or negative effect in kilotonnes per year. The results for reduced industrial CO_2 emissions resulting from decreased fertilizer manufacture are not included, because, although the demand for fertilizer is influenced by agricultural policy, the CO_2 released by its manufacture is accounted for in the industrial inventories. Additionally, the decrease will occur partly outside the UK, which is a net fertilizer importer. All the results have been converted to CO_2-equivalents, for ease of comparison, using the relative warming potential of 21 for CH_4 (i.e. one molecule of CH_4 will have 21 times the global-warming effect of one molecule of CO_2).

Methane

Table 10.3 shows an overall decrease in CH_4 emissions from agriculture of 13 kt year^{-1}, despite a predicted increase from dairy farming as a result of longer waste-storage times. Environmentally Sensitive Areas are likely to be responsible for a small amount of CH_4 as arable land is converted to pasture. These

Table 10.3. Summary of MAFF policy effects on agricultural methane and carbon dioxide fluxes.

Greenhouse-gas flux	Magnitude and direction of effect	CO_2-equivalent kt CO_2 year^{-1}
CH_4 emissions	−13 kt CH_4 year^{-1}	280
CH_4 sink	+ 2 kt CH_4 year^{-1}	47
CO_2 sink	+816 kt C year^{-1}	2,993
Net flux		3,320

increases are cancelled out, however, by the reduction in emissions caused by the declining beef herd. The largest reduction arises because of the ban on straw-burning, which is estimated to reduce emissions by 11 kt CH_4 year^{-1}.

The soil CH_4 sink is estimated to increase by 2 kt year^{-1}. It must be emphasized that, because of the large uncertainties and lack of data, this figure is speculative, intended only as a possible indicator of the direction and magnitude of any response. Not surprisingly, the arable sector has the largest potential impact on the CH_4 sink, as the greatest changes in fertilizer applications are estimated to take place here.

The UK Climate Change Programme (UK DoE, 1994) anticipates a 10% reduction in CH_4 emissions from 1990 levels by 2000, resulting from the use of CH_4 for energy generation and reduced industrial emissions. This amounts to 484 kt CH_4. The total effect of the policies analysed here is an effective reduction of 15.5 kt CH_4 (including both sources and sinks).

Carbon Dioxide

The CO_2 sink results are presented in kt C year^{-1} for ease of comparison, but it should be noted that these are long-term averages over 50 years, and a linear response cannot be assumed. Both set-aside and the ban on straw-burning are estimated to have significant effects on C sequestration, although this is a very small percentage of the total C sink. The effect of these two policies is relatively large, because they cover the whole UK arable area. More C is sequestered per hectare following land-use conversions to grassland in ESAs and NSAs, but the limited area extent of these policies means their overall impact is small. In fact, the overall increase in the C sink resulting from all the policies is less than 0.1% of the total soil and vegetation sink of 9.61 billion tonnes C (UK DoE, 1994).

The Climate Change Programme (UK DoE, 1994) states that the programme of measures to reduce greenhouse-gas levels includes reducing CO_2 emissions by 10 Mt C against projected emissions by 2000. The effect of the agricultural policies analysed here is an effective reduction of 3.0 Mt C. The estimated impact on industrial C emissions, as a result of fertilizer-use decreases, amounts to 1,552 kt CO_2 year^{-1} by 1995, the time the price reforms are phased in. This is 21% of the CO_2 emissions associated with agriculture in the 1990 inventory.

NOTE

1. The research reported in this chapter was funded by the UK Ministry of Agriculture Fisheries and Food under contract Nos CCO318 and CCO319, which is gratefully acknowledged.

11 Full Cycle Emissions from Extensive and Intensive Beef Production in Europe

Susan Subak

INTRODUCTION

Beef is a greenhouse-gas intensive food, and emissions of methane and other greenhouse gases related to its production vary considerably depending upon the feeding regime used and related technological inputs. Understanding the greenhouse-gas implications of different livestock-production modes is essential to an effective greenhouse-gas minimizing strategy. These implications are especially complex in the case of livestock systems because raising animals can involve a range of greenhouse gases and sources, including carbon dioxide from fossil-fuel inputs, nitrous oxide emissions from fertilizer use and biomass burning, alterations in carbon storage potential through alternative land uses for pasture and stock-feed crops and, finally, methane related to cattle digestion and manure. Worldwide, livestock are likely to be the greatest anthropogenic contribution to methane emissions, which, as a potent greenhouse gas, contribute to the risks of global warming (Watson *et al.*, 1992), and beef cattle contribute at least half of livestock-related methane emissions (Subak *et al.*, 1993). Recent research indicates that carbon dioxide emissions related to fossil-fuel inputs may be comparable in heating-equivalent terms to a quarter or more of the methane contribution from enteric fermentation and manures (see Löthe *et al.*, Chapter 12, this volume). For these reasons, it is important to understand how different livestock management systems, such as intensive feeding versus extensive pasture, contribute to different mixes of methane, carbon dioxide and nitrous oxide emissions. One of the main strategies for reducing total methane emissions is to increase animal productivity, e.g. production of beef per head, while another relates to changing the composition of animal feed.

For several decades, European agronomic research has investigated possible dietary interventions that may inhibit methane production in the animal rumen (see, for example, Czerkwaski, 1969; Demeyer and van Nevel, 1975). These experiments predate concern over the greenhouse-gas consequence of livestock production and were aimed at improving efficiency by preventing the 2–15% loss of gross energy in animal feed due to methane production. While some incremental reduction in methane emissions is expected to develop from the feed-supplement research, major reductions in methanogenesis are unlikely to be achieved without unacceptable negative effects on overall digestion and animal health (van Nevel and Demeyer, 1996). Accordingly, methane-abatement strategies related to livestock production have, for the most part, centred on increasing the productivity of the livestock industry (US EPA, 1993). Emissions per unit of beef or milk product vary greatly in different regions and have been reduced inadvertently over time in Europe and North America through the use of more productive breeds, careful management of age-cohort structure and improvements in feeding quality. These management strategies have tended to shorten the life span of a methane-generating animal, for it is then readied for slaughter at a younger age.

None the less, the issue of productivity improvements relative to greenhouse-gas emissions in agricultural systems is highly complex. Far more emissions, sources and sinks are relevant to the analysis than are conventionally included in national assessments of livestock emissions. In addition to the 'direct' emissions – methane released from enteric fermentation and manures and the nitrous oxide released from manures – carbon dioxide from fossil fuels is 'embodied' in most livestock-production systems, chiefly because of the mechanization of cropping used to raise stock feeds and because of the fuel inputs needed for fertilizer production. Moreover, land used for grazing or stock-feed production can represent displacement from higher-carbon-density land cover, such as forests or woodlands. As each bovine in Europe requires, on average, 0.5 ha of land under grazing and/or stock feed (European Commission, 1987) and there are more than 80 million beef cattle in Europe (FAO, 1996), the land-use implications of beef production are enormous.

Generally, there is a trade-off between land-intensive systems and energy-intensive systems and, with the intensification of agriculture, increased stocking densities have been achieved at the cost of added fossil-fuel consumption. The substitution between land and energy are most extreme between, on the one hand, feedlot beef systems in North America, Europe and Japan and, on the other, pastoral beef production in Sahelian Africa. The land required to produce 1 kg of beef protein in a high-energy-input feedlot in upstate New York has been estimated to be 136 m^2, compared with

25,700 m² for the same amount of beef protein produced through traditional herding in arid Africa (Giampietro *et al.*, 1992).

While there is tremendous variation in livestock-production systems within Europe, beef production is believed to be generally more energy-intensive in Europe than in any of the other world regions. An assessment completed in the early 1980s estimates gross energy requirement for beef production at 9–102 GJ t^{-1} of beef in Europe. This range compares with 4–79 GJ t^{-1} in the USA and 4–32 GJ t^{-1} for beef produced in Brazil for export (Slesser and Wallace, 1982). Unlike other agricultural products that have been compared for their energy intensiveness in Europe compared with North America, the higher-energy inputs in European beef production do not appear to be offset by higher product outputs. European production was found to be more energy-intensive at a given level of beef production. These findings are not surprising given that Europe is more densely populated than North America and lacks the vast expanse of rangelands found in the American West.

The question then arises, how does the greenhouse-gas intensity of beef production in temperate regions that are relatively energy-intensive compare with systems that are relatively land-intensive? Is higher-energy intensity compensated for by the shorter animal lifetime made possible by higher-quality feed inputs? These questions are difficult to answer given the complex boundaries of beef-production systems in Europe and elsewhere. None the less, in this chapter an assessment of greenhouse-gas emission intensity has been completed for two beef-production systems, characterized as 'intensive' and 'extensive'. In the first instance, the relationship between direct greenhouse-gas emissions and intensification is described based on methane-emission analysis derived from published estimates of average European feeding regimes for 'extensive' and 'intensive' systems. Secondly, sensitivity analysis is used to describe the trade-off between land and energy inputs in beef-production systems.

ENTERIC FERMENTATION

Intensive and extensive agricultural systems can involve a variety of energy inputs and land uses, but for this analysis the intensive systems involve larger daily intakes at a ratio as defined by the Commission of the European Communities studies. In their research on feed requirements for European beef cattle, they derive feeding amounts based, in the intensive case, on 20 months of feeding (610 days), compared with 30 months of feeding (915 days) in the extensive case (Janssens, 1990). Methane emissions can be calculated based on intake of metabolizable energy. Metabolizable energy is gross energy intake by an animal less combustible gas and urinary and faecal losses. From their records of metabolizable energy required for cattle in each system, both annual and lifetime methane emissions can be calculated and compared per unit of beef produced. Using established relationships between feed intake, energy requirements for animal energy expenditure and feed quality (IPCC/OECD, 1994), methane emissions from enteric fermentation have been estimated for both systems. The assumptions and results are listed in Tables 11.1–11.3.

Table 11.1. Feeding and growth characteristics of intensive and extensive beef systems in Europe (from Jansens, 1990).

	Intensive systems			Extensive systems			
	< 1 year	1–2 years	Mean	< 1 year	1–2 years	> 2 years	Mean
Average live weight (kg)	220	500		170	405	555	
Daily gain (kg day^{-1})	0.98	0.82		0.71	0.57	0.49	
Days in phase	365	245		365	365	215	
Metabolizable energy (ME)							
(MJ ME head^{-1} year^{-1})	17,626	21,346		11,547	23,637	15,707	
(average MJ ME							
head^{-1} day^{-1})			64				56
Digestible energy (DE)							
(MJ head^{-1} day^{-1})			78				68
Days in year			365				365
Beef yield (kg head^{-1})			217				217
Conversion factor			55.6				55.6
Lifetime (days)			610				915

DE = ME × 1.22.

Table 11.2. Methane emissions from enteric fermentation: uniform feeding and feeding effort.

	Intensive	Extensive
Feed excreted (%)	40	40
Gross energy (GE) intake (MJ day^{-1})	130	113
Methane yield (Ym)	0.06	0.06
Beef yield (kg head^{-1})	217	217
Lifetime methane (kg CH$_4$)	85	111
Emissions yield^{-1} (g CH$_4$ kg^{-1} beef)	0.39	0.51
CO$_2$ eq yield^{-1} (kg CO$_2$ eq kg^{-1} beef)	9.2	12.1
Carbon eq yield^{-1} (kg C eq kg^{-1} beef)	2.5	3.3

GE = DE + feed excreted; emissions = GE × Ym × days CF^{-1}.
DE, digestible energy; CF, conversion factor; CH$_4$, methane; CO$_2$, carbon dioxide; eq, equivalent; C, carbon.

Table 11.3. Methane emissions from enteric fermentation: improved feeding for intensive system.

	Intensive	Extensive
Feed excreted (%)	30	40
Gross energy (GE) intake (MJ day^{-1})	112	113
Methane yield (Ym)	0.05	0.06
Beef yield (kg head^{-1})	217	217
Lifetime methane (kg CH$_4$)	61	111
Emissions yield^{-1} (g CH$_4$ kg^{-1} beef)	0.28	0.51
CO$_2$ eq yield^{-1} (kg CO$_2$ eq kg^{-1} beef)	6.6	12.1
Carbon eq yield^{-1} (kg C eq kg^{-1} beef)	1.8	3.3

GE = DE + feed excreted; emissions = GE × Ym × days CF^{-1}.
DE, digestible energy; CF, conversion factor; CH$_4$, methane; CO$_2$, carbon dioxide; eq, equivalent; C, carbon.

Methane emissions from each system were derived first assuming the same quality of feed and energy expenditures by animals in both systems. These assumptions resulted in lifetime methane emissions that were 30% greater in the extensive system. Note that the animals raised in the extensive system live 50% longer than the intensively raised cattle, but it is assumed that the beef yield per animal is the same. If we assume that the animals in the extensive system were eating lower-quality feed and were expending more energy on grazing, the divergence in lifetime methane emissions becomes much greater. These assumptions, which are consistent with the lower biodigestibility of feeds expected in the more extensive system, result in 80% higher methane emissions in the extensive system.

MANURE MANAGEMENT

On a worldwide basis, methane emissions from cattle manure are believed to be about one-quarter of the scale of methane emissions from enteric fermentation. Most of the emissions from wastes are related to disposal methods used in feedlot systems. In Europe, liquid slurry systems are in widespread use for cattle. These systems involve storing the manure in water in lined tanks until it is applied to fields. Since the disposal is 'wet', considerable methane emissions may be involved and, for some feedlot systems, the manure emissions will surpass those from enteric fermentation. Methane emissions from the intensively raised cattle are estimated to release as much as 77 kg of methane over the animal's lifetime under the liquid slurry method. Manure from intensively raised animals can also involve dry disposal, such as 'daily spread', in which manure is collected and applied to fields regularly, in which case emissions from manure can be even lower than the extensive emissions, because of the animals' shorter life span.

Table 11.4. Manure-management emissions for cattle.

Manure management	Intensive		Extensive pasture
	Liquid slurry	Solid storage	
Emission factor (kg CH_4 head^{-1} year^{-1})	46	2	2
Emissions per lifetime (kg CH_4 head^{-1})	77	3	5
Emissions (kg CH_4 kg^{-1} beef)	0.35	0.01	0.02
Emissions (kg C eq kg^{-1} beef)	2.3	0.1	0.2

CH_4, methane; C eq, carbon equivalent.

Emissions from extensively grazed animals, however, tend to involve lower emissions because the manure is allowed to lie as it is and mainly decomposes aerobically. Methane emissions from extensively grazed cattle systems, therefore, can involve a fraction of the methane emissions – as little as a few kilograms of methane over the animal's lifetime. These calculations are based upon emission factors developed by Safley and colleagues (1992) and involve large uncertainties. None the less, the range of methane emissions from manure disposal in intensive systems, 3–77 kg of methane over the animal's lifetime, as summarized in Table 11.4, could encompass a variety of disposal systems that are intermediate in emission level.

ENERGY EMBODIED IN AGRICULTURE

Intensification in Europe and elsewhere has been achieved at the cost of added energy inputs for the production of crops raised for stock feed, such as wheat, maize, barley and oilseed rape. These crops require greater inputs of nitrogen fertilizer and mechanization in the processing of the grains than is required for the maintenance of most grazing land. Gross energy requirements for beef production to the 'farm gate' encompass a wide range in Europe. Energy requirement in gigajoules per tonne beef for production in Italy, for example, is estimated to vary from 29 to 59 GJ t^{-1}, compared with 67 to 102 GJ t^{-1} for France and 9 to 60 GJ t^{-1} in the UK (Slesser and Wallace, 1982). An increase in energy inputs, which usually takes the form of fossil fuel, has allowed greater vegetable and animal energy production for a given area of land. The relationship between energy inputs and beef production in Europe has been modelled by Slesser and Wallace (1982). They included in their analysis energy expended directly 'on farm' and for raising stock feed and estimated energy use per unit of land area as well as per unit of animal product. The regression equation developed by Slesser and Wallace (1982) is then used to derive estimates of carbon dioxide from both systems as related in Table 11.5 and Fig. 11.1.

Table 11.5. Land–energy trade-offs in European beef production.

Energy input (MJ ha⁻¹ year⁻¹)	Beef product (kg ha⁻¹ year⁻¹)	Energy input per product (MJ kg⁻¹ beef)	Fossil-fuel carbon emissions (kg C kg⁻¹ beef)	Land-use storage potential foregone (kg C kg⁻¹ beef)	Total C 'cost' (kg C kg⁻¹ beef)
500	15	5	0.1	10.3	10.4
726	16	7	0.1	9.4	9.5
1,078	19	9	0.2	7.9	8.1
1,639	23	11	0.2	6.6	6.8
2,555	28	14	0.3	5.4	5.7
4,091	34	18	0.3	4.4	4.7
6,738	43	24	0.5	3.5	4.0
11,435	54	32	0.6	2.8	3.4
20,033	69	44	0.8	2.2	3.0
36,296	89	61	1.2	1.7	2.8
68,147	118	87	1.6	1.3	2.9
132,877	158	126	2.4	0.9	3.3

C, carbon.

The relationship between energy input and land requirement is as modelled by Slesser and Wallace (1982):

$$P \text{ (kg ha}^{-1}\text{ year}^{-1}) = 0.88 \text{ (ES)}^{0.44}$$

$R^2 = 0.69$ for 41 observations, where P is the output of animal product in kilograms per hectare per year and ES is the energy input. Energy requirement for European beef production is estimated to range from 9 to 102 MJ kg⁻¹ beef (Slesser and Wallace, 1982). Fossil-fuel emissions for beef production are assumed to be 40% petroleum, 40% natural gas and 20% coal, based on studies of fossil-fuel use in US agriculture (D. Pimentel, Cornell University, Ithaca, New York, 1995, personal communication). In the medium case, carbon-storage potential foregone because of cropping or pasture is assumed to be 1.0 t carbon ha⁻¹.

In this study, the area of land used for grazing or to provide stock feed for cattle is considered as part of the environmental expenditure of beef production. In some regions of Europe, either grazing lands were previously woodlands or forests or the rainfall in the cropping or grazing area is sufficient to permit reversion to richer habitats or carbon-storing forests. The opportunity cost of beef production is posed in terms of carbon storage potential foregone. The indirect carbon cost is assessed for several moderate rates of annual carbon uptake – 0.5 t carbon ha⁻¹, 1.0 t carbon ha⁻¹ and 1.5 t carbon ha⁻¹. For example, higher energy inputs into agriculture usually mean that a smaller area of land is required to raise stock feed. Therefore, higher energy requirements represent a lower opportunity cost for carbon storage. If the uptake potential of the land is not very high because of low rainfall or other

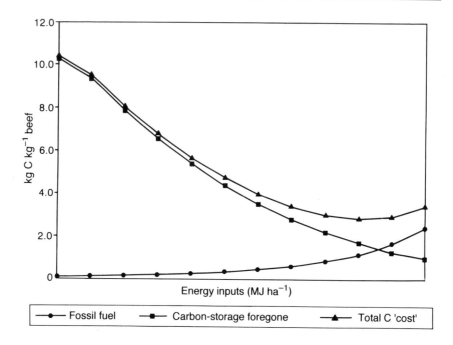

Fig. 11.1. Carbon dioxide emissions as a function of fossil-fuel inputs and carbon-storage opportunity foregone. Net carbon dioxide emissions decrease as fossil-fuel inputs increase (substituting for land) up to a point and then begin to rise with energy intensity greater than about 30 MJ ha^{-1} year^{-1}. Units: fossil fuel in MJ ha^{-1} year^{-1} as $x^{1.06}$. Land use carbon storage foregone: kg C kg^{-1} beef. Total C 'Cost': kg C kg^{-1} beef.

reasons, e.g. only 0.5 t carbon ha^{-1} year^{-1}, the carbon-storage potential foregone would remain low. The relationship between the carbon 'cost' of alternative energy inputs is illustrated with three different carbon uptake-potential rates, as shown in Tables 11.6 and 11.7 and Fig. 11.2. Of course, it is not realistic to expect that forage or stock-feed areas will always, or even usually, be converted to forests. Indeed, in some regions, traditional habitats rely on grazing cattle. Alternative options for converting the arable or grazing land also have implications for net carbon dioxide emissions. Rural areas currently used for grazing or to raise arable crops for stock feed could be used to produce a range of sources of renewable energy, including energy crops, forest residue, straw and farm waste. In addition, the land could be used to support wind systems, small-scale hydroelectricity, municipal waste incineration and landfill gas (Grubb, 1995b). None the less, the land-use implications of alternative feeding arrangements are very large and the opportunity cost of the land involved is never zero.

Table 11.6. Carbon dioxide emissions as a function of different carbon (C) offset potentials.

Fossil-fuel C emissions (kg C kg^{-1} beef)	Land intensity: C storage potential foregone (kg C kg^{-1} beef)			Total C 'cost' (fossil fuel + land inputs) (kg C kg^{-1} beef)		
	(0.5 t C ha^{-1}) (1)	(1.0 t C ha^{-1}) (2)	(2.0 t C ha^{-1}) (3)	(1)	(2)	(3)
0.1	5	10	21	5.2	10.4	20.7
0.1	5	9	19	4.8	9.5	18.9
0.2	4	8	16	4.1	8.1	15.9
0.2	3	7	13	3.5	6.8	13.3
0.3	3	5	11	3.0	5.7	11.1
0.3	2	4	9	2.5	4.7	9.1
0.5	2	4	7	2.2	4.0	7.5
0.6	1	3	6	2.0	3.4	6.2
0.8	1	2	4	1.9	3.0	5.2
1.2	1	2	3	2.0	2.8	4.5
1.6	1	1	3	2.3	2.9	4.2
2.4	0	1	2	2.9	3.3	4.3

Table 11.7. Indirect greenhouse-gas emissions for beef (kilogram carbon (C) equivalent per kilogram of beef).

		Intensive (kg C kg^{-1} beef)	Extensive (kg C kg^{-1} beef)
6 MJ kg^{-1} beef	0 t C ha^{-1} foregone	n/a	0.1
	0.5 t C ha^{-1} foregone	n/a	4.8
	1.0 t C ha^{-1} foregone	n/a	9.5
	1.5 t C ha^{-1} foregone	n/a	18.9
10 MJ kg^{-1} beef	0 t C ha^{-1} foregone	n/a	0.2
	0.5 t C ha^{-1} foregone	n/a	3.5
	1.0 t C ha^{-1} foregone	n/a	6.8
	1.5 t C ha^{-1} foregone	n/a	13.3
30 MJ kg^{-1} beef	0 t C ha^{-1} foregone	0.6	n/a
	0.5 t C ha^{-1} foregone	2.0	n/a
	1.0 t C ha^{-1} foregone	3.4	n/a
	1.5 t C ha^{-1} foregone	6.2	n/a
120 MJ kg^{-1} beef	0 t C ha^{-1} foregone	2.4	n/a
	0.5 t C ha^{-1} foregone	2.9	n/a
	1.0 t C ha^{-1} foregone	3.3	n/a
	1.5 t C ha^{-1} foregone	4.3	n/a
Indirect emissions range		0.6–6.2	0.1–18.9

n/a, not applicable.

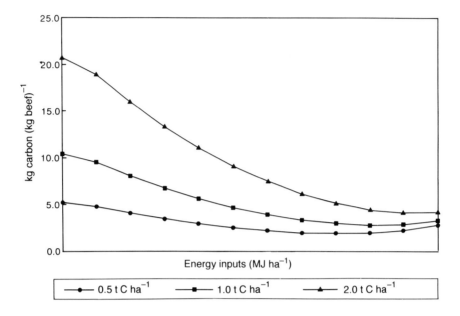

Fig. 11.2. Carbon dioxide emissions under different rates of carbon-uptake potential foregone. Net carbon dioxide emissions at a given level of fossil-fuel inputs (MJ ha^{-1} year^{-1} x$^{1.06}$) and alternative net carbon-storage change in grazing and feedcrop land of 0.5 t C ha^{-1}, 1.0 t C ha^{-1} and 2.0 t C ha^{-1}.

RESULTS AND DISCUSSION

Emissions deriving directly from enteric fermentation and manures are similar in level in intensive and extensive systems in Europe. Extensive systems, which involve longer-lived animals relying more heavily on pasture, give rise to higher levels of methane per unit of beef than do intensive systems, but methane emission levels for both systems may be similar if the intensive systems make use of a number of anaerobic disposal approaches, which lead to methane emissions. Emissions from manures are currently more uncertain in Europe than enteric-fermentation sources – because of the absence of record keeping and reporting related to waste disposal in many regions. None the less, available information on methane emissions from waste disposal indicates that the high- and low-emitting anaerobic waste-disposal systems span a range that has as its midpoint approximately 1 kg carbon kg^{-1} beef, which, combined with the enteric-fermentation emissions, is comparable to methane emissions from extensive systems.

Greenhouse-gas emissions from intensive systems will be higher than emissions from extensive systems if emissions from fossil-fuel inputs are considered in addition to direct methane emissions. This will not be the case, of course, for agricultural systems that rely on renewable energy rather than fossil fuels or that use organic rather than nitrogen fertilizers. But it appears to be true for more intensive beef systems, which consume 30 MJ or more of fossil fuel per kilogram of beef produced.

In practice, different cattle forms consume a wide range of energy. The range considered here is taken from a survey of energy use in agriculture published in the 1980s. The trend in the USA during the 1980s was towards decreased energy use per hectare (Cleveland, 1996). As much of the decrease in agricultural energy use is believed to be related to energy-price increases (Stout and Nehring, 1988; Bonny, 1993), this is likely to also be true in Europe. None the less, the examples calculated in this and other studies indicate that emissions from fossil-fuel use can be comparable in total heating-equivalent terms to those from direct methane and even conversion to intensive systems with relatively low energy inputs can significantly magnify total emissions from beef farms.

Beef production always involves a certain investment in land, which has an opportunity cost in carbon terms as well as in other agricultural uses. Land currently used for grazing and stock-feed production could be put to alternative uses. Conversion to forestry is one plausible use of grazing and arable land. For this study, the opportunity cost in carbon-storage terms of different carbon uptake rates on land has been considered. The carbon-storage potential foregone is relevant both in determining the optimal level of energy input at a given level of carbon opportunity cost and in comparing the relative emissions levels of extensive and intensive systems. In the case of a carbon-storage potential of 1.0 t carbon ha^{-1} $year^{-1}$, net carbon dioxide emissions were lowest at an energy input of about 60 MJ kg^{-1} beef. For any significant level of carbon-storage potential foregone, i.e. greater than 0.5 t carbon ha^{-1} $year^{-1}$, greenhouse-gas emissions from extensive systems surpassed emissions from intensive systems, as shown in Tables 11.8 and 11.9. These results suggest that the land-use implications of alternative beef-production modes make a greater difference to greenhouse-gas emissions than any of the other major factors – animal longevity or fossil-fuel inputs – given systems that are already relatively productive. In other words, if there is even the potential for a modest level of carbon uptake on pasture land, net greenhouse-gas emissions would be reduced by increasing the energy intensity of production.

If beef-production systems in Europe changed from one mode to another, as defined here, the implications for greenhouse-gas emissions can be roughly calculated. This analysis has several important caveats. The definitions of intensification and extensification used here are derived from mean estimates of feeding levels, energy inputs and land requirements. Information on the proportion of beef raised under different degrees of intensification is not available, and the data on energy inputs to beef systems span

Table 11.8. Direct greenhouse-gas emissions for beef (kilogram carbon equivalent* per kilogram of beef).

	Intensive systems		Extensive systems	
Enteric fermentation	kg CH_4 kg^{-1} beef	kg CE kg^{-1} beef	kg CH_4 kg^{-1} beef	kg CE kg^{-1} beef
Same feed	0.39	2.5	0.51	3.3
Higher-digestibility feed for intensive systems	0.28	1.8	n/a	n/a
Subtotal		1.8–2.5 (2.2)		(3.3)
Manures				
Liquid slurry	0.35	2.3	n/a	n/a
Dry storage	0.01	0.1	0.02	0.2
Subtotal		0.1–2.3 (1.2)		(0.2)
Direct emissions	0.29–0.74	1.9–4.8(3.4)	0.53	(3.5)

* The carbon equivalent (CE) unit for methane (CH_4) emissions used compares the warming potential of methane with that of carbon dioxide over a 100-year time horizon (IPCC, 1994).
n/a, not applicable.
Figures in parentheses indicate mean of range.

Table 11.9. Total greenhouse-gas emissions for beef (kilogram carbon equivalent per kilogram beef).

	Intensive	Extensive
Direct emissions (enteric fermentation and manures)	1.9–4.8 (3.4)	(3.5)
Indirect emissions		
Fossil fuels	0.6–2.4 (1.5)	0.1–0.2 (0.15)
Land use	0.2–5.6 (2.9)	3.3–18.8 (11.05)
Total excluding land-use potential	2.5–7.2 (4.9)	3.6–3.7 (3.7)
Total emissions	2.7–12.8 (7.8)	6.9–22.5 (14.7)

Figures in parentheses indicate mean of range.

a wide range. If we assume that the largest holdings of cattle in Europe involve intensive production and the remainder extensive production, we can surmise that about half (55%) of cattle in the current European Union are extensively raised (European Commission, 1995). These are holdings of from 1 to 99 cattle (European Commission, 1995).

The savings in greenhouse-gas emissions if the extensive holdings are converted to intensive systems can be calculated, based on beef consumed in Europe. As summarized in Table 11.10, if there is no carbon opportunity cost on the land used for grazing, conversion to intensive production represents an increase in emissions of about 7 Mt (6,776 kt) carbon dioxide as carbon each year. If the extra land commitment implicated in extensive production, however, has a carbon-uptake opportunity cost and this carbon-uptake potential can be exploited, calculated reductions in greenhouse-gas emissions range from 8 to 54 Mt carbon dioxide as carbon per year, based on a carbon opportunity cost of 0.5–1.5 ha^{-1} year^{-1}. It should be noted that a carbon opportunity cost considerably higher than 1.5 t carbon ha^{-1} year^{-1} is plausible at specific locations, but that national mean uptake rates for forests are usually below this level, as they encompass mature as well as growing stands. The results derived here are also sensitive to the choice of carbon dioxide equivalent used. These aggregate greenhouse-gas emission estimates have been calculated based on a global-warming potential integrated over a 100-year time frame. Use of a global-warming potential for a shorter time horizon would weight the importance of both the land-use and the fossil-fuel inputs less heavily, but the apparent advantages of intensification would be diminished.

Table 11.10. Theoretical savings in greenhouse-gas emissions in Europe after conversion of extensive beef production to intensive systems.

Opportunity cost in carbon of grazing and stock-feed cropping (t C ha^{-1} year^{-1})	Current greenhouse-gas emissions from beef production in Europe (kt CO$_2$-C eq)	Estimated greenhouse-gas emissions after intensification (kt CO$_2$-C eq)	Greenhouse-gas emissions difference (kt C eq)
0.0	35,704	42,480	6,776
0.5	50,424	42,480	−7,944
1.0	65,844	42,480	−23,364
1.5	96,217	42,480	−53,734

C eq, carbon equivalent.

The conclusions reached here – that intensive systems, involving higher fossil-fuel inputs, are better as a greenhouse-gas-reducing strategy if the alternative involves foregoing carbon-storage opportunities on land – should not be interpreted as an argument in favour of intensification of agriculture. When considering pasture conversion to forests or other uses, competing demands for land use need to be considered, especially the ecological suitability of land that has traditionally been used for grazing. The goal of reducing greenhouse-gas emissions on agricultural land represents only one of these competing demands.

12 Reduction of Emissions in Farming Systems in Germany

Konrad Löthe, Clemens Fuchs and Jürgen Zeddies

INTRODUCTION

Emissions of greenhouse gases (carbon dioxide (CO_2), methane (CH_4) and nitrous oxide (N_2O)) into the atmosphere are one cause of the greenhouse effect. Agriculture is responsible for a significant share, particularly in CH_4 and N_2O pollution. In this chapter, the methodology for comparing the balance between these emissions is discussed. The sources and sinks of greenhouse-gas emissions in different farming systems are incorporated in linear farm models. Based on extensive data collection in two research areas in Germany, the calculations are based on the possibilities for existing farms. Technical means for reducing the emissions are given by a reduction of intensity in grassland systems, by increasing milk yields and by biogas production. The ecological effects and the economic impacts of these techniques are analysed. The use of renewable resources in crop-land farms is discussed. It can be shown that in a crop-land farm the gas emissions can be reduced significantly, if a CO_2 tax is introduced.

To a large extent, emissions of CO_2, CH_4 and N_2O are connected with the intensification of food production. Methane is produced in agriculture from ruminants and while manure is stored and N_2O from fertilizer application on cultivated land. The contribution to emissions from agriculture in Germany is shown in Table 12.1. Ammonia (NH_3) is not a primary greenhouse gas but fertilizes forest growth, thereby enhancing emissions of N_2O (Papen *et al.*, 1993).

© CAB INTERNATIONAL 1997. *Climate-change Mitigation and European Land-use Policies* (eds W.N. Adger, D. Pettenella and M. Whitby)

Table 12.1. Agriculture as a proportion of greenhouse-gas emissions in Germany (from Trunk and Zeddies, 1996).

	NH_3 ('000 t)	CO_2 (million t)	CH_4 ('000 t)	N_2O ('000 t)
Emissions by agriculture	384	17	1,400	59.1
Sum of emissions	402	705	4,700	164
Share of agriculture (%)	96	2.4	30	36

METHODS

Comparable Balancing and System Boundaries

In order to create CO_2 balances, a reference system is needed. This reference system is an ecosystem undisturbed by humans. It is naturally conditioned, and sinks and sources of the different gases are almost in balance. Therefore, the system is limited in carbon sequestration by its total biomass.

The feed of a dairy cow has taken CO_2 out of the atmosphere and part of this uptake is emitted as CH_4 from the ruminant cow. Because of the higher greenhouse-gas potential of CH_4, this transmission process causes a net emission. These emissions are aggregated with the global-warming potential (GWP) to CO_2 equivalents, which provide the equal warming effect of the gas, depending on the residence time (Table 12.2).

Table 12.2. Lifetime and greenhouse-warming potential of different trace gases (from Deutscher Bundestag, 1993).

	CO_2	CH_4	N_2O
Lifetime (years)	120	10.5	132
GWP 20 years	1	35	260
GWP 100 years	1	11	270

For an exact calculation of the amount of greenhouse-gas emission, it is necessary to define boundaries for the system being analysed. Criteria are needed to define what is in the analysed system and what is out of the system. In this system, the complete nitrogen (N) cycle and the production and consumption of intermediate goods (e.g. fertilizer and pesticides) are included. Production of investment goods (machines and buildings) is not taken into account. The production of renewable resources is part of the system and can reduce emissions of CO_2 through the substitution of fossil energy.

Linear Programming Models

Generally, the aim of linear programming models at the farm level is to obtain information on the best farm organization. The objective of the model is to maximize the gross margin of the farm. The model derives from activities and constraints and shows the possibilities and the ability to adapt to different economic conditions, e.g. the changed relation between prices and costs. In using models in farm planning, it is assumed that the model is a realistic representation of the farm system considered. The database for the models used contains data from 122 grassland farms and 1,600 field records from 80 arable farms over some years. From the data production, functions for crops and grassland have been estimated (Schanzenbächer, 1994; Trunk and Zeddies, 1996) and activities for agricultural mechanization have been built up. Other model components are the livestock component, including a nutrition component, the manure and fertilizer component, the N cycle and the use of renewable resources. In the model, environmental parameters are calculated, such as N input–output balances and greenhouse-gas emissions. One of these balances is the 'farm-gate balance', which is the sum of N inputs minus the N leaving the farm in products for the market.

Sources in the Farm Models

Carbon dioxide emissions are caused by direct and indirect consumption of fossil energy. Direct use of energy means the consumption of diesel fuel and electricity. Each consumed litre of diesel fuel on the farm causes a direct CO_2 emission of 2.88 kg. Indirect consumption means the consumption of inputs, such as fertilizer and pesticides, which are produced with the use of fossil energy. The use of 1 kg N as calcium ammonium nitrate causes a CO_2 emission of 2.37 kg. Table 12.3 shows the CO_2 emissions of some goods, including direct and indirect consumption.

Table 12.3. CO_2 emissions from direct and indirect energy consumption (based on Reinhardt, 1993).

Energy consumption	kg CO_2
1 l diesel fuel	2.88
1 kW electrical power	0.618
1 kg N as calcium ammonium nitrate	2.37
1 kg phosphate as P_2O_5	1.00
1 kg potash as K_2O	0.86
1 DM* pesticide	0.138

* 1 Deutschmark (DM) = US$0.67 = £0.43.

Nitrous oxide emissions are caused with and without agriculture. Bouwman *et al.* (1992) distinguish between natural 'background emissions' and emissions caused by N-fertilizer. Eichner (1990) describes the effect of N-fertilizer on N_2O emissions. Because of the uncertainty in the data of N_2O emissions, Mosier (1993) assumed a fixed relation of 1% of the N-fertilizer used, which is also used in the linear programming model. A second source of N_2O in the farm model is the emission from leached nitrate. Nitrate leaching depends on crop rotation and therefore N use is estimated in the farm models (Table 12.4).

Table 12.4. N_2O emissions from N-fertilizer use (from Eichner, 1990).

Type of fertilizer	N_2O-N emissions (%)
Ammonia fertilizer	0.86–6.84
Ammonium nitrate	0.04–1.71
Ammonia type	0.02–0.90
Urea	0.07–0.18
Calcium nitrate	0.001–0.5

The sources of CH_4 considered are rumen fermentation in cattle and digestion in pigs. The amount emitted depends on the diets of the animals (Kirchgessner *et al.*, 1991a, b), which is assumed in the linear programming model to be set at a level that optimizes growth and yield.

For cows, CH_4 emissions are given by:

$$CH_4 = 62.5 + 77.2\ CF_G + 82.5\ CF_T + 10.8\ NfE + 18.5\ CP - 194.6\ EE \quad (1)$$

where CH_4 is the CH_4 emission in g day^{-1}, CF_G is the crude fibre from dried-grass rations (kg), CF_T is the crude fibre from corn-silage rations (kg), NfE is the N-free extract (kg), CP is the crude protein and EE is the ether extract (kg). Depending on the type of farm and the different milk-yield level of the cows, the CH_4 emissions are calculated.

Although monogastric animals were thought to have only a limited fermentation capacity, pigs also emit CH_4 emissions. These emissions depend on the amount of consumed bacterially fermentable substrates (BFS) (kg day^{-1}). Growing pigs emit between 14 and 19 g CH_4 kg^{-1} BFS. The CH_4 emissions per sow (g day^{-1}) are given by (Kirchgessner *et al.*, 1991b):

$$CH_4 = 2.85 + 13.0\ BFS \quad (2)$$

An additional source of CH_4 from manure emissions is during manure storage. Part of the organic matter in the manure ferments to CH_4 when it is stored anaerobically (Heyer, 1994).

Sources of NH_3 in farming systems are livestock (housing, manure storage and manure spreading) and the use of N-fertilizers. Emissions from housing vary with the kind of livestock husbandry (Table 12.5). Manure loses NH_3 during storage. A cover on the store can reduce the emission significantly. Manure spreading is another source of NH_3 emission. The emission rate depends on the kind of manure, the weather at spreading and the spreading technique (Horlacher and Marschner, 1990). Injection and incorporation of manure are possibilities to reduce emission and to increase the N content of the manure. These effects would substitute for more N-fertilizer, as it would be substituted without emission reduction. Ammonia emissions are also caused by using N mineral. Depending on fertilizer type, between 2 and 15% of the N is lost as NH_3 (ALB, 1989).

Table 12.5. Ammonia emission from housed livestock (from DLO, 1993).

Kind of animal	NH_3-emissions (kg head^{-1} year^{-1})
Cows	3.9
Sows with piglets	8.1
Pigs (fully perforated floor)	3.0

RESULTS

Grassland Farm

Reduction of intensity in grassland use
One possibility for reducing greenhouse-gas emission in agriculture is to reduce the intensity of grassland use. In Table 12.6, a 34 ha dairy farm, with a milk quota of 300 t, 50 cows, and silage feeding, reduces the intensity significantly, using less N-fertilizer. In farm model 2, with intermediate intensity, the gross margin is decreased by only 1,000 Deutschmark (DM) year^{-1}. Young stock is reduced and any feed deficit is compensated by concentrates.

Overall, the reduction of intensity has an influence on the emission. Ammonia emission per livestock unit (LU) and per kilogram of milk is decreased by about 10%, because the protein content is lower and the cows excrete less urea. Nitrate leaching and N_2O emission are decreased more than proportionally. The N emission decreases by one-third. The CH_4 emission is reduced, because the share of concentrates in the diet is higher and the emission caused by digestion of concentrates is lower than the emissions from roughage. The sum of the greenhouse gases is reduced by 18% LU^{-1} and 15% kg^{-1} milk.

A further reduction in the intensity (variant 3) of the grassland decreases the gross margin to a higher extent (5,000 DM year^{-1}). The number of young

Table 12.6. Impact of reducing the intensity of grassland use (from Trunk and Zeddies, 1996).

	Representative farm no.			
	1	2	3	4
Intensity of grassland use	High	Intermediate	Low	Low
Grassland (ha)	34	34	34	44
Gross margin (DM year^{-1})	153,259	152,289	148,100	156,657
Labour (h year^{-1})	3,985	3,736	3,431	3,905
Cows (head)	50	50	50	50
Young stock (head)	42	37	27	42
N-fertilizer bought (kg ha^{-1})	85	42	0	0
Net yield grassland (t dry matter)	94	83	75	72
Concentrate bought (t)	56.5	70.9	76.4	62
Protein surplus in ration (%)	19	9	4	5
Nitrogen segregation cow^{-1} (kg N cow^{-1})	112	100	94	95
Sum of NH_3 loss (kg LU^{-1})	29.5	26.8	24.2	22.1
NH_3 loss kg^{-1} milk (g)	5.9	5.4	4.9	4.4
NO_3 leaching (kg ha^{-1})	38.7	11.1	1.8	1.8
N_2O emission (kg ha^{-1})	5.6	3.3	1.9	1.9
Sum of N emissions (kg)	4,193	2,815	2,066	2,171
'Farm-gate balance' (kg N)	1,909	718	−347	−742
CH_4 emission (kg)	9,846	8,777	7,980	9,280
CH_4 emission kg^{-1} milk (g)	21.7	20.3	20.3	20.7
CO_2 emission fossil energy (kg)	37,946	30,822	34,923	31,901
CO_2 emission concentrate production (kg)	19,499	20,684	22,331	18,159
CO_2 emission from fertilizer production (kg)	7,525	3,676	0	0
CO_2 equivalent (100 years) (kg)	331,113	269,040	235,317	260,395
CO_2 equivalent (100 years) (kg kg^{-1} milk)	0.73	0.62	0.59	0.58

stock is reduced to a lower level and the diminished roughage production requires more concentrates to be purchased. Ammonia and N_2O emissions and NO_3 leaching are sharply decreased. Compared with the highly intensive variant 1, the N emission is reduced to one-half. Only CH_4 emission kg^{-1} milk stays constant, because the roughage contains more crude fibre.

In variant 4, the farm has 44 ha of grassland, instead of 34 ha. As a result, the farm can work with the lowest intensity. The gross margin is 3,000 DM year^{-1}, higher than it is in the highly intensive variant with 34 ha. The ecological parameters are again low.

Increase of milk yield
When the milk yield rises from 5,000 to 6,000 l cow^{-1}, the gross margin increases by 4,500 DM year^{-1} and the NH$_3$ emission decreases by 13% and the CH$_4$ emissions by 9%, as shown in Table 12.7. The CO$_2$-equivalent output of the whole farm is reduced by 10%. Because with increasing milk yield the share of the food needed for maintenance diminishes, an increase of the milk yield has positive effects on the income and on the environmental parameters, up to a certain level. Above a yield of 6,000 l cow^{-1}, the gross margin can only be increased with additional resource inputs. However, the net environmental benefits continue to be positive.

Table 12.7. Impacts of different milk-yield levels on greenhouse-gas emissions (from Trunk and Zeddies, 1996).

	Representative farm no.		
	5	6	7
Milk yield (l)	5,000	6,000	7,000
Grassland (ha)	34	34	32
Gross margin (DM)	140,038	144,551	143,080
Labour (h year^{-1})	3,948	3,564	3,073
Cows (head)	55	46	39
N-fertilizer bought (kg ha^{-1})	33	40	35
NH$_3$ loss during spreading (kg)	1,805	1,264	524
Sum of NH$_3$ loss (kg LU^{-1})	24.8	26.2	27
NH$_3$ loss kg^{-1} milk (g)	6.1	5.3	4.7
NO$_3$ leaching (kg ha^{-1})	32.8	33.0	32.0
N$_2$O emission (kg ha^{-1})	3.9	3.9	3.6
Sum of N emissions (kg)	3,552	3,419	3,074
CH$_4$ emission (kg)	8,849	8,583	7,706
CH$_4$ emission kg^{-1} milk (g)	23.1	21.0	19.1
CO$_2$ emission fossil energy (kg)	31,779	29,194	25,420
CO$_2$ emission concentrate production (kg)	15,976	14,003	15,249
CO$_2$ emission from fertilizer production (kg)	2,923	3,564	2,925
CO$_2$ equivalent (100 years) (kg)	275,221	266,570	240,080
CO$_2$ equivalent (100 years) (kg kg^{-1} milk)	0.72	0.65	0.59

Biogas production on a dairy farm
With a biogas fermenter, 30–40% of the organic matter in the manure is transformed into biogas. It can be used for production of heat or electricity and there are several investment subsidies for new fermenters. Biogas fermenters are important in two ways. The first is the prevention of CH_4 and NH_3 emissions from the manure store. The other is that the production of biogas can substitute for fossil-fuel energy.

Table 12.8 shows the effect of biogas production on a dairy farm. In variants 9 and 10, an on-farm fermenter is included. The fermenter in variant 9 is less productive than the central fermenter in variant 10. Both variants decrease the amount of greenhouse-gas emissions significantly. When manure is transformed in the fermenter, fixed N is mineralized. Therefore,

Table 12.8. Biogas production on dairy farms (from Trunk and Zeddies, 1996).

	Representative farm no.		
	8	9	10
Biogas fermenter	Off farm	On farm	Central
Investment subsidy (%)		30	30
Gas production kg^{-1} organic matter ($l\ kg^{-1}$)		220	300
Use of heat energy (%)		50	80
Gross margin (DM)	144,551	138,532	140,347
Labour (h)	3,564	3,554	3,728
N-fertilizer bought (kg ha^{-1})	40	26	26
NH_3-N loss kg^{-1} milk (g)	5.3	4.9	4.9
NO_3-N leaching (kg ha^{-1})	33	32	32
N_2O emissions (kg ha^{-1})	3.9	3.8	3.8
N emissions total (kg)	3,419	3,258	3,259
'Farm-gate balance' (kg N)	450	−28	−26
CH_4 emission manure storage (kg)	1,574	0	0
CH_4 emission total (kg)	8,583	6,998	6,996
CH_4 emissions kg^{-1} milk (g)	21.0	17.2	17.2
CO_2 emissions fossil energy (kg)	29,194	29,048	30,733
CO_2 emissions from concentrate production (kg)	14,003	14,309	14,259
CO_2 emissions from fertilizer production (kg)	3,564	2,308	2,305
Electricity from biogas (kW)	0	45,848	62,645
Heating energy (oil equivalent) (l)	0	2,709	5,923
CO_2 equivalent (100 years) total (kg)	266,572	266,225	208,356
CO_2 equivalent (100 years) (kg kg^{-1} milk)	0.65	0.55	0.51

the amount of N-fertilizer bought is reduced from 40 to 26 kg ha^{-1} and the farm-gate balance is also decreased.

Arable-farm Carbon Dioxide Tax

One possibility for reducing greenhouse-gas emissions from farming systems could be an introduction of a CO_2 tax. The direct effects of the tax are price increases for the goods, according to their CO_2 potential. The price of diesel fuel at farm level rises from 0.60 DM l^{-1} to 0.94 DM l^{-1} (+44%) if a CO_2 tax of 0.10 DM kg^{-1} CO_2 is introduced. The price of N-fertilizer rises from 1 DM kg^{-1} to 1.32 DM kg^{-1} (+22%). Pesticides have a low CO_2 emission, according to their price, with a change in price of +1.4% if a CO_2 tax of 0.10 DM kg^{-1} is introduced. In the farm model, there is no CO_2 tax on any other greenhouse gas.

Table 12.9 shows the economic and physical parameters of a representative arable farm of the Kraichgau region (Baden-Württemberg, Germany), with 42.4 ha of arable land and 3.5 ha of grassland. The enterprise profile includes straw, whole barley and rape-seed, which all can be grown on set-aside, as long as they are used for biomass crops or, in the case of oil-seed rape, for biodiesel.

One effect of the CO_2 tax is that the special intensity of the crop production is reduced. In the model, the amount of N-fertilizer used in winter wheat is decreased from 175 kg N ha^{-1} to 165 kg N ha^{-1}. There is also a change in the crop rotation. The CO_2 tax reduces the production and intensity of maize, because the drying of maize causes high CO_2 emission. The share of sugar beet and wheat increases, while their intensity decreases. The substitution of fossil energy, in this case fuel oil for heating purposes, by renewable resources causes a reduction of emission in the farming system.

Another effect is that renewable resources become more profitable as the prices of direct energy increase. With a CO_2 tax of 0.10 DM kg^{-1} CO_2, the price of fuel oil increases from 0.40 DM l^{-1} to 0.69 DM l^{-1}. At this level, 68 t straw is used to produce heating energy and 5,480 l biodiesel is manufactured on EU set-aside land (variant 12), which substitutes 114 t CO_2. At the level of 0.20 DM kg^{-1} CO_2, 8.3 ha winter barley is additionally produced for biomass use. Altogether 165 t biomass is used for energy production and 267 t CO_2 are substituted (variant 13).

Without taking into account the substitution effect, the CO_2 equivalents (20 years) are decreased from 143 t to first 133 t and then 122 t. Including the substitution effect, the CO_2 equivalents (20 years) are decreased from 143 t to first 18 t and then −145 t. This net calculation shows that this farming system could be a sink for greenhouse gases.

With the introduction of a CO_2 tax, the agricultural output, measured as 'grain equivalents', is decreased. As a coefficient of the efficiency of

Table 12.9. Arable farm and CO_2 tax.

	Variant		
	11	12	13
Arable land (ha)		42.4	
Livestock (head)		3.5 cows; 5.8 sows; 107 fattened pigs	
CO_2 tax (M kg^{-1}CO_2)	0	0.10	0.20
Winter wheat (ha (kg N ha^{-1}))	19.2 (175)	18.8 (165)	20.4 (165)
Sugar beet (ha (kg N ha^{-1}))	6 (190)	6.2 (145)	6.2 (130)
Maize (ha (kg N ha^{-1}))	11.9 (155)	8.5 (120)	
Winter barley (ha (kg N ha^{-1}))	0.5 (90)	4 (135)	2.8 (140)
Winter barley biomass (ha (kg N ha^{-1}))			8.3 (140)
Rape-seed 'biofuel' (ha (kg N ha^{-1}))		4 (190)	4.7 (190)
Land set-aside (ha)	4.7	0.7	
Grassland (ha (kg N ha^{-1}))	3.5 (100)	3.5 (75)	3.5 (75)
Gross margin (DM)	107,223	102,880	113,542
Biomass (t)		68	165
Biodiesel (l)	–	5,480	6,100
N_2O from fertilizer (kg N_2O)	86	83	87
N_2O from groundwater (kg N_2O)	195	180	154
Sum of N_2O (kg N_2O)	281	263	241
CH_4 from manure store (kg CH_4)	454	212	548
CH_4 from pigs (kg CH_4)	10	10	11
CH_4 from cows (kg CH_4)	384	298	297
Sum of CH_4 (kg CH_4)	850	520	557
CO_2 from fertilizer (kg CO_2)	21,325	21,695	23,363
CO_2 from pesticides (kg CO_2)	905	1,103	1,308
CO_2 from drying (kg CO_2)	11,839	10,109	773
CO_2 from fuel and electricity (kg CO_2)	4,134	13,038	13,832
Sum of CO_2 (kg CO_2)	40,152	45,945	40,722
CO_2 substitution (kg CO_2)	–	114,203	267,409
Without substitution			
$\quad CO_2$ equivalents 20 years (kg CO_2)	142,962	132,525	122,877
$\quad CO_2$ equivalents 100 years (kg CO_2)	125,372	122,675	111,919
With substitution			
$\quad CO_2$ equivalents 20 years (kg CO_2)		18,322	−144,532
$\quad CO_2$ equivalents 100 years (kg CO_2)		8,472	−155,490
Sum of NH_3 (kg NH_3-N)	1,058	1,191	1,193
Grain equivalent (GE)	3,441	3,391	2,719
Grain equivalent kg^{-1} CO_2 equivalents 20 (kg CO_2 GE^{-1})	41.5	5.4	−53.2
Grain equivalent kg^{-1} CO_2 equivalents 100 (kg CO_2 GE^{-1})	36.4	2.5	−57.1

'greenhouse-gas productivity', the ratio between gas emissions and grain equivalents can be used. In variant 13, the ratio is negative. This is the case because the net CO_2 emission of the farm model is negative.

CONCLUSIONS

Despite uncertainty as to the exact quantity of greenhouse-gas emissions, there are several technical possibilities to reduce emissions of greenhouse gases from agriculture in the short and medium term. Technical possibilities for emission reduction are energy-saving, the production of renewable resources, rising milk yields and biogas production. Yet the economic constraints of any option being economically beneficial – at least, in terms of the ratio between costs and product prices, for example – ensure that only rising milk yields fit the criteria of reducing emissions and increasing margins, given the present situation.

Political possibilities to reduce emissions of greenhouse gases include the introduction of a CO_2 tax and subsidies for protection measures and for the use of renewable resources, such as the production of energy from agricultural wastes. These subsidies could be part of other agri-environmental programmes. With the necessary conditions created, this chapter has demonstrated that agriculture can substantially contribute to reduced environmental stress, in general, and reduced greenhouse-gas emissions in particular.

The Effects of the Dutch 1996 Energy Tax on Agriculture

13

Marinus H.C. Komen and Jack H.M. Peerlings[1]

IMPLEMENTING THE DUTCH ENERGY TAX

Like the other signatory countries of the Framework Convention on Climate Change (FCCC), the Netherlands has recognized the importance of its national contribution to the global enhanced greenhouse effect. National policy in the Netherlands goes further than the FCCC, in establishing a carbon dioxide (CO_2) emission target of a 3–5% reduction of 1989/90 levels by 2000. To achieve this target, the Dutch government decided to introduce a unilateral energy tax in 1996. This chapter discusses the effects for the Dutch economy of the introduction of this energy tax, using applied general equilibrium (AGE) analysis, and as such can be considered as a first attempt to determine the effects at a disaggregated level of a unilateral energy tax which has already come into force.

The government implemented the policy for reducing emissions of greenhouse gases in 1996 by taxing the use of natural gas, other gas, gas oil and electricity by households and 'small' energy users (Energie Beheer Nederland, 1995). Setting up the tax base in this manner implies that horticulture (glasshouses) and large industrial users are exempted from the tax. Meanwhile, the revenues generated by the energy tax have been used to reduce its adverse effects by lowering the pre-existing distortionary taxes related to labour. In doing so, the policy as implemented seems to take into account four important criteria in designing acceptable environmental taxes, as now briefly reviewed.

First, the Netherlands chose to introduce the energy tax unilaterally where the European Union (EU) has thus far failed to design an acceptable

© CAB INTERNATIONAL 1997. *Climate-change Mitigation and European Land-use Policies* (eds W.N. Adger, D. Pettenella and M. Whitby)

171

multilateral tax system. A fundamental problem for a small open country is that foreign countries may choose not to use taxes and often pursue a less ambitious environmental policy. Sectors which are particularly energy-intensive and which face international competition bear a high cost burden in these circumstances, and it is argued that this could potentially also harm the domestic economy as a whole (Bovenberg, 1993). There are, however, studies arguing the contrary case (see, for example, Welsch, 1995a, b), complete unilateral taxation does not seem very promising. Second, related to the exemption of energy-intensive industries competing internationally, there is the question whether global emissions of CO_2 will fall because of the energy tax since production of energy-intensive commodities may relocate to other countries. International reallocation of production is less of a threat if only households and small users are taxed (Bovenberg, 1993; Hoel, 1996).

The third issue concerns the recycling of the energy-tax revenues in a second-best world. Some have argued that such a revenue-neutral reform might offer a 'double dividend' of: (i) improved environmental outcome; along with (ii) reduction of the distortionary costs of the tax system (see, for example, Pearce, 1991b); or (iii) reduced unemployment (Bovenberg and de Mooij, 1993; Bovenberg and van der Ploeg, 1994a; de Wit, 1995). In other studies, however, a second dividend seems not to occur, due to the fact that the gains from using pollution-tax revenues to substitute for other distortionary tax revenues tend to be more than offset by the distortionary effects that result from an interaction of the pollution tax and pre-existing taxes (Bovenberg and de Mooij, 1994; Bovenberg and van der Ploeg, 1994b, c; Parry, 1995). In general, the term 'double dividend' is often vaguely defined and used in several different ways and its occurrence very much depends on specific assumptions in the models used. This leads to some confusion (for a clear discussion, see Goulder, 1995b). General agreement, however, seems to exist about the fact that 'there exists a "double dividend" in the sense that a welfare cost reduction can be achieved by using revenues from pollution taxes to cut distortionary taxes rather than returning these revenues in a lump-sum fashion' (Bovenberg and de Mooij, 1994).

The fourth issue raised by the Dutch unilateral energy tax concerns the fact that, being part of the EU, the Netherlands is not allowed to tax those energy sources which can easily be transported across national borders, such as petroleum. The rationale of this regulation is the prevention of widespread cross-border shopping in an internal market without any border controls. Moreover, the EU does not allow environmental levies on inputs to be refunded at its internal borders. This implies that the Netherlands cannot refund taxes on inputs to domestic producers when final goods are exported (Bovenberg, 1993).

It appears that, in designing the energy-tax scheme, the Dutch government takes these issues into account. Energy-intensive industries competing on international markets are exempted; only 'non-transportable' energy sources are taxed; the revenue recycling is in accordance with EU regulation; and the tax seeks a double dividend to take into account the fact that total distortionary costs can be reduced by reducing pre-existing labour taxes.

The purpose of this chapter is to examine the effects of the Dutch 1996 energy tax, and its possible extension to horticulture and other large users, on production, income formation and employment in Dutch agriculture and the economy as a whole. Different modelling approaches can be used to determine the effects of reducing carbon emissions (for an overview, see Gaskins and Weyant, 1993; Weyant, 1993). Given the economy-wide effects of an energy tax, AGE analysis seems the most appropriate tool (Bergman, 1991; Conrad and Schröder, 1991, 1993; Jorgenson and Wilcoxen, 1993; Breuss and Steininger, 1995; Zhang, 1996). The use of an AGE model is even more promising, as it facilitates the explicit consideration of tax-revenue recycling issues (Bovenberg and Goulder, 1994; Capros *et al.*, 1995; Goulder, 1995a). The Dutch Central Planning Bureau (Netherlands Central Planning Bureau, 1992, 1993) have analysed the effects for the Dutch economy, using ATHENA, an AGE model for the Netherlands. The disadvantage of this study is that it does not explicitly contain the commodities taxed. Moreover, agriculture, the sector which is the main focus of this chapter, has not been previously disaggregated, as it is here.

In this chapter, the effects are determined using WAGEM (Komen and Peerlings, 1996), an AGE model of the Netherlands. This model gives a complete description of the Dutch economy, with a significant level of detail with respect to agricultural industries and commodities; allows for factor mobility between industries; incorporates both the Armington and product homogeneity assumption in international trade; and allows for endogenous total labour supply.

The next section provides information on the available data. An explanation of the model and assumptions used is then outlined, followed by policy simulations, with results being elucidated and discussed. General aspects of modelling energy-tax impacts and the impact of the Dutch tax are considered in the final section.

DATA FOR THE MODEL

In this section, the data sources and data aggregations are discussed. The data for WAGEM are presented in a social accounting matrix (SAM), showing the transactions and income flows in the Dutch economy in 1990 (see Appendix, Table 13.A1). In the SAM, the rows contain the receipts recorded by origin, while the columns contain the payments by destination. The totals of the rows (total receipts) equal the totals of the columns (total payments). A list of commodities and industries is given in Table 13.A2 in the Appendix. Because industries can produce more than one commodity, while the same commodity can be produced by several industries, a one-to-one relationship between categories of industry and commodities does not exist, as is usually the case in SAMs. In the SAM presented, the 38 commodities and 34 industries are aggregated into one column and one row, for neatness of presentation.

THE MODEL AND ITS CALIBRATION

A complete description of the AGE model and its calibration can be found in Komen and Peerlings (1996). In the model, depicted in Fig. 13.1, the outputs of an industry are produced with one hypothetical aggregate output, according to fixed proportions. The aggregate output is composed of two hypothetical aggregate inputs: an aggregate intermediate and a factor input. It is assumed that production takes place with a constant elasticity of

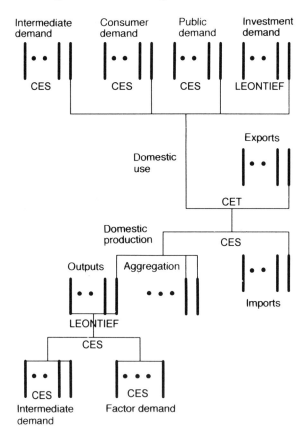

Fig. 13.1. Flow of commodities and assumptions in general equilibrium model of the Dutch economy.

substitution (CES) production function with constant returns to scale. The aggregate factor input is composed of labour and capital. The composition also takes place according to a CES production function with constant returns to scale. The aggregate intermediate input is composed of several commodities (intermediate inputs) according to a CES production function with constant returns to scale. Cost minimization yields the demand functions for the aggregate inputs, the intermediate inputs and capital and labour by industry.

In the model, the Armington assumption for modelling trade is used (de Melo and Tarr, 1992). The Armington assumption states that commodities imported and exported are imperfect substitutes of domestically produced and used commodities. This assumption is necessary to take into account two-way trade, while an unrealistically high degree of specialization is avoided. The imported (exported) commodity and the domestically produced (demanded) commodity are aggregated into a new composite commodity-using constant returns to scale CES (Constant Elasticity of Transformation; CET) functions. Cost minimization and revenue maximization result in import-demand and export-supply functions, where the levels of imports and exports depend on the ratio between domestic and world-market prices and the degree of substitutability between domestic and foreign commodities. The consequence of a high level of disaggregation for agriculture is that individual commodities can be distinguished that are internationally homogeneous. For such commodities, which have small import or export shares, the Armington assumption is not valid. In the model, the homogeneity assumption is applied for the trade in pigs (G2) and eggs (G12).

A commodity can be used as an intermediate input, a consumption good (consumed by private and public households) or an investment good. It is assumed that there is one representative private household that consumes commodities, saves and supplies factor services and therefore receives factor income. Total capital supply is perfectly price-inelastic. Labour supply is divided into immobile labour and imperfectly mobile labour. In the agricultural industries, supply of own labour is assumed to be fixed and therefore perfectly immobile. Total imperfectly mobile labour supply, however, is assumed to be price-elastic, which implies that more people will enter the labour market when the real wage rate increases. Both capital and labour are imperfectly mobile between industries. This implies that factor rewards can differ between industries. Complete factor mobility would equalize factor rewards between industries (for a discussion, see Peerlings, 1993). Factor supply per industry is modelled, using CET supply functions resulting from revenue maximization.

It is assumed that the household has CES uncompensated-demand functions, which implies that expenditure elasticities equal 1 (for discussion and an elucidation of a more flexible demand structure see Peerlings, 1993). Private-household income is formed by factor rewards, corrected for taxes and an exogenous income transfer from the rest of the world. Private-household expenditure, savings and paid non-product-related indirect taxes are assumed to form a fixed share of private-household income.

Indirect taxes and price-reducing subsidies are incorporated in the buyers' prices, creating, together with the market margins, a price wedge between suppliers' and buyers' prices. Ideally, one would like to know on which transaction the indirect taxes (positive or negative) are levied. However, data are only available for taxes on exports and total domestic demand (wholesale taxes).

Tax revenue (product-related, non-product-related and income taxes), together with net income transfers from the rest of the world (e.g. development aid), create government income. This income is divided into savings and expenditure on consumer goods, according to fixed proportions. Government demand is modelled by means of CES uncompensated-demand functions, implying expenditure elasticities of 1.

Total gross saving in the economy equals the sum of private and government saving, capital depreciation and the surplus on the balance of trade. Savings, corrected for non-product-related taxes on investment (e.g. investment subsidies), equal investment. If saving is seen as buying a capital good by households, the government and the rest of the world to store wealth, investment demand for individual commodities is the input demand of a hypothetical capital-goods industry that produces this (aggregate) capital good. Investment demand is modelled, using Leontief input-demand functions. Leontief instead of CES input-demand functions are used, because in the initial situation demand is sometimes negative (reduction of stocks).

Trade and transportation services (G38: market margins) are produced by different industries. The use of these services is incorporated in the buyers' prices of exports, total domestic use (wholesale margins) and household demand (retail margins). Because all equations in an AGE model are homogeneous of degree zero, a price *numéraire* is required to determine the actual price level. The Laspeyres index in output prices is used in the present simulations.

A calibration procedure is used to select the parameter values of the AGE model. The parameters of the model are chosen such that the model can reproduce the SAM of 1990 as an equilibrium solution: the 'benchmark' equilibrium. In the model, CES production and CET product-transformation functions are used, which implies that substitution and transformation elasticities have to be specified exogenously (Shoven and Whalley, 1992). The substitution and transformation elasticities used in the production, consumption and foreign-trade specifications originate from Zeelenberg *et al.* (1991) and from own approximations (Komen and Peerlings, 1996). The price elasticity for labour supply originates from van Soest (1993).

POLICY SIMULATIONS AND RESULTS

In this section, the effects on production, income formation, prices and trade for the Dutch economy of the introduction of the 1996 energy tax are determined.

The 1996 Energy Tax

The major parameters of the 1996 energy tax are summarized in Tables 13.1 and 13.2. Table 13.1 shows that the first 800 m³ gas and 800 kWh electricity are excluded from taxation. Moreover, usage above 170,000 m³ gas and 50,000 kWh electricity is also excluded. The different thresholds imply that the relevant price of electricity and gas for small users excludes the energy tax and for large users includes the tax. Although in horticulture small users are dominant, firms are exempted from the energy tax on gas. The government considers taxing this energy-intensive industry to be a significant threat to the international competitiveness of this sector.

Table 13.1. Threshold levels for the Dutch 1996 energy tax.

	Gas	Electricity
Households	> 800 m³	> 800 kWh
Firms (small users)	800–170,000 m³	800–50,000 kWh
Firms (large users)	800–170,000 m³	No tax
Horticulture	No tax	800–50,000 kWh

Table 13.2. Energy-tax rates for electricity and gas, 1996–1998.

	1996		1997		1998	
	Tax rate (cents)	As % of 1990 prices	Tax rate (cents)	As % of 1990 prices	Tax rate (cents)	As % of 1990 prices
Electricity per kWh	3.47	19.1	3.47	19.1	3.47	19.1
Gas per m³	3.76	9.3	7.52	18.5	11.2	27.6

100 cents = 1 guilder.

Table 13.2 shows that the introduction of the energy tax will take place in three steps. The *ad valorem* energy tax on electricity is 19.1%, while the *ad valorem* tax on gas will increase from 9.3% in 1996 to 27.6% in 1998. Tax revenues are returned through detailed fiscal measures, most of which lower labour costs for firms or increase net income of the working force (Energie Beheer Nederland, 1995).

Policy Simulations

In the first policy simulation, the 1996 energy tax is simulated by raising the gas price (G23 and G24) and electricity (G25) in Table 13.A2 in the Appendix. Consumers and industries where small users are dominant (mainly agricultural and services industries) have to pay the tax. Domestic and small commercial users do not have to pay the tax if they remain under the threshold of 800 m^3 gas and 800 kWh electricity. This is modelled by taxing total use and a partial tax return. For those industries where large users are dominant (mainly food-processing, chemical and metal industries), it is assumed that the relevant prices exclude the tax. However, these industries have to pay the energy tax over the range of 800–170,000 m^3 taxable gas use. This is modelled by means of a transfer from the industries to the government. As a result, this policy measure gives no incentive for large users to reduce gas use (Netherlands Central Planning Bureau, 1993). Horticulture is totally exempted from the energy tax on gas.

In the second simulation, gas use by horticulture is also taxed, to assess the consequences of the exemption. In the third simulation, the taxation is extended to all industries, with the exception of the gas- and electricity-distribution industries. Under this general taxation, there is no distinction between large and small energy-using industries.

In all simulations, the proposed tax rates for 1998 are used and tax revenue is used to lower average total labour costs. The higher real-wage rate that results from this policy increases labour participation, which is modelled by means of a supply function of total mobile labour, where the real wage is the only labour supply-determining variable.

Results

The effects of the energy tax in the three different simulations are summarized in Tables 13.3 and 13.4. The simulations show that the energy tax has an insignificant effect on agriculture (except horticulture under glass). The main focus in this section will therefore be on horticulture and the economy as a whole.

The results show that the 1996 energy tax of 19.1% on electricity and 27.6% on gas for households and 'small' energy users results in higher prices of electricity and gas for small users (14.1% and 17.6%, respectively) and lower prices for large users (–4.2% and –7.8%, respectively). Being exempted from an energy tax on gas, horticulture profits from the lower price of gas.

Table 13.3. Effects of three energy-tax scenarios: the 1996 energy tax; extension to horticulture; and extension to all industries.*

		Effect of tax scenarios (% change)†		
Variable	Industry, consumer, commodity‡	1996 energy tax (small users)	Including horticulture under glass	All industries (general tax)
Gas and gas-oil use	Consumer (4,825)	−4.4	−4.1	−3.3
	Total (9,065)	−3.8	−4.4	−4.2
Natural-gas use	Fertilizer industry (617)	7.0	8.0	−10.3
	Gas distribution (6,042)	−3.8	−4.3	−3.3
	Electricity supply (1,567)	−2.2	−2.7	−3.1
	Total (10,031)	−1.9	−2.1	−4.4
Electricity use	Consumer (2,699)	−3.6	−3.6	−3.0
	Fertilizer industry (56)	6.3	7.1	−10.1
	Total (9,813)	−3.1	−3.1	−4.5
Price gas and gas oil	Selling price gas distribution	−7.3	−8.3	−10.6
	Consumers	17.4	16.2	13.0
	Small-user industries	17.6	16.3	13.1
	Large-user industries	−7.8	−8.8	13.1
Price electricity	Selling price electricity supply	−4.4	−4.5	−6.6
	Consumers	14.1	13.9	11.6
	Small-user industries	14.1	13.9	11.6
	Large-user industries	−4.2	−4.4	11.6
Production	Fertilizer industry (2,344)	5.1	5.9	−7.9
	Oil and chemical industry (93,473)	−1.3	−1.5	−3.4
Mobile labour	Fertilizer industry (343)	3.2	3.7	−4.7
	Oil and chemical industry (10,475)	−0.8	−0.8	−1.7
Income	Fertilizer industry (676)	9.6	11.1	−13.9
	Oil and chemical industry (34,842)	−2.5	−2.7	−5.5
Net export	Fertilizer (1,184)	7.7	8.8	−11.6
Miscellaneous	Exchange rate	0.06	0.18	0.57
	National income (516,267)	0.50	0.49	0.47
	Total mobile employment (267,742)	0.07	0.08	0.12
	Equivalent variation	376	293	−181
	Energy-tax industries	1,284	1,505	2,707
	Energy-tax consumers	968	965	952

* Energy tax on electricity 19.1%; energy tax on gas and gas oil 27.6%.
† Equivalent variation and tax revenues, in million guilders (1990).
‡ Base-year quantities in parentheses, in million guilders (1990).

This results in a 2.8% higher use of gas, which enables horticulture to increase production by 1%. Similar results occur (7% more gas use, 5.1% more production) for the fertilizer industry, which is highly energy-intensive. On the national level, electricity and gas use decrease by 3.1% and 1.9%,

Table 13.4. Effects of three energy-tax scenarios on horticulture.*

	Effects of tax scenarios (% change)†		
Variable‡	1996 energy tax (small users)	Including horticulture under glass	All industries (general tax)
Gas and gas-oil use (959)	2.8	−5.9	−4.3
Electricity use (101)	−0.4	−5.6	−4.1
Production (7,382)	1.0	−2.3	−1.5
Mobile labour (1,169)	0.8	−1.5	−0.8
Income (2,946)	1.8	−4.1	−2.4
Net export flowers/plants (5,721)	0.7	−1.9	−1.9
Net export vegetables/fruit (2,041)	−2.8	−2.8	−1.4

* Energy tax on electricity 19.1%; energy tax on gas and gas oil 27.6%.
† Equivalent variation and tax revenues, in million guilders (1990).
‡ Base-year quantities in parentheses, in million guilders (1990).

respectively. The reduction of labour costs, by means of recycling the energy tax revenue of 2,252 million guilders (1990), results in a 0.07% larger labour force. Moreover, the changing tax basis decreases welfare losses due to existing tax distortions (second-best welfare improvements). Both the higher labour participation and the second-best welfare improvements increase national welfare, measured by the equivalent variation, by 376 million guilders (1990).

If the energy tax on gas is also applied to horticulture, production and income in horticulture fall by 2.3% and 4.1%, respectively, while electricity and gas use in horticulture decrease by 5.6% and 5.9%, respectively. Smaller horticultural production results in lower net exports of flowers and plants of 1.9% and vegetables and fruits of 2.8%. Welfare is 83 million guilders lower than under the 1996 energy tax. The exchange rate goes up by an extra 0.12%, compared with the situation under the 1996 energy tax. This illustrates that a unilateral energy tax negatively affects industries, such as horticulture, which are not protected domestically and which compete on world markets.

The latter result is confirmed by the third simulation, in which all industries are taxed. The fertilizer industry, for example, uses 10% less energy, which results in a 7.9% lower production and 13.9% lower income. Moreover, net exports of fertilizer decrease by 11.6%. Total use of gas and electricity decrease by 4.2% and 4.5%, respectively. Although the labour force increases by 0.12%, through an increase of the real wage, welfare decreases by 181 million guilders (1990). Therefore, a general taxation of energy does indeed hurt the economy as a whole more than an energy tax for small users does.

The results of the three simulations in terms of reduced emissions of CO_2 are presented in Table 13.5, comparing these with 1990 national emissions of greenhouse gases (Simons *et al.*, 1994; de Haan and Keuning, 1995). Electricity is primarily generated from natural gas. Hence, only the emissions of gas associated with electricity, rather than electricity and directly distributed natural gas, are presented to avoid double counting. Moreover, the effects of the energy tax on CO_2 emissions of other, non-taxed, energy sources are not calculated. In future development of the model, such analysis will be possible when emission factors by energy source and industry are incorporated into the current model. In Table 13.5, it is estimated that total CO_2 emission will decrease by 1.370 million tonnes as a direct result of the 1996 tax, of which most is due to emission reductions by consumers. If horticulture is not exempted from taxation of gas, total emissions decrease by 1.514 million tonnes CO_2. Thus, including horticulture would only result in an additional reduction of 0.144 million tonnes compared with the 1996 energy-tax case, while the total emission by horticulture would decrease by 0.574 million tonnes. The reason for this modest impact on total emissions from the inclusion of horticulture is that, due to the lower demand by horticulture, prices for both gas and electricity will decrease and hence demand by the other energy users will increase. If a general energy-tax rate is introduced, total emissions would decrease by 3.173 million tonnes CO_2, which is a 2% reduction of total CO_2 emission or a 1.5% reduction of total greenhouse-gas emission.

Table 13.5. Changes in CO_2 emissions as a result of the imposition of different energy taxes.

	Base-year (1990) quantities (kt CO_2)	1996 energy tax (small users) (kt CO_2)	Including horticulture (small users) (kt CO_2)	All industries (general tax) (kt CO_2)
Natural gas*	72,100	–1,370	–1,514	–3,173
horticulture	6,600	185	–389	–284
consumers	19,200	–845	–787	–634
other	46,300	–710	–338	–2,255

* Total CO_2 emission in 1990 was 158.5 million tonnes CO_2 (de Haan and Keuning, 1995), representing 73.4% of total annual greenhouse-gas emissions (Swart *et al.*, 1993).

SUMMARY AND CONCLUSIONS

This chapter quantifies the effects for the Dutch economy of the introduction of the 1996 energy tax for small users. The simulations give a detailed insight into the effects of the 1996 energy tax, and its possible extension to horticulture and other large energy users. An AGE model is used to analyse the effects of the energy tax, such that different levels of taxes can be simulated for different industries, while the effects of the destination of the tax revenue can simultaneously be taken into account. Moreover, the model is very detailed with respect to relevant commodities and industries.

There are, however, drawbacks of the model relevant to the analysis of energy taxes and revenue recycling. First, unilateral energy taxation might give 'first-mover advantages' if it is expected that other countries will subsequently introduce their own unilateral taxes. Due to the static nature of the model, such factors cannot be incorporated. Secondly, possible technological changes are not modelled. High energy prices will affect the speed and direction, for example, towards energy saving, of technological change (Bovenberg, 1993; Jorgenson and Wilcoxen, 1993; Carraro et al., 1995). Thirdly, the labour-supply function is exogenous in the present model. Theoretically, it would be more consistent to model labour supply as the result of utility optimization, with the consumer choosing between leisure and labour. If such an extension were incorporated, welfare results would take into account welfare reductions associated with increased time in work or reduced leisure time.

A further limitation of the present model is that emission coefficients per energy source and per industry are not incorporated in the model, making it difficult to judge to what extent overall emissions of greenhouse and non-greenhouse pollutants are affected by the energy tax. Hence, it is difficult to judge to what extent the environmental situation has improved due to the energy tax. Further, the impact of emissions on relocation of production to other countries is ignored. This applies especially to the third simulation. A fifth drawback stems from the fact that, within industries, there are large and small energy users. This implies that the exact effects of an energy tax are difficult to calculate, because in the model the energy price for an industry either includes or excludes the tax. However, the heterogeneity within industries is taken into account when the tax returns and transfers are calculated above. Finally, the comparison of reduced emissions with total national emissions is calculated using 1990 data, while the tax was introduced in 1996.

Given these caveats and possible extensions, the present analysis shows that the 1996 energy tax of 19.1% on electricity and 27.6% on gas for households and 'small' energy users is relatively insignificant for agriculture. Horticulture benefits from the lower prices of energy (excluding taxes), because it is exempted from the energy tax on gas. On a national level, electricity and gas use decreases by 3.1% and 1.9%, respectively, while total emission of CO_2 decreases by 1.370 million tonnes (0.8%). The emission reduction is less than the 3–5% reduction targets. National welfare, measured by the equivalent variation, increases by 376 million guilders (1990) due to the tax, which is partly due to increased employment (0.07%) and partly due to second-best welfare improvements. The tax therefore realizes a double dividend, in the sense that CO_2 emissions reduce and employment increases. However, from a welfare economic perspective, it is indeterminate whether this employment effect raises total welfare, as the calculated welfare growth should be corrected for the welfare reduction caused by the decrease in leisure.

If the 1996 energy tax on gas were extended to horticulture, production and income in horticulture would be reduced by 2.3% and 4.1%, respectively, while electricity and gas use in horticulture would decrease by 5.6% and 5.9%, respectively. The equivalent-variation measure of aggregate welfare shows a fall of 83 million guilders (1990) in these circumstances and the exchange rate would be 0.12% higher than under the 1996 energy tax. These results illustrate that a unilateral energy tax is more costly for those industries competing on world markets, such as horticulture. The latter results are confirmed by a policy simulation where all industries are taxed.

NOTE

1   The authors thank the Netherlands Organization for Scientific Research and the Foundation for Economic Social-Cultural Geographical and Environmental Sciences (NWO-ESR), for research funding.

APPENDIX: SOCIAL ACCOUNTING MATRIX FOR THE NETHERLANDS

Table 13.A1. Social accounting matrix for the Netherlands in 1990 (in million guilders) (calculated from consumption and production tables for 1990 in basic prices, Netherlands Central Bureau of Statistics, 1992, 1993).

	Commodities	Production	Income formation			Income distribution		Institutions					
	1.	2.	3.	4.	5.	6.	7.	8.	9.	10.	11.	12.	13.
Account	Commodities	Industries	Labour income	Capital income	Non-prod. taxes	Households	Govern.	Households	Capital	Govern.	R.o.w. current	R.o.w. capital	Total
Commodities													
1. Commodities	0	471,829						280,134	105,640	74,795	279,746		1,212,144
Production													
2. Industries	948,964												948,964
Income formation													
3. Labour income		275,312											275,312
4. Capital income		191,354											191,354
5. Non-prod. related taxes		10,469											10,469
Income distribution													
6. Households			275,032	133,124									408,156
7. Government	7,352				10,469	19,606		22,962	8,818				69,207
Institutions													
8. Households						391,320							391,320
9. Capital				58,230				88,224		-12,238			134,216
10. Government							62,557						62,557
11. R.o.w. current	255,828		280			-2,770	6,650					19,758	279,746
12. R.o.w. capital									19,758				19,758
13. Total	1,212,144	948,964	275,312	191,354	10,469	408,156	69,207	391,320	134,216	62,557	279,746	19,758	

R.o.w., rest of the world.

Table 13.A2. Industries and commodities in the Netherlands social accounting matrix.

Industries		Commodities	
B1	Dairy farming	G1	Cattle
B2	Pig farming	G2	Pigs
B3	Poultry farming	G3	Poultry
B4	Arable farming	G4	Flowers and plants
B5	Horticulture under glass	G5	Grain
B6	Other horticulture	G6	Grain substitutes
B7	Forestry and agricultural services industry	G7	Other arable-farming products
B8	Fishery	G8	Vegetables and fruits
B9	Beef and other meat products manufacturing	G9	Milk
B10	Pork and pork products manufacturing	G10	Grain-mill products
B11	Poultry and poultry products manufacturing	G11	Other agricultural and food products
B12	Dairy products manufacturing	G12	Eggs
B13	Fish products manufacturing	G13	Raw materials dairy and dairy products
B14	Canned and preserved fruits and vegetables manufacturing	G14	Bakery products
B15	Grain-processing industry	G15	Fish and fish products
B16	Compound feed-processing industry	G16	Beef and other meat products
B17	Flour mills	G17	Pork and pork products
B18	Other food products manufacturing	G18	Poultry and poultry products
B19	Wool products, cotton products, textiles, clothing, leather, footwear, wood, paper and paper-processing industry	G19	Beverages and tobacco
		G20	Raw materials for leather, textiles, footwear and paper
B20	Fertilizer industry	G21	Oil products, building materials, glass, synthetic products
B21	Chemical pesticides manufacturing		
B22	Oil and other chemical industry	G22	Fluid fuel
B23	Stone, clay, glass, metal products and fabricated metal products manufacturing	G23	Gas and gas oil
		G24	Natural gas
B24	Machinery, tools and electrical products manufacturing	G25	Electricity
		G26	Water
B25	Transport equipment industry	G27	Fertilizer
B26	Electricity supply	G28	Pesticides
B27	Gas distribution	G29	Rubber, ores, other minerals, metal, other materials and semimanufactured products
B28	Water supply		
B29	Construction		
B30	Wholesale and retail trade	G30	Equipment and machinery
B31	Repair services industry	G31	Transport equipment and parts
B32	Transport and communication services	G32	Home and office furniture and equipment packing, commodities in work, etc.
B33	Veterinary services		
B34	Other services (banking, insurance, public services, etc.)		
		G33	Garden services and provision of agricultural services
		G34	Construction and ground work, painting, building, public utilities, etc.
		G35	Repair and maintenance commodities
		G36	Veterinary services
		G37	Other services
		G38	Market margins

Policy Instruments for Environmental Forestry: Carbon Retention in Farm Woodlands

14

Robert Crabtree[1]

INTRODUCTION

The contribution of forestry to national policies for enhancing carbon sinks depends primarily on the rate of expansion of the afforested area. Carbon retention in forests only occurs in young plantations (in the pre-equilibrium development phase), and the rate of additional (non-replacement) planting ultimately determines the rate of carbon retention. In the UK, the state forestry service no longer engages in significant new planting, with the consequence that the private sector is the major investor in new forests and woodlands. However, the investment characteristics of forestry are such that very little new planting would take place without government provision of financial or fiscal incentives. Forestry produces low returns to capital, reflecting high investment costs, particularly for land, and the long time period to harvest. This leads to a dependency on state intervention, with the scale and type of private investment largely determined by policy and regulated by conditions attached to the policy instruments (Crabtree, 1995). Within the UK, most of the new planting is located in Scotland, where land prices are lower, reflecting a more limited set of alternative opportunities for land use.

UNITED KINGDOM FORESTRY POLICY

With planting almost entirely in the private sector, government is faced with a principal-agent problem (Kreps, 1990). This occurs where the principal (in this case government) does not have complete control of resources and depends for achievement of its objectives on modifying the behaviour of agents (in this context, landowners), who are motivated by private goals. Specifically, in the absence of intervention, new planting by landowners would not result in the quantity, quality or mix of forestry outputs desired by society. The problem for the principal is how best to design a set of policy instruments that deliver the objectives of policy through the indirect action of private landowners and investors. Following the removal of tax incentives in 1988, government has sought to deliver public benefits from afforestation by offering financial incentives to landowners, coupled with constraints on planting in areas of high environmental sensitivity.

This raises the question of what intervention is trying to achieve: that is, what the national forestry objectives are. This is an issue not categorically resolved by a recent review of policy objectives (Scottish Office Environment Department, 1994). The review of the sector identified a multiobjective role for forestry, with specific objectives to increase the output of timber and extend the amenity and environmental benefits from forests. The importance of environmental benefits in the mix of forestry outputs contrasts markedly with cost–benefit analyses undertaken a decade earlier (e.g. National Audit Office, 1986) and reflects the advances made in valuation of non-market outputs and the acceptance of their validity by government for inclusion in forestry-policy appraisal. The environmental benefits identified in the policy review are numerous, and include benefits to public health from recreation, to amenity from improved landscapes, to biodiversity from habitat preservation or enrichment, and to sustainability from the creation of carbon sinks. One explanation for a lack of clearly defined quantitative objectives is that policy develops incrementally, with the emphasis on policy adjustment. Each change in policy, rather than reflecting a reappraisal through rational policy-making, tends to be a response to changing circumstances and previous achievements and failures (Gregory, 1989).

The UK forestry-policy objectives for planting on agricultural land have much in common with those underpinning European policy under European Community (EC) Regulation 2080/92. Although priorities differ between countries, policy objectives include the provision of environmental benefits from forests and diversification opportunities for farmers, together with some contribution to resolving the problems associated with 'surplus' farm production (Anz, 1993; Pearse, 1995). Success in expanding farmland afforestation has been variable, depending on the level of incentives, opportunity costs (reflecting Common Agricultural Policy (CAP) support), institutional barriers and the willingness or otherwise of farmers to engage in forestry (e.g. OECD, 1994b; see also Lippert and Rittershofer, Chapter 15, this volume). The expansion of woodland planting does not have major implications for agricultural output, given that output control (set-aside and livestock quotas) is already embedded in policy, and the location of much of the new planting is on relatively unproductive land.

In order to deliver policy aims in the UK, incentives are provided to stimulate private-sector planting. These take the form of a payment per hectare planted, which varies depending on the type of planting (conifers, broadleaves), the quality of the land and the size of plantation. A number of additional targeted incentives are given for the creation of recreational forests and those in locations where environmental benefits are especially high. For planting on farmland, a set of additional annual payments have been made under special schemes, which have operated since 1988. This chapter analyses aspects of the current scheme (Farm Woodland Premium Scheme, FWPS), which was specifically established to support the planting of woodlands on farmland (MAFF, 1992). In this chapter, the contribution of the scheme to carbon sequestration is calculated, the scope for enhancing the rate of carbon retention examined, and the trade-offs between carbon retention and other environmental benefits from farm planting identified.

NEW FORESTRY PLANTING IN GREAT BRITAIN

In the 1990s new planting in Great Britain (GB) averaged around 15,000 ha year^{-1}, increasing to 17,012 ha in 1994/95 (Forestry Commission, 1995). Around two-thirds of this was in Scotland, where 60% of the planting is in conifers (mainly Sitka spruce) and 40% in mixed broadleaves. The contribution of farm woodlands planted under specific schemes for farmers has increased in recent years, such that they now form a substantial part of the total planting. During the 1988–1992 period, 13,613 ha were planted on GB farmland and, under the present FWPS scheme, 7900 ha were planted on average in the 1994 and 1995 years. This comprises 46% of the total planting, and most of it was in Scotland (Table 14.1).

Table 14.1. Mean (1994/95) new annual private forestry planting, Great Britain (from data supplied by the Forestry Commission and the Scottish Office Agriculture and Fisheries Department).

	Great Britain (ha)	Scotland (ha)
Farm Woodland Premium Scheme	7,868	5,080
Other planting excluding the above	9,144	5,798
Total	17,012	10,878

Voluntary incentive payments are available on any new forestry planting that meets the design specifications laid down by the Forestry Commission. Under the FWPS, additional annual payments over 10–15 years are available, but only to farmers planting on farmland. These payments are designed to take into account the higher (private) opportunity costs of farm forestry, as compared with afforestation on non-farmed land. Table 14.2 indicates the present cost (in public expenditure terms) of the total package of financial incentives for different types of farm woodland. Incentive costs are highest for mainly broad-leaved planting on lowland sites (arable, improved grassland), at £4,078 ha^{-1}. The financial cost for such planting is nearly four times that for conifers on unimproved land – a reflection of the variation in the opportunity costs for land and differences in planting costs for different species. The mean cost per hectare to the public exchequer for the mix of planting that has taken place under the scheme is £2,190 ha^{-1}, in present cost terms.

Table 14.2. Farm woodland incentive payments, UK. Incentive payments available for plantations exceeding 10 ha.

	Conifers (> 50%)		Broadleaves (> 50%)	
	Present value (£† ha^{-1})	50-year annuity* (£ ha^{-1})	Present value (£ ha^{-1})	50-year annuity (£ ha^{-1})
Less favoured areas				
Improved land	2,057	130	2,912	184
Unimproved land	1,141	72	1,632	103
Lowland (non-less favoured areas)	2,940	185	4,078	257

* Annuity estimated at 6% discount rate.
† 1 European currency unit (ECU) = £0.813.

The time period over which an unthinned commercial plantation is a net retainer of carbon is the period to clear felling. This will vary with the species and management system, but approximates to 50 years for high-yielding Sitka spruce. Forests with lower growth rates and longer time periods to felling will continue to fix carbon over a longer period. If the 50-year period is used for comparing the average annual rates of carbon retention from different woodlands, the corresponding public-expenditure costs, calculated in Table 14.2 as annuities at a 6% cost of capital, vary between £72 and £257 ha^{-1}, depending on species and land quality.

FARM FORESTRY PLANTING

The FWPS has the objective of providing farmers with an alternative productive use for land and delivering environmental benefits. The aims can be interpreted as partly the provision of an adjustment mechanism for farmers to progressive agricultural policy reform, and partly the generation of environmental services from farming through improvements to biodiversity and landscape. There is no specific aim to encourage farm forestry as a carbon sink, although this appears to be a broad policy goal in publicly assisted forestry planting (Scottish Office Environment Department, 1994). In a recent evaluation of the FWPS undertaken by the author (Crabtree, 1996), a sample of farmers planting under the scheme were asked what their objectives were in entering the scheme. Nearly all entrants had amenity, landscape or wildlife objectives in planting, and 80% claimed that benefits from shelter were important. Only 35% had commercial timber production as an objective. In unprompted questioning, no respondents mentioned carbon fixing as an objective, although 76% did indicate that benefits for their family (but not future generations of the public in general) were important in their decision to plant trees. Hence, the private decisions of farmers do not apparently take into account benefits from carbon fixing, a fact hardly surprising if it is assumed that behaviour is determined largely by self-interest. Any carbon-retention benefit can thus be taken as an external benefit from forestry planting. Much the same conclusions probably apply to forestry investors other than farmers. Since such investors generally plant plantations for commercial motives, a greater importance would be attached to timber production and less to environmental benefits – suggesting that the impact of the environmental benefits from enhanced carbon retention on investment decisions would be extremely small.

CARBON-RETENTION ESTIMATES

Any estimate of the impact of forestry on carbon sequestration faces the problem of uncertain technical coefficients associated with the different rates of carbon storage and loss in components of the carbon cycle. The forestry management system can also have an important impact (Cannell, 1995) and this presents greater problems with farm woodlands as compared with commercial forests, since the future management of many woodlands is not well defined. The mixture of species commonly present implies a more complex management policy than that for single-species commercial planting. For estimation purposes, it was assumed that conifers, planted as a monoculture or in mixtures with broadleaves, would be unthinned but clear-felled and replaced, according to normal commercial practice. Broadleaves were assumed not to be harvested. Carbon-fixing estimates were calculated for a random sample of 100 plantings under the FWPS (an 11% sample) by applying growth and carbon-retention coefficients derived from the literature. The procedure was as follows: yield-class estimates for each sample planting were made on the basis of their geographical location, using Allison *et al.*'s (1994) yield-class map of Scotland for Sitka spruce (the predominant conifer species planted) and data from Tyler (1994) for broad-leaved species. Scotland exhibits a wide range of growth conditions and this, coupled with a substantial variation over space in the planting under the scheme (from good arable land to unimproved land in the less-favoured areas), produced a range in the estimated Sitka spruce yield class from 4 to 21 m^3 ha^{-1} year^{-1}. The mean figure of 17 is high for Scottish planting and is greater than observed on state plantations (National Audit Office, 1986). This is to be expected, since much of the farm woodland planting is on more productive land with less harsh climatic conditions. Even so, given the uncertainty over future management, the yield estimates should be regarded as at the upper end of those likely to be attained.

Carbon-retention estimates for different species by yield class were taken from Dewar and Cannell (1992) and Cannell (1995), who used Dewar's (1991) model of carbon retention by forests. These estimates take account of the temporal patterns of fixation and release, not only in the timber but also in the wood products, litter and soil (Table 14.3). The retention rates are higher and more reliable than those used by Pearce (1994) in a recent cost–benefit analysis of UK forestry. The latter accounted only for retention in timber and release through end use.

Table 14.3. Carbon (C) storage estimates for plantation species, UK (from Dewar and Cannell, 1992).

Species	Yield class (m^3 ha^{-1} year^{-1})	Rate of storage* (Mg C ha^{-1} year^{-1})	Equilibrium carbon storage (Mg C ha^{-1})
Picea sitchesis (unthinned)	24	5.6	254
	16	4.5	229
	8	2.9	169
Fagus sylvatica	6	2.4	200
Quercus spp.	4	1.8	154

* Total C storage at end of first rotation divided by rotation length.

The rate of carbon storage changes over time, depending on the species, growth rate and management regime for the woodland. Fast growth is associated with rapid storage, but earlier felling and an associated carbon release that depends on product use. Slower-growing broadleaves accumulate carbon over a much longer time period but at a slower rate, with the result that differences in equilibrium storage levels between species and growth rates are less marked. The drainage of wetlands and deep peatland releases carbon, thus reducing both the net rate of retention through afforestation and the time period over which net retention is positive (UK Department of Environment, 1994; Cannell, 1995). Although very large areas of Scotland have peat soils, very little of the FWPS was on peatland and none on wetlands. No adjustment to retention rates for increased carbon loss is thus made.

The mean annual rate of carbon fixing for all planting under the FWPS, weighted by area, was estimated from the sample data as 3.2 t ha^{-1} (Table 14.4). There was little difference between the rate of carbon fixing on the different types of land planted. On poorer-quality (unimproved) land, the higher proportion of coniferous planting compensated for the generally lower productivity of that land. On arable land, the higher proportion of broad-leaved planting limited the mean annual rate of carbon retention achieved. Total carbon retention was estimated at 53,000 t year^{-1} for a planted area of 16,500 ha in Scotland, and applying the same growth and retention estimates to the total GB planting gives an estimate of around 85,000 t year^{-1} over at least the next 50 years.

Table 14.4. Carbon (C) retention by farm woodlands planted under the FWPS (Scotland).

	Percentage area planted	Mean rate of carbon retention (t C ha^{-1} year^{-1})*
Less favoured areas		
Improved land	38	3.3
Unimproved land	46	2.9
Lowland	16	3.2
Mean (weighted by area)		3.2

* Calculated from yield-class estimates, as in Table 14.3, by dividing the total C storage at the end of the first rotation by the rotation length.

The annual rate of planting during the 2 years following the launch of the FWPS averaged 5,100 ha for Scotland and is estimated at around 7,900 ha for GB. This gives a rate of retention of 25,300 t year^{-1} for each year's new planting, which compares with an estimate of the total carbon currently being removed from the atmosphere by British plantations of 2.5 Mt year^{-1} (Cannell and Dewar, 1995). The contribution of recent farm woodland planting to this overall level of carbon retention appears trivially small. However, this is a comparison between the stock of all non-mature forests (covering planting over the last 50 or more years) and annual planting rates under the recent farm woodland scheme. In terms of current planting, the proportion of forests planted under farm and non-farm schemes suggests that at least one-third of carbon retained in the forest stock as a result of new planting is derived from on-farm afforestation associated with farm woodland grant schemes (see Table 14.1). This is explained by the higher proportion of conifers in commercial forests compensating for the poorer site conditions, as compared with farm woodlands. Farm planting will therefore make a significant contribution to retention in the medium term, so long as the incentive structure is progressively modified to keep farm woodland planting attractive to farmers and to maintain planting rates. Over a longer-term horizon of, say, 100 years, farm woodlands will contribute an increased proportion of the total retention from current planting, because of the higher proportion of broadleaves planted.

VALUING THE BENEFITS FROM CARBON RETENTION

A number of different approaches have been used to ascribe a value to increased carbon retention. Studies based on a global cost–benefit approach have been reviewed by Pearce (1994), Fankhauser (1995a) and Fankhauser in Chapter 5 in this volume. In general, they adopt a scenario in which carbon dioxide (CO_2) concentration is doubled. Valuing retention in terms of avoided damage costs led Pearce (1994) to adopt a value of $US14 t^{-1} carbon, based on the estimates of Nordhaus (1991) and Ayres and Walter (1991). Fankhauser's (1995a) estimate of the marginal damage costs to society of CO_2 emissions was $US20.3 t^{-1} carbon, rising to $US27.8 after 30 years. Other cost estimates reviewed by Fankhauser vary from $US5.3 to $US10 t^{-1} over the years 1991 to 2000, increasing to $US10–18 in 30 years' time. These are lower-bound estimates; upper-bound estimates were substantially higher. Both the annual estimates of the marginal social costs of CO_2 emissions and their distribution through time are highly sensitive to the discount rate used, an aspect discussed in detail elsewhere (Price and Willis, 1994; Price, 1995).

An alternative approach to the valuation of carbon sequestration is to identify the alternative public project specifically directed at reducing carbon emission with the highest social cost per unit net emission reduction (e.g. Anderson, 1991; van Kooten *et al.*, 1992). In the UK, government has invested little in such projects (House of Lords, 1995): the shadow price of sequestration thus appears to be very low. Where carbon taxes are implemented, as, for example, in Norway, the rate of taxation can be used as an indication of society's shadow price of carbon fixation. Despite the case for a carbon tax (Pearce, 1991b; Peck and Teisberg, 1993a), this has not been introduced in the European Union (EU), and therefore cannot provide a guide to the pricing of benefits from measures that enhance carbon sequestration.

Given the very small contribution (1.5%) that forestry makes to UK net emission reduction (Cannell and Dewar, 1995) and the limits imposed by land availability for new planting, forestry can never play more than a minor role in emission control. The main thrust of policy has to focus on reducing energy consumption and emissions through the application of regulation, moral suasion and economic instruments. If these approaches prove cheaper in social-cost terms than mechanisms that enhance carbon retention, the value of forestry's contribution to reducing net emissions would be more appropriately valued on the basis of its ability to provide a limited substitute for emission-reduction mechanisms.

Nevertheless, if typical estimates of the global damage associated with present emissions (£3–20 t^{-1}, approximating to \$US5–30 t^{-1}) are applied to carbon retention achieved through farm woodland planting, the mean annual cost saving is £10–65 ha^{-1} planted. This relates to the estimated annual rate of carbon retention of 3.2 t ha^{-1} for planting under the scheme as a whole. How are such benefits to be evaluated? One approach is to consider woodland planting as a single objective policy instrument, in which public expenditure is allocated to farmers in order to increase the size of the carbon sink. Using a 50-year horizon, the cost to the exchequer of financing incentive payments varies from £72 to £257 ha^{-1}, depending on land quality and species planted (see Table 14.2). For the FWPS planting in Scotland, the mean annual expenditure is £138 ha^{-1}, a figure which is substantially greater than the public benefits derived from carbon retention. Admittedly, the exchequer cost could be lower if savings from any reduction in the cost of agricultural support were included, but such cost-savings are likely to be small. With the introduction of livestock quotas, afforestation will displace rather than reduce much of the existing agricultural production. In this type of 'value for money' appraisal, public expenditure on farm forestry can only be justified in terms of producing other environmental benefits, including landscape improvement and increases in biodiversity.

Applying a cost–benefit appraisal to farm woodlands is not practicable, since there is an absence of complete benefit information. Most of the economic valuation research on forestry has concentrated on quantifying the recreational benefits to the public from free access to forests. However, recreation is not a benefit that will flow from farm woodlands, where public access is usually prohibited. It could be, however, that the sort of locational differentials found in recreational analysis (Willis, 1991) would also apply to the visual 'use' values of farm woodlands. If so, this would indicate higher benefits from planting near population centres (Bateman *et al.*, 1996). Nevertheless, research on the public's willingness to pay for agri-environmental improvements suggests that the non-use benefits from farm woodlands will be most significant (Hanley *et al.*, 1996). It is known that the Forestry Commission is currently attempting to value the biodiversity benefits from British forests, but no estimates are as yet available.

It is of interest to note that carbon retention by forests is not treated as a benefit in 'official' UK forestry cost–benefit analyses and does not appear to influence decision-making on public investment in forestry. There seem to be a number of possible explanations. One is the considerable uncertainty surrounding the costs associated with global warming in the UK context and an unwillingness to impose additional social costs until the situation becomes clarified. A further reason relates to the context of commitments under the framework Climate Change Convention (UK Department of Environment, 1994). The Convention operates in a framework of physical targets for emission control, a context in which cost-efficient, rather than cost–benefit, approaches to policy formulation are most relevant. There are a number of aspects of the UK commitment that also bear on the valuation ascribed to reductions in net emissions. The UK has separate commitments for reductions in emissions and the enhancement of carbon sinks (OECD, 1994a). It anticipates meeting its emission targets largely through the adoption of natural gas, rather than coal-fired, technology in privatized electricity generation (House of Lords, 1995, and Chapter 1 this volume). This is supported by a partnership approach with industry, business and the public, which includes a number of regulatory and fiscal measures. In so far as this will produce the required emission-reductions profile, it can be argued that the short-term benefits from additional measures to reduce emissions are small or zero.

Much the same argument can be applied to the benefits from increasing carbon sinks. Existing forestry is estimated to fix carbon at a rate of 2.5 Mt carbon year^{-1}, a level that will be sustained at least until the year 2000 (UK Department of Environment, 1994). If the commitment to enhance carbon sinks can be attained without requiring increased carbon retention from new planting, the shadow price of carbon fixing from new forestry is again zero in the short term. Thus, in a context for evaluation where the objective is to increase carbon sinks, a reluctance to value marginal gains in carbon retention from new planting can easily be explained. In the longer term, decisions about how to value carbon retention in forestry policy will depend on the extent to which planting rates determined by other policy objectives, such as recreation and timber supply, satisfy national commitments for carbon-sink provision. It is likely that carbon retention from forestry will assume increasing importance over time, as damage costs rise and the marginal benefits from other forestry outputs (such as recreation) decline with an increase in the planted area.

POLICY-INSTRUMENT ANALYSIS

Despite the limited extent to which carbon retention currently impinges on public investment decisions in the UK forestry sector, it is worth considering how incentives to farmers and landowners might be modified to enhance the role of woodlands as a carbon sink. Most obviously, the rate of planting could be increased through higher incentive payments. Under 3% of farmers have participated in woodland planting, so the scope for additional planting is substantial. A 'willingness to accept' analysis of non-entrant farmers to increased levels of incentive payment suggested that planting rates would respond markedly to increased rates of payment, particularly on poor-quality land (Crabtree, 1996). Within the arrangements for farm woodlands, the least-cost carbon-fixing option (in both financial and economic terms) is to concentrate planting on low-quality land. At the margin, both the financial payments required to encourage participation and the resource costs would be lower on land with limited potential for agriculture. Exchequer costs could also be reduced by encouraging coniferous rather than broad-leaved planting, and this would maximize carbon retention over the medium term. Broad-leaved planting is a higher-cost option, because of higher planting costs and slower rates of growth. Planting on arable land also requires a high level of incentive and is an extremely costly route for increasing the rate of carbon retention.

Nevertheless, a strategy that concentrated planting on poor-quality land and minimized planting on arable land would require a trade-off of biodiversity and landscape benefits in favour of carbon fixing. This is so because potential gains from enhanced biodiversity are greatest when arable land is planted. The loss of existing flora and fauna is small and planting is taking place in areas with relatively little existing afforestation. A shift to planting on unimproved land would therefore tend to reduce net biodiversity benefits at both farm and regional scales, and such benefits are a prime objective of current woodland planting. Landscape benefits from new planting may also be lower on unimproved land. Much of this land in Scotland is remote from centres of population and already has some forest cover, so public use benefits would be lower. A shift to conifer planting would itself involve a biodiversity loss, since long-term ecological benefits are greater from mixed and broad-leaved woodlands.

The overall conclusion is that current farm woodland planting is making a sizeable contribution to enhancing the carbon sink in the UK. The incentive structure could be changed both to increase the annual rate of retention and reduce the associated exchequer cost but not without trading off other environmental benefits. However, afforestation on farms is a high-cost route for increased carbon retention, compared with commercial afforestation on less productive land. Special high-incentive schemes to encourage planting can only be justified in terms of a wider mix of environmental benefits. If the sole policy objective were to be carbon retention, non-farm commercial plantations on mineral soils are the least-cost approach.

NOTE

1 The author thanks Neil Chalmers for assistance in data analysis.

The Role of the Common Agricultural Policy in Inhibiting Afforestation: the Example of Saxony

15

Christian Lippert and Michael Rittershofer[1]

INTRODUCTION

One of the major possibilities of reducing the atmospheric concentration of carbon dioxide (CO_2), as discussed in this book, is the sequestration of carbon to the biosphere by afforestation (Winjum *et al.*, 1992). Although the possible contribution of countries in the temperate zones to afforestation may not be as great as that of tropical countries, afforestation measures should be promoted, at least to improve the credibility of the developed countries (Brabänder *et al.*, 1992), who ask for forest programmes in tropical countries. If further afforestation in the European Union (EU) is considered desirable for many reasons, certain conditions must exist. As well as the obvious constraint of the availability of suitable land, there is a need for an appropriate social, economic and administrative framework. In the absence of such a framework, national and regional afforestation programmes (transforming the Directive (EU) No. 2080/92 into national law, for example) will have no real effect on the extension of forest.

This chapter analyses problems of the political implementation of afforestation measures in two different areas of Saxony in eastern Germany. After a discussion on the property rights associated with the policy context and a description of the relevant geographical and political conditions, the theory is applied to the specific situation in Saxony. From a political and economic point of view, the issues of the lack of afforestation in the past 5 years and the mechanisms for the promotion of afforestation in this region are discussed, given an apparent social demand for more forest. There are two dilemmas inhibiting afforestation in the areas under study: a profitability-related constraint, resulting from the EU-caused distortions; and an institutional constraint, manifested by the property conditions, which are somewhat unique in a German and Western European context.

PROBLEMS OF POLITICAL IMPLEMENTATION OF AFFORESTATION: THEORETICAL REFLECTIONS

Marginal Benefit and Cost Theory to Explain Profitability-Related Obstacles

From an economic point of view and assuming that all benefits and costs are known and can be attributed to the user of the land, afforestation will be economically desirable and will take place if the annual benefits of future forest exceed the annual costs of afforestation. The major constraint to such afforestation is that some social benefits, such as the CO_2 sequestration function of forests, are not directly remunerated to the afforesting land user, who has to bear the entire costs, including the opportunity costs, of land use, which are, moreover, raised artificially by agricultural policy measures. The result is a distortion in the allocation of land: too much agricultural land at the expense of forest land. In the following paragraphs, a simple theoretical model is used to describe the factors promoting or inhibiting afforestation. These factors, associated with property rights and policy failures, take place in the context of the economic costs and benefits now briefly outlined.

The benefits of afforestation can be divided into two elements: direct economic benefit obtained by the land user (revenue from production of wood); and externalities or benefits that have the character of public goods, such as recreation, mesoclimate, soil protection, water protection and regulation and sequestration of carbon. The external benefits are not part of the prices obtained for wood products (there is no market for these externalities). Their value depends on a number of factors: location of the forest, proximity to settlements and markets, type of forest function considered and the preferences and tastes (and monetary proxies) for those to whom externality accrues. For example, the CO_2 sequestration externality accrues globally, whereas the local inhabitants are the relevant population for the mesoclimate imports of afforestation.

The costs of afforestation include the costs of forest management: afforestation measures, maintenance and cultivation; and its opportunity costs: profit foregone of alternative land uses such as agriculture, industrial use, or for housing.

The main aspects of the land-allocation problem (forest use versus agricultural use) are illustrated by a marginal benefit and cost analysis in Fig. 15.1. The distance between E and A represents the potential farm land and forest land – declining soil quality along the abscissa, in a given region, with

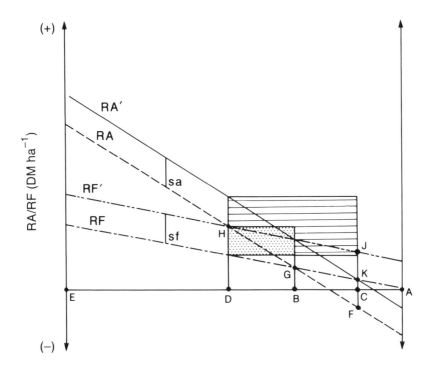

Fig. 15.1. Soil-dependent rents from agriculture and forest. RA, soil-dependent agricultural rent, excluding subsidies; RA′, soil-dependent agricultural rent, including subsidies; RF, soil-dependent rent of forest use; RF′, soil-dependent rent of forest use, including forest subsidies.

agriculture increasing from the left-hand side and forestry from the right. In this formulation, all land must be allocated to either agriculture or forestry.

RF is the soil-dependent (private) rent of forest use, which is the highest above E and the lowest above A. It amounts to the difference between direct economic benefits and the costs of forest management, including planting costs. This approach corresponds to the land-rent theory of Ricardo. Another approach would be to consider soil of unchanging quality according to the location-rent theory of von Thünen (von Alvensleben, 1995). In this case, the abscissa would represent the distance to the next market or urban settlement (increasing from E to A).

RA represents the soil-dependent agricultural rent (excluding subsidies). In this model, RA represents the opportunity costs of forest use, neglecting the fact that there may be other possibilities of land use. In this hypothetical case of free-market conditions, without government taxes or subsidies, there will be an equilibrium, with EB hectares of agricultural land and BA hectares of forest.

In reality, agricultural production has always been subsidized, through price support or by direct land-related payments (after the 1992 reform of the Common Agricultural Policy (CAP)). In Fig. 15.1, the impact of agricultural support can be accounted for by a payment per hectare (sa Deutschmark (DM) ha^{-1}), which shifts the curve from RA to RA', so that only CA ha will be allocated to forest use in the considered region. This loss of forest is an indirect result of agricultural protection (higher opportunity costs of forest use). The area FGK represents the corresponding welfare loss. Notice that the true rent (CF) at point C is negative.

What impact does government policy, responding to a recognized social demand for afforestation, have on forest area? A subsidy of sf DM ha^{-1} paid to land users who want to change land use is considered to remunerate the positive externalities of 1 ha of forest, so that, in Fig. 15.1, RF shifts to RF', leading to the additional BC ha of forest. Note that the amount of sf also depends on the number of externalities in the considered region. The externalities might be more significant in close proximity to large cities (additional mesoclimate and recreation function of forest) than in an area with a low population density. The intersection of RF' and RA represents the social optimum (including positive forest externalities, such as the reduction of CO_2 and recreation functions): ED hectares should be used for agriculture whereas DA ha should be reserved for forest. Previously, agricultural policy in the German Democratic Republic (GDR), which subsidized agriculture heavily for self-sufficiency purposes, probably led to an equilibrium point close to C. This situation has not changed since reunification and the reform of the CAP: in our model, the state, through taxes and subsidies, would still have to pay an amount corresponding to the stippled and hatched surface of (sf + sa) × (DB + BC) in Fig. 15.1 if the optimum is to be attained. Without the EU direct payments for agricultural land use, only sf × DB DM (stippled surface in Fig. 15.1) would have to be spent for afforestation measures. The triangle FHJ stands for the entire welfare loss which would occur if agriculture is subsidized (FGK) and positive externalities of forest are not internalized (KGHJ).

For methodological and practical reasons, the exact quantification of the amount and of the value of the externalities mentioned is rather difficult. In the case of CO_2, for example, there are serious difficulties in explaining the carbon cycle, hence leading to uncertainty in the cost and impact of CO_2 mitigation policies (Edmonds, 1992). Moreover, there is the problem of valuation of the benefits of carbon sequestration, as dealt with in the earlier chapters. What damage (e.g. future loss of agricultural yield) will be avoided if 1 t of carbon can be sequestered? In our simple model, we assume that government knows the value of positive forest externalities, so that the correct amount of the necessary payment per hectare (sf) can be determined. For most of the externalities, this value also depends on the share of forest in the region (declining marginal benefit). In areas with a high share of forests (such as north-eastern Saxony), it may become even negative in some places, because of the aims of nature protection. Conservationists claim that, in these areas, afforestation has negative effects on the variety of species and on the landscape. In contrast to the other externalities, the benefit of integration of carbon in the biosphere has a constant marginal profit in Saxony.

Apart from the uncertainty surrounding the costs and benefits of CO_2 sequestration, there is a perceptible social demand for additional forest areas in the forest-scarce regions of Saxony. Therefore, the political aim of the Saxon government is to raise the average forest share by 3% (Sächsisches Staatsministerium fuer Umwelt und Landesentwicklung, 1994). It seems quite clear that the attainment of this fixed objective would lead nearer to the social optimum (from C towards D in Fig. 15.1). Nevertheless, there have been only low rates of afforestation in the last few years under the new economic conditions. This fact indicates that payments for afforestation have been too low or, put in other words, that rents for agricultural land use, including set-aside payments (RA') have been too high. In the following section, we shall apply this marginal benefit and cost analysis to specific situations in Saxony.

Property-rights and Transaction-costs Theory to Explain Institutional Obstacles to Afforestation

The theoretical optimum of land allocation (e.g. point B in Fig. 15.1) is not stable over time. It changes according to shifts of the depicted rent functions, which are caused by changing price–cost ratios, as a result of technological progress and changing preferences (Henrichsmeyer, 1988, p. 179). For example, it is obvious that declining agricultural-product prices (shifting RA downwards) and the new social preference for multiple-use amenity forest, which is ultimately reflected in afforestation subsidies, have the effect of shifting the optimum to the left. It should be noted that, in the future, such a shift may also be caused by climate change (see van Kooten and Folmer, Chapter 9, this volume). In any case, at any point in time, the actual spatial distribution of forest and utilized agricultural area (UAA) will differ more or less from the theoretical optimum, since the land allocation observed today is the result of yesterday's prices and costs. Here the question arises on which conditions, apart from the price- and cost-related profitability analysed above, land users will reallocate their land. In trying to answer this question, we shall use a property-rights and transaction-costs approach.

Property rights
According to property-rights theory, it is not goods or resources themselves which are possessed and/or traded but the property rights related to these goods or resources (Coase, 1960). In common parlance, to own a resource, such as a plot of land, means to enjoy of a bundle of certain well-specified

property rights, which define: (i) the manner in which this resource can be used (i.e. *usus*); (ii) the possibilities of a change of use (i.e. *abusus*); (iii) the enjoyment of yields and profits resulting from the use (i.e. *usus fructus*); and (iv) the possibilities to sell the resource (Richter, 1990, p. 575). In this sense, ownership – 'the right to possess, to use, to manage, to benefit, to be secure and to alienate' – does not mean complete control of a resource, disregarding the interests of others (Bromley, 1991, p. 159). A further major aspect of the property-rights theory is that the way in which a resource will finally be used – the allocation of resources – depends not only on the profitability of different ways to use the resource but also on the specification of the property rights related to it (Furubotn and Pejovich, 1972, p. 1,139).

Property rights and duties related to land use have become rather complicated within the EU: the owner (or tenant) of a plot of agricultural land has to observe a great number of environmental regulations; on the other hand, he/she is entitled to benefit not only from crop yields but also from the agricultural-subsidy schemes, including the rights to take set-aside subsidies or certain fees for extensive farming. Bromley (1991, p. 160) suggests two types of rights transfers. If a farmer sells a plot of land to another farmer, with all rights and duties related to it, an explicit rights transfer takes place, without affecting the rest of the population. Implicit rights transfers consist in a modification of the legal entitlements attached to a plot of land. New environmental regulations, such as the German Fertilizer Act or the 1992 set-aside obligation of a certain part of the agricultural area, associated with the right of taking certain subsidies, are examples of implicit rights transfers. It should be noted that such a rights transfer affects directly the distribution of costs, for example, of certain environmental measures, between landowner and other parts of society.

Given high opportunity costs of afforestation resulting from subsidies, there is the possibility of changing land use by an implicit rights transfer, as well as the more usual route of subsidizing afforestation. For example, similar to the so-called compliance programmes in the USA, where certain farm-programme benefits are coupled to the farmers' observation of soil-conservation measures (Bromley and Hodge, 1990, p. 210; Petersen, 1993, p. 259), the granting of certain CAP subsidies could be linked to afforestation. In the same manner as farmers are obliged to set aside part of their UAA in order to be entitled to the EU subsidies, they could be enjoined to afforest part of their land in the case of a recognized social demand. Such an implicit rights transfer and redistribution of property rights would entail a change of the property-rights status quo and thus of the status quo of income distribution. By such a modification of the existing property-rights structure, land users would have to bear higher charges, to the advantage of the rest of society. In other words, part of the rent RA' (Fig. 15.1) caused by the EU subsidy would be redistributed. In so far as this agricultural-rent reduction amounts to part of the welfare loss FHJ in Fig. 15.1, such a 'property-rights redistribution policy' would increase social welfare, unless the redistributed rents were not seen as socially or environmentally motivated transfers. Although Bromley and Hodge (1990, p. 211) point out, giving examples, that redefinitions of rights in land are easier to carry out in situations of strong structural change ('upheaval'), it should be noticed that such a policy would certainly entail strong opposition from the farmers concerned. In any case, it would cause transaction costs in the form of costs of institutional change at the state level.

Transaction costs
The exchange and modification of implicit and explicit property rights between individuals and institutions cause transaction costs, which, besides the production-costs aspects analysed above, under the assumption of a 'zero-transaction-cost world', directly affect the individuals' choices concerning resource allocation (Richter, 1990, p. 579). Thus, it seems to be

> a logical solution to extend the utility maximization hypothesis to all choices and to consider explicitly the constraints imposed by the system of property rights and by transaction costs, broadly defined as the costs of acquiring information about alternatives and of negotiating, monitoring, and enforcing contracts.
>
> (De Alessi, 1990, p. 7)

Relevant transaction costs of land-use change in the case of afforestation that have to be considered in a land-users decision-making process include the following.

1. Costs of acquiring knowledge concerning the afforestation programmes and relevant legal constraints (i.e. acquiring the information necessary to exercise the right of land-use change and taking the corresponding subsidy).
2. Costs of dealing with administration.
3. Costs of acquiring knowledge concerning the planting and forest cultivation, etc. (i.e. acquiring specific human capital, e.g. by hiring new workers or making actual workers learn forestry matters).
4. Costs of purchasing the property rights needed for a land-use change, especially the right of *abusus*. These negotiation costs depend on the existing property-rights structure, which determines, for example, how many contracting parties are involved.
5. Risk-dependent opportunity costs caused by the irreversibility of an afforestation and hence loss of the opportunity for future agricultural use.

A proportion of these transaction costs may be borne by the state, such as costs of information and legal advice, and may be lowered because of degression effects. It is important to note that, if society wants to change land use in a given area, institutional arrangements should be modified in order to lower at least part of the corresponding transaction costs. For example, changing forest laws in a way that future clearance will not be strictly forbidden will lower the costs mentioned under item 5 above. Some of the transaction costs at least partly have the character of fixed charges, such as knowledge of forestry practices, while others, such as costs associated with changes in actual forest area, vary more or less with the area finally afforested.

Figure 15.2 shows the influence of different types of transaction costs on the allocation of land. First, let us suppose that the considered agricultural land is used by one single farm. Here the rent of land-use change RLUC represents the difference between the theoretical rent of forest use after the introduction of sf (RF′) and the agricultural rent (RA′), including subsidies. The consideration of transaction costs per hectare shifts RLUC downwards (RLUC-TCvar). The curve ATCfix shows the average fixed transaction costs (TCfix ha^{-1} afforestation). As in the case of allocating pollution entitlements, the optimum (B) is not affected by the fixed transaction costs unless they exceed a certain amount, but it moves to the right (B′) due to the variable transaction costs (Henrichsmeyer and Witzke, 1994, p. 279). If the fixed charges (area B′LNC) are higher than the corresponding profits, including the variable transaction costs (B′MC), they will have a prohibitive effect and there will be no change at all.

If the new optimum (B) were not so far away from the present equilibrium (C) and if the charges TCvar and TCfix were relatively high, afforestation

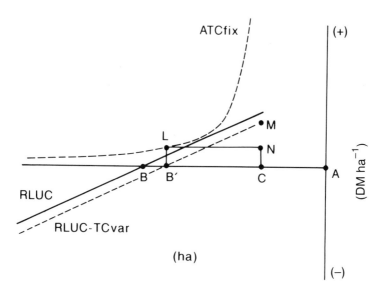

Fig. 15.2. Transaction costs and rents associated with land-use change. RLUC, rent of land-use change (RF′ – RA′; see Fig. 15.1); RLUC-TCvar, rent of land-use change, including variable transaction costs; ATCfix, average fixed transaction costs.

would not be reasonable, even from a social-welfare point of view, because transaction costs would exceed the possible welfare gains. But, as the institutional framework influences these costs, efforts should be made by the administration to lower them and to make the land user benefit from possible degressions. Land users willing to afforest could be assisted in setting up enterprises – for example, through forest associations (*Forstbetriebsgemeinschaften*), which encourage appropriate forest management and cultivation (Niesslein, 1992, p. 82). Through such mechanisms, specific knowledge and human capital would be integrated and centralized, thus especially lowering the costs mentioned above under items **1** and **2**. In the following discussion, we shall apply property-rights and transaction-costs theory to the specific situation in Saxony.

GEOGRAPHICAL CONDITIONS AND AFFORESTATION POLICY IN SAXONY

Geographical Characterization

Saxony is a 'Land' in the Federal Republic of Germany and borders on Poland in the east and on the Czech Republic in the south. Compared with the German average (30%), Saxony has a low share of forest land (27%). Moreover, the distribution of forest is rather heterogeneous. Its territory is roughly characterized as three distinct areas.

- Region I Low mountain range (Ore mountains) and their foothills (along the Czech border), characterized by a high share of grassland and marginal-site areas.
- Region II Neighbouring in the north a large area of very productive loess soil between the Polish border in the east and the town of Leipzig in the west. In this region, agriculture has been predominant for hundreds of years. Because of the high soil quality, there has been a low share of forest since the Middle Ages. Today, this region is also characterized by a high population density in the three congested areas of Leipzig, Chemnitz/Zwickau and Dresden.
- Region III The northern part of Saxony is a region with poor sandy soil, with many marginal areas and a lower percentage of utilizable agricultural area. The share of forest land is high compared with the German and Saxon average. Population density in this region is low.

The problems of political implementation of afforestation measures in Saxony will be illustrated by two examples of different Saxon regions: the agglomeration area of Leipzig (within region II) and north-eastern Saxony (within region III).

The area under study around Leipzig (1,147 km^2) shows a population density of 229 inhabitants km^{-2} and a share of forest of 7%, whereas the population density in north-eastern Saxony (1,714 km^2) amounts to 113 inhabitants km^{-2}. Forest covers 44% of its surface. In both areas, lignite mining is a significant land use.

Present Afforestation Instruments and Their Acceptance

Within the scope of Directive (EU) No. 2080/92 common payments are granted for forestry measures by the German government and the governments of the German Länder in order to support the forest economy and to promote a greater share of forest in Germany. Afforestation is supported by subsidies for capital expenditures, such as planting, depending on the species of tree, and annual payments for afforestation, to compensate farmers' losses in income from former agricultural production, dependent on soil quality.

Previously, afforestation rates were very low, leading to increases in subsidy after the reform of the CAP in 1992 (Brabänder et al., 1992, pp. 6–7). The major increases included, the payment for afforestation was raised from a maximum of 500 DM ha^{-1} to 1,400 DM ha^{-1} for farmers; non-agricultural landowners are entitled to a lower payment of 350 DM ha^{-1}. The share contributed by the EU amounts to 600 European currency units (ECU) (1,230 DM) ha^{-1} year^{-1} for farmers and 150 ECU (307 DM) for non-agricultural landowners. The subsidy rates for capital expenditures were raised. The share contributed by the EU for capital expenditures was raised from 25% to 50% in order to encourage the member countries of the EU to implement profitable measures of afforestation.

As already mentioned, the Saxon government intends to increase the share of forest land by 3%, or 55,000 ha, during the next 15–20 years. Although the subsidies have been raised, the rate of afforestation has not come up to the expectations in the years following the increases. According to data from the Saxon Ministry for Agriculture, Nutrition and Forestry between 1991 and 1994, only 2,437 ha were afforested in total, made up of 1,426 ha recultivation and 1,011 ha afforestation of agricultural land. Hence, the average afforestation per year amounts to 609 ha, which is less than 25% of the minimum target of the Saxon government.

Between 1992 and 1994, only 38 ha were afforested in the region of Leipzig, excluding the afforestation for the purpose of reclamation of former coalmines (Marsch, 1995a). This area is very small, compared with the intended increase of forest land at the expense of agricultural area in this region (about 3,400 ha) in the next 15–20 years (Marsch, 1995b). Since 1996, afforestation is accepted under the land set-aside commitment of the CAP reform. This new instrument increases the likelihood that farmers will be willing to afforest agricultural land with poor soil quality. However, such an increase depends on several other factors, which will be discussed in the following section.

OBSTACLES TO AFFORESTATION IN TWO DIFFERENT SAXON AREAS

Profitability Obstacles

Our explanation of obstacles to afforestation is based upon the model developed earlier in the chapter. As the area surrounding Leipzig is characterized by a high soil quality, forests were converted to agriculture many hundred years ago. As a result of highly productive agricultural structures and increasing suburbanization, the opportunity costs of an implementation of forest (RA) in this area are high.

In comparison, the opportunity costs for afforestation measures in the north-east of Saxony are low, because of a low agricultural productivity and a lack of alternative land uses (RA in Fig. 15.1 is shifted downward for northeast Saxony). Yet the increase of the forest share in this region is not much higher than in the surroundings of Leipzig.

A major reason for the low afforestation rates in both areas are the economic distortions resulting from agricultural policy. The high subsidies improve the income from agriculture: because of these subsidies, agricultural production is profitable, even on marginal sites in the north-east of Saxony or on former mining land in both areas. This is the primary explanation of why farmers are not interested in afforestation.

To reduce these distortions caused by the CAP, afforestation payments would have to be increased. However, given a limited budget, there is the danger that the declared objective of the Saxon government cannot be fulfilled (the amount of (sf + sa) × (DB + BC) not being available). The observed demand for agricultural use on restored coalmining land is understandable in view of the high agricultural subsidies, ranging between 660 and more than 1,200 DM ha^{-1}. These payments (corresponding to sa in Fig. 15.1) maintain agriculture in the sense of turning even a negative agricultural rent (RA) into a positive one, as in the case of rape and sunflower on sandy soils in north-eastern Saxony.

Compared with these agricultural subsidies, remuneration of positive externalities of forest land use is not sufficient, especially in the congested area of Leipzig with its high population density, high pollution and lack of recreation areas. By way of explanation, it is possible that the payments granted for afforestation in the area of Leipzig have been insufficient to internalize all the listed externalities. On the other hand, even if subsidies were set at the level of estimated externalities, they were partly offset by high opportunity costs, taking into account that, for example, 836 DM ha^{-1} were paid for set-aside of agricultural land. A higher amount of the afforestation payment (sf + sa) seems to be necessary.

Seen from the angle of policy analysis, the main problem is that politicians try to attain several social aims, such as economic stabilization of farms, and remuneration of positive externalities through inappropriate instruments, such as land subsidies, with the result of growing economic losses (Ahrens and Lippert, 1994). Despite surpluses of agricultural products and a need for other land uses, agriculture is maintained even on the poorest sites, often in the form of set-aside in the case of the area of Leipzig (Labudda, 1995, p. 26). Moreover, the increased rent of marginal crop land will entail higher tenancy payments and market prices for land, so that, as we discussed above, alternative land use, such as forestry, cannot be realized to the extent desired by society.

Institutional Obstacles

There are several other reasons for agricultural farms to refuse afforestation, which are explained here through a property-rights and transaction-costs approach.

Lack of information
Farmers may not be well informed by the administration concerning the possible promotion measures (Plochmann and Thoroe, 1991, p. 32). Although no study details this issue for Saxony, there are signs indicating such a lack of information. Most of the executive land managers interviewed about afforestation described below were not well informed. In addition, Labudda (1995, p. 5) describes the lack of information in the Saxon forest administration, as a result of the quickly changing conditions after German reunification. These include problems of adapting the different administrations to the new situation and of the still ongoing restitution of property, so that the forest administration cannot establish contact with the landowners. These factors and that of the hierarchical organization of the forest administration limit the information flows and act as an obstacle to afforestation.

Specific human capital
Agricultural enterprises in the region generally have no or little experience in forestry. If they decide to afforest, they would have to hire, or at least to instruct, labour in these enterprises. Such an acquisition or reallocation of labour to this specific task incurs transaction costs. Here the establishment of forest associations could be helpful in lowering these charges.

Every change from agricultural land into forest land, even if the land user can earn more money than before, bears the danger of a loss of employment of the labour still employed on the farms. In Saxony, where the agricultural labour force diminished by 40% between 1991 and 1993 (Statistisches Landesamt des Freistaates Sachsen, 1994, p. 184), this is a significant factor. It explains, at least partly, the low acceptance of afforestation measures and some of the pressure against the delimitation of local afforestation areas which seems to be exerted by interest groups, as well as by officials dealing with agricultural affairs.

Property conditions and tenancy contracts
Less than half of the agricultural land in western Germany is leased land. The average leased-land share of 6,292 selected full-time farms in the former federal states comes to 47.0%, whereas the corresponding share of 680 selected full-time farms (284 selected legal persons) in the new federal states amounts to 88.7% (99.6%) (Bundesministerium fuer Ernaehrung, Landwirtschaft und Forsten, 1995, p. 45, p. 48). So, in former GDR, usually the bundle of property rights related to a plot of land is not in the hands of one person: *usus* and *usus fructus* (including the right to take afforestation subsidies) are given to the tenants (who are often legal persons, such as cooperatives, in our case), whereas the possibility of *abusus* (i.e. change of use) and the right to sell the plot remain in the hands of the lessors. In an agriculture mainly based upon family holdings, as in western Germany, these four property rights are held by the farmer, who is to a great extent the owner of the land he/she is cultivating. Such a farmer can take independently all decisions concerning the future land use on his/her own, which means that the transaction costs of land-use change are much lower than in Saxony.

The existing afforestation programme is based upon the agricultural ownership structure in western Germany. It does not take into account the completely different institutional situation in eastern Germany. For example, even long-term tenancy contracts do not cover the possibility of a change from agricultural land to forest land. On the other hand, the afforestation subsidy is given to the agricultural cooperative or closed corporation that plants the trees, not to the owners. Hence a multitude of owners and tenants will have to negotiate in order to share the subsidies, negotiations which inevitably mean higher transaction costs. This is an example of an already existing political instrument introduced in eastern Germany after reunification without sufficient adaptation to the specific circumstances. As in other cases, a loss of efficiency is unavoidable (Ahrens and Lippert, 1994).

Loss of flexibility
First of all, forest clearance is generally not allowed (Sächsisches Staatsministerium fuer Landwirtschaft, Ernährung und Forsten, 1992), which means that afforestation on agricultural land is often irreversible. This loss of flexibility (i.e. loss of the right of *abusus*) seems to be unacceptable for many farms, especially with regard to an uncertain development of the CAP.

Empirical Evidence

In order to better understand the low afforestation rate observed, the executive managers of 20 agricultural legal persons (cooperatives and closed corporations, with an average size of 1,202.8 ha of UAA) in north-eastern Saxony were interviewed. These farms represent together 62.9% of the farmland in the area under study. Notice that there are very few family holdings in this less favoured area. A list of some possible reasons for the rejection of afforestation were given, such as property conditions, decreasing soil value, lack of profitability, loss of flexibility, liquidity difficulties, lack of experience concerning forest cultivation, and loss of employment possibilities. The managers were asked to evaluate the weight of these reasons by points (from 1 = very important to 5 = irrelevant). Furthermore, they could express their objections against afforestation in their own words. First of all, they were asked if afforestation would be a viable activity for their enterprise. Many of them (12) said it would not, three managers had no opinion and only five managers thought of afforestation as a viable activity. Until 1995, none of them had taken subsidies for afforestation measures. Only three farmers thought they were able, within their present capacities, to carry out afforestation and cultivation measures. Most of the farms did not have enough (10) and/or suitable (five) workers (no opinion four, other obstacles two).

Some of the managers stated that, even if profitability of forest use was given, they would not be able to change the land use, because they would have to deal with a multitude of owners (lessors). It became quite clear that the most important reasons for their lack of interest – besides or even before the low profitability – were institutional ones. Property conditions turned out to be the most important obstacle to afforestation in the managers' opinion (very important 17, important one, no opinion two), even before the lack of profitability (very important seven, important three, relevant two, no opinion eight). The other reasons (e.g. liquidity) mentioned in the questionnaire seem to be of less importance.

CONCLUSIONS

In summary, there are two constraints inhibiting afforestation in the areas under study: a profitability-related constraint, resulting from the distortions associated with CAP, and an institutional constraint, mainly caused by the property conditions. As the first-best solution, the reduction of the agricultural subsidies, is not possible under the present political circumstances and as the second-best solution, the general economic enhancement of afforestation in order to counterbalance the elevated opportunity costs and transaction costs, will remain limited, because of the given budget and the legal constraints, other solutions have to be sought.

The acceptance of the classification of afforestation as agricultural land under set-aside is a step in the right direction, in order to reduce the negative economic effects for the farms. But, as long as the annual payment for afforestation is lower than the payment for fallow set-aside land, there will be little economic interest in afforesting set-aside land. Such interest will continue to be constrained by the institutional obstacles described here.

A new instrument is being promoted by the Saxon government to increase afforestation, as alternatives to the integration of forestry matters in regional planning and supplementary subsidies in areas with a high demand for afforestation. A private foundation has been established to buy up land for afforestation purposes. So the transaction costs of land-use change will no longer have to be borne by the agricultural enterprises. Moreover, these costs will be lower, because there will be one specialized negotiating partner for land afforestation. Agricultural enterprises have the possibility to rent the land of the foundation at a low fee, provided that it will be afforested. In comparison to the afforestation of the farms' present set-aside land, there will be no competition between agricultural and forest land use. Furthermore, by this means, the farms will be capable of fulfilling part of their set-aside obligation.

NOTE

1   The authors are indebted to H. Ahrens and M. Whitby for critical comments.

Forest Management and Policy Options for Emission Mitigation in Finland

16

Heikki Seppälä and Kim Pingoud

INTRODUCTION

Despite extensive industrial use of wood, forests in Finland, when managed sustainably, provide great opportunities for slowing down the increase in atmospheric carbon. The aim of this chapter is to assess the effects of potential mitigation policies on the national carbon budget. In addition, the economic impact of the present carbon tax, as it relates to wood as fuel, is considered. Alternative wood-production scenarios are analysed in respect to their effect on carbon balance for the years 1990–2040.

The overall infrastructure for using the forest sector as a management tool in mitigating greenhouse-gas (GHG) emissions in Finland is favourable. Carbon could be sequestered most effectively by increasing the growing stock as much as possible. This might, however, contradict other policy objectives, such as those to diminish foreign debt and unemployment via markedly increasing production of forest-industry products. Increased use of wood, based on sustainable forestry, to substitute for fossil fuels and materials, could in many cases decrease the overall carbon emissions. However, when considering the introduction of carbon taxes, society should concurrently account for the risk to forest owners' investments in forestry.

Forest-industry products account for nearly 40% of Finland's current export income. In this context, carbon stored in wood-based products with a long life span could provide additional storage in mitigating the increase in the concentration of atmospheric carbon dioxide (CO_2). In carbon-balance assessments, the sequestration potentials of stemwood, ground vegetation, litter and soil organic matter should all be accounted for in a comprehensive manner. As forests are extensively utilized to produce fuel wood or raw material for forest industry, the utilization of forests and the retention of carbon in wood-based products should be included in the assessment.

Large uncertainty in climate warming projections implies that we should not only concentrate on long-term (100–200 years) analysis. A relevant complementary approach is to study what is feasible, for example, within 10–40 years, in order to slow down the build-up of GHGs in the atmosphere, until we have more precise information and knowledge about the possibilities of global climate changes. As time passes, we shall have better information about the probability and efficient actions, based on new or improved technology developed during such a time frame. Given this need for consideration of the short and medium term, the analysis here focuses on the period 1990–2040.

In this chapter, we first quantify stocks and fluxes of carbon in the Finnish forest sector, including also forest ground vegetation, litter and soil organic matter and forest-based products in Finland in 1990. Based on these quantity estimates, our aim is to assess the effects of potential mitigation policies on carbon balance. Two alternative wood-production scenarios are analysed in respect to their effect on carbon balance for the years 1990–2040. The budget of carbon in wood products is assessed with regard to its sensitivity to changes in recycling, in the life span of products and in the rate of landfill decay. The impact of carbon tax related to wood as fuel is considered in the concluding section.

THE FORESTRY SECTOR IN FINLAND

Individual countries may choose among a diverse range of forest practices to implement plans for reducing and limiting GHG emissions. This range is based upon types of policy instruments, characteristics of land-use and forestry practices, and other characteristics of countries (see, for example, Richards *et al.*, 1995). Each country customizes its own policies, based upon economics, political and cultural conditions, and its forestry practices and resources. A policy that will be effective in one country may not be possible or workable in another.

Finland, located in the boreal coniferous zone, has 66% forest cover. One-third of the land is classified as peatland or bogs, of which a sizeable proportion has been naturally forested and about half has been converted into fields or productive forests by large-scale drainage. Silviculture is simplified by the limited number of commercially important tree species; logging conditions are also comparatively easy. Timber markets, however, are complicated by the pattern of forest ownership. About 80% of timber is supplied by non-industrial private owners, mostly families, with an average holding of less than 35 ha.

Finland was the first country in the world to complete a systematic field-sample-based national forest inventory, with pilot results published in 1923. Since then, seven more inventories have helped to develop a timely, valid and reliable monitoring system, which also takes into consideration biodiversity and carbon accounting. In the 1960s, the Finnish Forest Research Institute began to develop a comprehensive system of forest statistics. The comprehensive and reliable forest-sector statistic, along with high-level forest research and education, enables intensive national level forestry planning and policy-making. It also allows consideration of forest conservation, biodiversity, endangered species and other forest-related services, including also global-warming concerns.

In the economy, Finland has recently been fighting the most serious recession in her history. Foreign debt has grown and unemployment is the most severe problem. In agriculture, Finland has been decreasing the level of regulation that enabled it to be self-sufficient in food production. Energy policy, too, is in transition.

These salient characteristics of Finland enable us to understand some of the infrastructure and institutional facts that will affect the choice of forestry practices to manage carbon. Activities such as rehabilitation of degraded forest lands, agroforestry, anti-erosion plantations and fire-control intensification are not relevant to Finland, because of socio-economic and natural circumstances. However, other practices, such as preservation and conservation, biomass for energy, increased solid-wood forest products, changes in rotation and field afforestation, may be more relevant.

FINNISH FOREST-SECTOR CARBON BUDGET IN 1990

Forest Ecosystems

The total area of forest and scrubland in Finland is 23.2 million hectares, of which 16.3 million hectares are mineral soil lands and 6.9 million hectares are peatlands. At the beginning of 1990, the total volume of the growing stock (stemwood) was 1,880 million m^3, of which 1,363 million m^3 was on mineral soil lands and the rest, 517 million m^3, on peatlands. The annual volume increment was about 80 million m^3 and the drain over 55 million m^3 (Yearbook of Forest Statistics, 1995).

Since the 1950s, the growing stock has increased by 22% and its annual increment by 44% (Kauppi *et al.*, 1995). To obtain the total carbon amount in tree biomass, the carbon storage of other parts than the trunk (branches, foliage, needles and roots) was estimated to be 75% of stemwood storage. The carbon storage in trees in 1990 was about 710 Mt carbon (Fig. 16.1). Based on an average estimate (1.2 t carbon ha^{-1}) and total forest area (23.2 million hectares), the carbon storage in ground vegetation was 28 Mt carbon. Thus

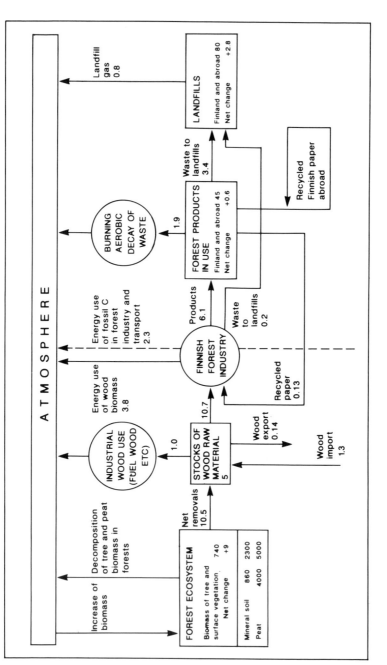

Fig. 16.1. Carbon reservoirs (Mt C) and fluxes (Mt C year^{-1}) from the Finnish forest sector in 1990 (from Pingoud et al., 1995).

the total carbon reservoir of tree and surface vegetation in 1990 was about 740 Mt carbon. The annual gross increase was 29 Mt carbon year^{-1} and the net increment (sink) 9.0 Mt carbon year^{-1}. The carbon storage of living trees in Finland has increased at an accelerating rate over the period 1960–1995.

Using the results measured by Westman *et al.* (1994) of carbon storage on mineral soils in 30 stands in southern central Finland (uppermost 1 m layer 4–12 kg m^{-2} carbon and between the depth of 1 m and the groundwater 1.3–2.4 kg m^{-2}), one can arrive at an estimate of 860–2,300 Mt for the carbon reservoir in mineral soils in Finland. Estimates of carbon accumulation on soil over the whole country are still lacking. Finnish forest soils are young on the geological time-scale. Only 10,000 years have elapsed since the last glaciation and mineral forest soils may still be actively accumulating carbon (Westman *et al.*, 1994).

Natural peatlands bind CO_2 from the atmosphere and emit methane. Bogs that have not been drained or are already abandoned are a sink of carbon. The original mire area in Finland is 10.4 million hectares (Lappalainen and Hänninen, 1993). The present peatland area is about 9 million hectares, of which virgin peatlands account for some 4.3 million hectares, the rest being almost entirely drained for forestry. Draining peaked in the 1960s but has now practically ceased. On the basis of peatland areas and amounts of peat, the estimate for the amount of carbon in peatland soils is 4–5 billion tonnes.

Forest Products

The carbon content of forest products in use is only 40–50 Mt carbon, which is only 5–7%, compared with that in living vegetation and 10–12% with that of stemwood. Only less than 30% of the carbon in Finnish-made products in use is located in Finland and the rest abroad, because of exports of forest products.

Landfills

After their primary use, wood and wood-based products are usually either burned or disposed as waste in landfills. The carbon content of wood and wood-based products in landfills is 70–90 Mt carbon or 10–12% compared with that in living vegetation. From the point of view of GHGs, landfill dynamics are of great importance, because part of the decaying wood in landfills generates methane, which is a much more effective GHG than CO_2.

Therefore, based on our calculations, the net GHG impact of landfills on the Finnish forest sector's total GHG balance was negative. However, most of the forest products made in Finland were exported and consequently methane emissions were realized in landfills outside Finland.

Net Greenhouse Impacts

In summary, we estimate that the overall greenhouse budget of the Finnish forest sector (including foreign trade of forest products) showed a surplus rather than a deficit in 1990, even though fossil-fuel emissions were about 2 Mt carbon (equivalent) (Figs 16.2–16.4).

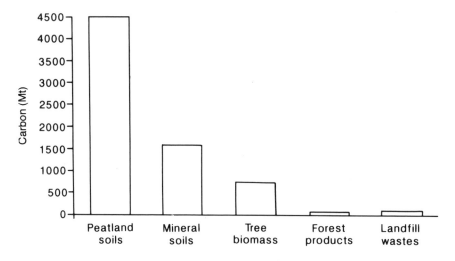

Fig. 16.2. Profile of carbon storage in the Finnish forest sector in 1990 (from Pingoud *et al.*, 1995).

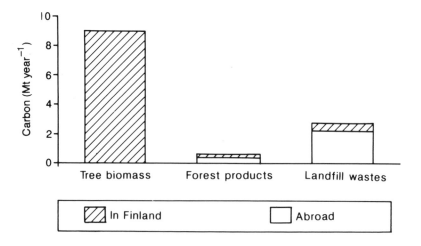

Fig. 16.3. Estimated net increase in rates of the carbon storage in the Finnish forest sector in 1990 (from Pingoud *et al.*, 1995).

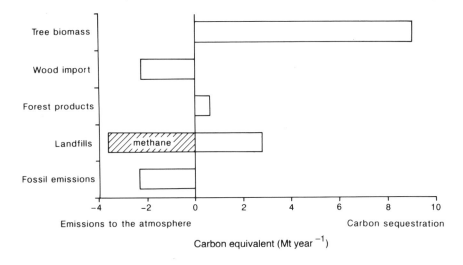

Fig. 16.4. Net GHG fluxes from the Finnish forest sector in 1990 (from Pingoud et al., 1995).

MANAGEMENT OPTIONS

To evaluate the management of the forest sector for mitigation of CO_2 emissions, we compare here two different management and utilization options for forest resources over the period 1990–2040. These options (scenarios) are based on the work of Karjalainen et al. (1995). In the first scenario, cutting remains at the level of the late 1980s (Table 16.1). We call this the

Table 16.1. Outlines of management options for Finnish forestry: growth, removal and standing volume of stemwood (million m³) (from Karjalainen et al., 1995).

	1990–1999	2010–2019	2030–2039
Growth			
Low-growth scenario	85	106	136
Maximum-growth scenario	83	88	97
Removal			
Low-utilization scenario	59	60	56
Maximum-utilization scenario	76	93	87
Volume			
Low-utilization scenario	2,049	2,850	4,290
Maximum-utilization scenario	1,860	1,750	1,720

'low-utilization scenario'. Under this alternative, the amount of wood lost due to cutting remains less than the increase due to growth, which results in an increase in growing stock. The second scenario outlines the maximum sustainable harvest, assuming that cutting is gradually increased, with the aim of balancing growth and removal within the period 2000–2040. Furthermore, it is assumed that the capacity of the forest industry is not limiting, i.e. all the harvested timber can be processed. This is called the 'maximum utilization scenario'. Differences in growth between the scenarios are due to differences in growing stock by management. Calculations in scenarios concern only stemwood.

Under the low-utilization scenario, the amount of wood lost due to cutting remains less than the increase due to growth, which results in an increase in growing stock. To realize the maximum-utilization scenario, cuttings and growth are balanced in the long run. Furthermore, it is assumed that the capacity of the forest industry is not limiting and all harvested timber can be processed. Differences in growth between the scenarios are due to differences in growing stock affected by management. The effects of changing climate on tree growth have been excluded. For comparison, results both with and without the export and import of timber and wood-based products are expressed. We also refer to sensitivity analyses of the carbon balance in wood products with regard to changes in product life span and changes in final use of products from being recycled to being used as an energy source.

Under both management scenarios, the average annual carbon balance is positive in the forest sector (Table 16.2). More carbon is accumulated than

Table 16.2. Forest-sector carbon reservoir 1990–2040 (Mt carbon) (based on Karjalainen et al., 1995).

	1990	2015	2040	2040*
Low-utilization scenario				
Growing stock	375	540	870	870
+ Products in use	55	120	145	35
+ Products in landfills	80	85	115	30
= Total storage	510	745	1,130	935
Maximum-utilization scenario				
Growing stock	375	370	355	355
+ Products in use	55	175	245	60
+ Products in landfills	80	95	140	35
= Total storage	510	640	740	450

* Export and import of timber and wood products included.

released. There are, however, substantial differences between the scenarios. The net annual sink with the low-utilization scenario is over 2.2 times as high as with the maximum-utilization scenario. The annual gross differences are much lower than the net differences, which implies that carbon is stored more efficiently in forests than in products.

At the end of the study period, carbon storage in the forest sector with the low-utilization scenario (1,130 Mt carbon) is over 50% higher than with the maximum-utilization scenario (740 Mt carbon). Within the low-utilization scenario the carbon storage increased by a factor of 2.3 or 620 Mt carbon from 1990 to 2040 and in the maximum-utilization scenario by a factor of 1.5 or 230 Mt carbon. Taking into account the export and import of timber and wood products, the carbon storage within Finland increased, however, by only 425 Mt carbon in the low-utilization scenario and in the maximum utilization scenario decreased by 60 Mt carbon between 1990 and 2040.

The sensitivity analysis showed that lengthening the life spans of wood-based products has only a small effect on the amount of carbon stored in products. In the case of the terminal use of products, the options for enhanced energy production and enhanced disposal to landfills substantially affect the amount of carbon in products, but the effect on the Finnish carbon balance is small, especially when the export and import of products is taken into account.

The results clearly indicate that it is easier to store carbon in the forest than in wood products, even in the case where the life span of the products or terminal use of the products would drastically increase. In other words, the current trend in the use of Finland's forest resources, with increasing growing stock, produces the largest carbon reservoir. This may, however, conflict with the need to strongly increase wood utilization in the forest industries in order to reduce foreign debt and unemployment, especially in the case of absence of international implementation of emission taxes or markets for emission permits.

GREENHOUSE-GAS TAXES AND WOOD AS A SOURCE OF ENERGY

International cooperation to slow down CO_2 and other GHG accumulation in the atmosphere may require emission taxation or markets for emission permits. According to the concept of net national emissions, individual countries should pay for permits or should be taxed according to their net emissions. This has led to a discussion of how carbon emissions arising from the burning of wood should be taxed at the national level. The most common argument is that, in contrast to fossil fuels, CO_2 emissions caused by the burning of wood should be neglected in carbon taxation, because sustainable forestry guarantees carbon accumulation in growing forests. Accordingly, it is argued that forest owners need not be subsidized for sequestration of carbon. The analysis by Tahvonen (1994), based on a dynamic general equilibrium model, with productive capital and the stock of forests as state variables, shows, however, that the reverse is true.

The ability of forests to decrease national expenditure on CO_2 taxation increases wood production of unharvested forests. As a consequence, the steady-state level of the forest stock increases. In a decentralized economy, this means that private forest owners should produce positive externalities. This is not possible without subsidies to forest owners. An optimal subsidy equals the (annual) amount of carbon stored by a given forest stand, multiplied by the international price of emission permits. In addition to this, optimality requires that all CO_2 emissions must be taxed, independently of whether their origin is in fossil fuels or wood. However, when wood is used in durable commodities, the optimal tax per carbon content may be lower than in the case where wood is burned. The simplest solution, and also following the above-mentioned general arguments in Finland, is to put a tax on fossil fuel. Its main effect would be to change from production with high direct and indirect use of fossil-fuel products to production with relatively lower use of fossil-fuel products.

This would increase the production of most of the forestry-based products, including wood-based energy. In the short run, this could lead to higher total net emissions of CO_2 from forest biomass and in the total economy. In the longer run, there would be a balance here, but on a higher level of consumption of wood products. This would, however, depend on how the GHG taxes are channelled back to the economy (e.g. Seppälä, 1994). If the GHG tax regime is high, we may get a general economic recession, which again would reduce the emission.

The higher timber prices, caused by the increase in wood demand, would also lead to increased investments in forest management (including afforestation and regeneration), increasing the future growing stock of timber in the forest and giving higher sequestration of atmospheric CO_2 (Hoen and Solberg, 1995). This effect would, however, in most cases be less than what is optimal from a national or international point of view, if special measures are not taken. This is because of risks. Forest owners will not be sure of getting any benefits of their extra long-term investments in increased forest stocks. First, it is not certain today that increased emission of GHG causes climate change. Secondly, even in the case of future climate change, the future damage is uncertain, as well as the amount of marginal resources to spend to moderate the damage. And, thirdly, even in the case of severe damage by climate change, new renewable-energy technologies may make higher taxes on fossil fuel unnecessary in 30–100 years ahead, when the forest owner would be benefiting from present investments in forestry.

These risks are mainly caused by the fact that forest owners are carrying all forestry investment costs themselves, whereas the benefit of decreased concentration of atmospheric CO_2 is a true public good. Even if the climate change damage will not be as severe as we currently fear, the build-up of forest resources will benefit society in the form of forest products and many other uses. However, the forest owner will probably not receive all of the expected returns of his extra investments, because the increased supply of timber will produce lower-than-expected timber prices.

To enhance efficiency and equity, the government should, in addition to a tax on fossil fuel, introduce a subsidy for absorption of carbon. This subsidy should be as closely as possible linked to the fixation of carbon in the forest biomass over time. The size of the tax should equal the emission tax on CO_2 but less by the cost of emission of CO_2 from the future decay of the biomass fixed. An alternative to this net subsidy would be to introduce a gross tax, similar to that on fossil fuel, on all fixations and emissions of atmospheric carbon in forests, and to charge the end-user (paper, sawn-wood and bio-energy users) for emission. This might, however, be difficult to measure and control.

Another alternative could be to pay credits according to estimated net growth and impose a tax to account for decay when the forest biomass is harvested. This harvest tax should reflect the assumed decay profile of the forest biomass and should be varied according to end use. To reflect current decay profiles, sawn timber would be given a lower tax than pulpwood, while wood for bioenergy would be given a higher tax than for pulpwood.

SUMMARY AND CONCLUSIONS

The overall infrastructure for using the forest sector as a management tool in mitigating GHG emissions in Finland is favourable. Finland therefore represents an important case conflict in this respect. Effective carbon sequestration could be implemented by increasing the growing stock as much as possible. This may, however, contradict the policy objective of reducing foreign debt and unemployment through significantly increasing production of forest-industry products. It seems likely that foreign debt and unemployment will continue to be regarded by policy-makers as the most severe socio-economic problems.

However, an appropriate increase in the length of the rotation in forestry could be optimal for balancing the needs of forest resources for carbon sequestration and timber production. Increased use of wood, based on sustainable use of forest resources, to substitute for fossil fuels and materials, could in many cases decrease the overall carbon emissions. However, when considering whether to establish carbon taxes, society should concurrently take into account the forest owner's need for taking risks in investments in forestry.

Release of sequestered carbon back to the atmosphere can be delayed by prolonging product life spans, by increasing recycling, or by disposing of discarded products in landfills. To delay carbon release and offset the carbon-budget deficit, however, these changes should be substantial. From the point of view of the Finnish national economy, this policy tool is less effective, since the major part of wood-based products is exported.

German Forests in the National Carbon Budget: an Overview and Regional Case-studies

Klaus Böswald

GERMAN FORESTS IN THE CARBON CYCLE: STOCKS AND CHANGES

The objective of this chapter is to explain the present role of German forests and forestry in the national carbon budget and to introduce options for increasing the forest stock of carbon. An overview of the carbon stocks in forests and wood products is presented and changes in the carbon pools of forests and wood products are quantified, highlighting methodological issues in describing changes in the forests' carbon stock. The carbon mitigation obtained by substitution of high-energy raw materials by wood and the substitution of fossil fuels by wood as a source of energy are analysed. Additionally, the carbon-mitigation effects of various silvicultural measures, such as the modification of the rotation time, underplanting, change of tree species, afforestation of non-forest land and increased use of slash and wood from thinning operations for energy, are quantified. Finally, estimates of the costs of sequestering carbon due to the afforestation of non-forest land are presented.

Carbon Storage in German Forests

The stock of wood in German forests amounts to 2.68 billion m^3 on a total forested area of roughly 10 million ha. That implies a carbon storage of 888 Mt or 89 t ha^{-1} for the living biomass of the trees. Depending on the site characteristics, the carbon storage in the soil varies between 100 and 150 t ha^{-1}.

The higher values are due to more favourable conditions for microorganisms in the soil (elevated average temperature, greater share of deciduous trees), which lead to vigorous disturbance and consequently to an increase of carbon in the subsoil. If the carbon stock in the soil and in the ground vegetation is included, the total carbon stored in German forests amounts to 2.514 Mt (Burschel *et al.*, 1993a). This is 10.4 times the quantity emitted by burning fossil combustibles in Germany during 1991 (Enquete Commission, 1992) (Table 17.1).

Table 17.1. Carbon accumulation in German forests (from Burschel *et al.*, 1993a).

Compartment	Storage	
	Mt	%
Living biomass		
Trees	888	35.3
Ground vegetation	10	0.4
Dead wood	50	2.0
Humus		
Litter	392	15.6
Soil organic matter	1.174	46.7
Forest ecosystem	2.514	100

The quality of the database to assess carbon stocks in German forests is well founded for standing volume and the annual timber cut, as well as for the calculations of the biomass and their carbon content. It is more difficult to gain reliable values about the increment of the stands and the carbon storage in soils and ground vegetation. Further information about the methodology, indicating equations for the calculation of the carbon pool and the carbon-mitigation effects, is found in Burschel *et al.* (1993a).

Changes of the Carbon Storage in German Forests

The question whether the German forests are to be considered a carbon sink or a carbon source may be answered either by the difference between increment and timber harvest or by comparing two inventories. If increment and timber harvest are compared, it is only possible to estimate the carbon sink. Such an investigation has been done by Burschel *et al.* (1993a) for Germany. Based on the same methodology, data are available for the regions of Baden-Württemberg and Lower Saxony (Table 17.2). According to Burschel *et al.* (1993a), the sink effect of the living dendromass in Germany amounts to 5.5 Mt carbon year^{-1} or 0.55 t carbon ha^{-1} year^{-1}. This value is clearly below the results of regional studies, for example, in Baden-Württemberg,

Table 17.2. Estimated carbon (C) sink of forests in selected regions of Germany (from Böswald, 1996).

Region	Sink effect (Mt C year^{-1})	Sink effect (t C ha^{-1} year^{-1})	Forest area (million ha)
Baden-Württemberg*	0.9	0.69	1.31
Bavaria†	2.6	1.09	2.38
Lower Saxony‡	1.0	0.93	1.07
Bundesrepublik Deutschland§	5.5	0.55	10.03

* From Volz *et al.*, 1996.
† From Böswald, 1995.
‡ From Böswald and Wierling, 1995.
§ From Burschel *et al.*, 1993a.

with 0.69, and in Lower Saxony, with 0.93 t carbon ha^{-1} year^{-1}, which have been ascertained with comparable methods.

In contrast, the change of the carbon storage in the dendromass can be accurately estimated, if two comparable forest surveys are available with a suitable time interval. A calculation of change of carbon stock over time for Bavaria, for example, was possible, after the stock of wood and the forest area investigated by the Bayerische Waldinventur (Bavarian Forest Survey) (Franz and Kennel, 1973) and the Bundeswaldinventur (German Forest Survey) (BML, 1990) had been compared (Foerster and Böswald, 1994). Between 1971 and 1987, the increase of the carbon stocks in the living dendromass in Bavaria had been 1.09 t carbon ha^{-1} year^{-1}.

The findings presented indicate that the sink effect of the German forests is actually above the 5.5 Mt carbon evaluated by Burschel *et al.* (1993a). Assuming a medium carbon storage of 0.95 t carbon ha^{-1} year^{-1} (considered forest area – 4.76 million ha, sink effect – 4.5 Mt carbon year^{-1} (Table 17.2)), a sink effect of 9.5 Mt carbon year^{-1} can be calculated for the total forest area of Germany (Table 17.3).

The results presented by Burschel *et al.* (1993a) underestimate the actual carbon stock as they take a mean annual increment of only 5.7 m^3 ha^{-1} year^{-1}. In comparison with that, the increment of the living stock in Baden-Württemberg and Bavaria, which represents more than 35% of the forested

Table 17.3. Annual carbon sequestration of German forests (estimates in parentheses, see text).

Carbon sink	Germany (Mt year^{-1})
Trees	9.5
Humus in mineral soils	(1.0)
Ground vegetation	
Forest ecosystem	9.5 (10.5)

area in Germany, is greater than 10 m³ ha⁻¹ year⁻¹ (Foerster *et al.*, 1993; Schöpfer, 1993). Moreover, information about felling quantities is only partially suitable, because the felling statistics, especially for private forests, do not describe the actual circumstances (Kollert, 1990; Foerster and Böswald, 1994). Indeed, in Germany the increment in the last two decades has been nearly twice as high as the sum of fellings and mortality.

The amount of carbon in the forest soils in Germany is probably roughly in balance, because accumulation and decomposition of the biomass are in equilibrium. However, it is likely that soils are recovering from past destructive practices, which have led to an increase of the carbon storage in the soil. This effect covers approximately 1 Mt carbon year⁻¹ if it is cautiously assumed that on one-half of the forested area in Germany 0.2 t carbon ha⁻¹ year⁻¹ is accumulated (Burschel *et al.*, 1993b).

Consequently, German forests have an important carbon-sink effect. A comparison with the anthropogenic-initiated carbon emissions in the Federal Republic of Germany, which in 1991 were roughly 250 Mt, shows that each year about 4% of these emissions are additionally fixed in forests.

In several regions of Germany, the recent sequestration potential of the forests results from the age-class structure, as young stands are predominant in the forests. In these stands, the difference between increment and timber cutting is more favourable for a net carbon accumulation than in older stands. Given the probable change in the age-class structure (Cannell and Dewar, 1995; Kohlmaier *et al.*, 1995), a minor sequestration in German forests can be predicted in the future centuries (Böswald, 1996). It might even be possible that, on long-term considerations, the current sink effect in German forests could convert into a carbon source.

Carbon Dioxide: Ecological Importance of Wood Utilization

Unforested areas in potentially forested regions will again become forests, in the absence of human intervention. Normally, establishing forests accumulate biomass until the natural maximum volume of the site is reached (Borman and Likens, 1979). At this point, carbon storage in mature forest ecosystems is complete and assimilation and dissimilation are in equilibrium. In these forests, a carbon-sink effect can only be imagined if either soil humus accumulates or organic substances are withdrawn by draining and spread into the hydrosphere (Burschel *et al.*, 1993b).

However, because most of the forests are intensively managed, uninfluenced natural forest ecosystems are found in very few sites in Central Europe. Thus carbon storage of the dendromass is influenced primarily by logging.

However, for a comprehensive estimation of the role of forests and forestry in the carbon budget, it must be taken into consideration that the carbon ecological effects of wood utilization compensate to a large extent for the losses of carbon storage in the forests. Figure 17.1 shows the different carbon dioxide-related (CO_2) ecological effects for a spruce stand with a rotation period of 120 years and which is growing and thinned according to yield tables (Assmann and Franz, 1963). As a simplification, it has been presumed that the stand is finally used in a clear cut. Up to the age of 120, altogether 262 t carbon are accumulated in the woody biomass. Most of that is withdrawn at the final use, while the logging slash remains on the surface.

During the first selective logging, carbon storage in wood products is establishing, which lengthens the storage effect of the forests within the duration time of the wood products. The size of this stock settles at roughly 100 t carbon after two rotation periods (Table 17.4).[1] Additionally, carbon emissions are reduced if wood, as a raw material of low energy input, is used as a substitute of equal quality for high-energy raw materials. In the example shown, the reduced emission due to the substitution of high-energy raw material after three rotation periods amounts to 145 t carbon (Table 17.4).[2] Finally, sustainably produced timber, which is harvested and used as a source of energy, results in a considerable mitigation of CO_2 emissions, if by this the consumption of fossil combustibles is reduced. Due to the substitution of fossil fuels after three rotation periods, a carbon mitigation of 45 t can be calculated, as in Table 17.4.[3]

Consequently, a sustainably managed forest has the effect of CO_2 reduction and mitigation, provided that the harvested wood is used as a substitute

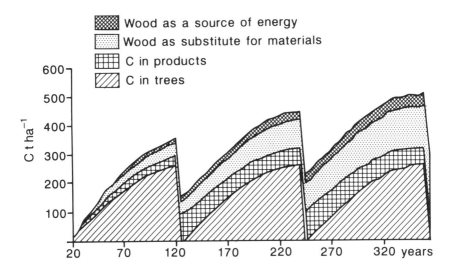

Fig. 17.1. Carbon (C) storage in wood products, carbon mitigation by wood as raw material of low energy input and wood as a source of energy in a spruce stand (from Böswald 1996). Grown and thinned according to yield table of Assmann and Farnz (1963), rotation period 120 years.

Table 17.4. Carbon (C) storage in wood products and carbon mitigation by material and energy substitution in a spruce stand (t C) (see Fig. 17.1).

Rotation length	Products	Wood as a substitute for materials	Wood as a source of energy
120	93	48	15
240	100	98	30
360	100	145	45

for high-energy raw materials or fossil fuels. While carbon storage in stocks and products is limited by the site conditions and the decomposition rate, the effects of material and energy substitution are adding up. In contrast, an unmanaged forest has no further effect of carbon mitigation if the maximum of carbon storage on the site has been reached.

The stock at the age 120 of the yield table, according to Assmann and Franz (1963), amounts to 958 m^3. The expansion factor, to calculate branches, brushwood and roots, is 1.45 (Burschel et al., 1993a). With a density of 0.377, a stock of 262 t carbon ha^{-1} can be assumed.

Carbon storage in wood products

In Germany, 340 Mt carbon are stored[4] in wood products, which averages out at 4.1 t carbon per capita. As expected, the greatest partial storage is to be found in residential and industrial buildings, where 220 Mt or 65% of the total is accumulated (Table 17.5). Surveys from Finland, assuming a carbon storage of 6 t per person in wood products in use, seem to confirm the results of Germany, as in Northern Europe the wood consumption per capita as well as the share of long-term utilization of wood are far higher (Karjalainen et al., 1995; Pingoud et al., 1995).

Table 17.5. Carbon (C) storage in wood products in use in Germany (from Frühwald and Wegener, 1993).

Products	C (Mt)
Residential buildings	170
Wood in durable use	35
Wood in public use, e.g. in playgrounds	40
Industrial buildings	50
Wrapping material	5
Wood in the form of paper	25
Wood stored while in production	15
Total	340

There is a discrepancy between the values given by Burschel *et al.* (1993a) and those of Bramryd (1982). Burschel *et al.* (1993a) ascertain a medium carbon storage in wood products of 1.6–1.7 t. The authors availed themselves of basic data evaluated by Kroth *et al.* (1991), who assume a lower use of wood in the building and construction industry than Frühwald and Wegener (1993). It cannot be explained why the value calculated by means of the information given by Frühwald and Wegener (1993) is even above the 2.1 t carbon per capita evaluated by Bramryd (1982). This is also astonishing, as the surveys of Bramryd (1982) have been made in Scandinavia and the USA, where traditionally a higher use of wood takes place.

Due to the carbon-related ecological effects of wood utilization, including wood in products, material and energy substitution, the annual carbon mitigation is presumed to be 5.1 Mt (Table 17.6). This carbon-mitigation effect has been calculated by Burschel *et al.* (1993a) for a product storage covering roughly half the size presented. If the surveys of Frühwald and Wegener (1993) are applicable (see Table 17.5), it might be possible that the real sink effect is above the data presented.

Table 17.6. CO_2-mitigation effect (Mt C) of wood products produced in Germany (from Burschel *et al.*, 1993a).

Effect	Annual mitigation
Wood utilization	1.1
Substitution of raw material	2.6
Wood as a source of energy	1.4
Total	5.1

C, carbon.

POTENTIALS FOR STRENGTHENING THE ROLE OF FOREST AND FORESTRY WITHIN THE CARBON BALANCE

Optimization of the Carbon-mitigation Potentials by Intensification and Extension of Forestry

In consideration of the described carbon storage and carbon-sink effect of the German forests and the wood industry, the question arises as to what extent their carbon-mitigation effect may be influenced or ameliorated through specific measures. Regarding this question Burschel *et al.* (1993a) have investigated several options for Germany. Among others, they enquired into silvicultural methods, such as lengthening the rotation period, alternation of tree species and underplanting, but also increased use of wood as an energy source and afforestation (Table 17.7). It is emphasized that, by means of realistic measures,

Table 17.7. Options of CO_2 mitigation by forests in Germany (Mt C) (modified from Burschel et al., 1993a).

	Period (years)		
	20	40	60
Options in existing forests			
Lengthening of rotation periods			
Medium option	14.7	3.0	−16.6
Maximum option	35.1	18.8	−44.7
Underplanting: 0.5 million ha year^{-1} in 20 years	0.2	1.7	6.0
Substitution of Scots pine by Douglas fir			
10,000 ha year^{-1}	1.5	5.0	9.2
5,000 ha year^{-1}	0.7	2.5	4.6
Afforestation			
5,200 ha year^{-1}	1.3	5.3	13.2
1.2 million ha in 20 years			
0.655 million ha shelter belts	20.1	89.0	189.8
4.5 million ha in 20 years			
0.5 million ha energy plantations	107.0	364.6	737.6
Increased use of wood as fuel			
Wood waste, thinning material	26.0	52.0	78.0
Wood waste, thinning material, slash	54.0	108.0	162.0
All options			
Medium (year^{-1})	61.2 (3.1)	154.4 (3.6)	253.2 (4.2)
Maximum (year^{-1})	197.0 (9.9)	495.6 (12.4)	865.5 (14.4)

C, carbon.

up to 2 Mt carbon annually can be fixed additionally in the forests in the short term. After 20 years, according to the model calculations, a reduction of up to 50% is feasible. In the longer term, carbon mitigation could even become negative, if, as a consequence of the age-class structure, age classes III and IV are harvested gradually.

The greatest carbon-mitigation effects could be achieved when non-forest land is afforested on a large scale. Roemer-Mähler (1992) estimated the extension of this area in Germany to be almost 5 million ha. The afforestation area is limited for ecological and aesthetic landscape reasons (Philipp, 1987). If these arguments are included in the determination, the afforestation area decreases to 1.2 million ha. This is roughly 7% of the total agricultural area in Germany. The lower value of the carbon sequestration (20–190 Mt carbon (Table 17.7)) through afforestation is due to that reduced area. The maximum mitigation occurs if 5 million ha are afforested, of which 0.5 million ha will be stocked with energy plantations (Table 17.7). Thus, within a period of 60 years, a maximum sequestration and mitigation of 737.6 Mt carbon could be achieved (Table 17.7).

Costs of Carbon Storage by Afforesting Agricultural Land

The cost calculation of carbon mitigation is made more difficult by the fact that the diverse effects of forests cannot exactly be assigned to the individual expenditures. Such a comparison becomes meaningful only if means are found to sequester and to mitigate carbon. An option for this is given by afforestation of agricultural land. Some estimates are now presented, based on an investigation made in Bavaria (Böswald, 1995, 1996).

For the calculation, expenses of planting and site preparation, as well as losses of income (based on European Union (EU) Regulation 2080/92), have been considered and have been expressed in terms of carbon-mitigation performance. Expenses for opportunity costs and soils, as well as any income from thinning, which could contribute to reduction of the costs up to the end of the 60-year period of observation, have not been taken into account. Moreover, the payment of interest on the different parameters has not been considered, nor has the carbon uptake been discounted.

The medium costs for the mitigation of 1 t carbon through afforestation are described in Table 17.7. With a great share of deciduous trees, the costs range between \$US386 and \$US112 t^{-1} carbon, depending on the observation period. With an increased share of conifers, the costs will be \$US86 within the observation period of 60 years (Table 17.8). The costs arising from a 100% share of conifers to obtain a maximum increment are not included in Table 17.8, as this option is completely unrealistic, for ecological reasons. Thus, for the mitigation of 1 t carbon, having a tree species share of 90% spruce and 10% Douglas fir, the sum of \$US117 t^{-1} carbon (20 years' observation interval) and \$US22 t^{-1} carbon (60 years' observation interval) would have to be spent (for further information, see Böswald, 1996).

If the costs necessary to sequester 1 t of carbon by afforestation are transferred to those responsible for the emissions, the price for 1 l of fuel oil would increase by \$US0.05 to \$US0.26, while the price depends on the share of tree species. In the calculation a medium carbon-emission of 0.666 kg l^{-1} fuel oil has been assumed (0.0185 kg carbon MJ^{-1}; 1 l of fuel oil is considered to have a heating value of 36 MJ; with the consumption of 1,500 l fuel oil 1 t carbon is released).

Table 17.8. Costs per tonne of carbon (C) sequestration through afforestation in Germany (\$ t^{-1} C) (from Böswald, 1996).

	Period (years)		
	20	40	60
Deciduous trees	386	214	112
Conifers	346	179	86

In contrast to the estimates in Table 17.8, the costs arising from substitution of conventional oil heating by a system with less emission are shown in Table 17.9. For the evaluation, it has been considered that heating plants in local latitudes mostly run for 1,000–2,000 h year^{-1} at their maximum capacity (Demmel and Alefeld, 1993). The costs of mitigating carbon with these options vary between $US260 and $US1,288 t^{-1}. Almost all of them are above the costs evaluated for afforestation. Only the carbon-mitigation costs that arise where the share of deciduous trees is highest are at a comparative level.

It can be concluded that the afforestation of agricultural land is also economically justified and represents a favourable option for carbon sequestration. Although, it is necessary to restate that, despite all technical innovation, heating plants run by fossil fuels still have a negative carbon balance. In contrast, the carbon balance is in equilibrium if wood is used as an energy source. A maximum carbon reduction, with reference to the produced kilowatt-hour of energy benefit, is, in particular, ensured when wood from sustainable production is used, as discussed.

One further reason for using wood as a source of energy is the current limitations on silvicultural management, due to the lack of a market for wood from thinnings and slash wood, despite it being suitable as a fuel wood.

Table 17.9. Costs of investment for the reduction of carbon (C) emissions for selected energy-production systems (from Demmel and Alefeld, 1993).

	$US t^{-1} C	
Facilities to produce thermic energy	1,000 h year^{-1}	2,000 h year^{-1}
Electronic heating due to nuclear power plant	1,288	644
Electronic heating due to coal power plant	729	364
Electronic heating due to gas power plant	520	260

h, power efficiency = 0.7.

CONCLUSIONS

The carbon storage in German forests is estimated as 2,500 Mt. This is roughly ten times the annual carbon emission in the Federal Republic of Germany. German forests annually sequester approximately 9.5 Mt, corresponding to approximately 4% of the annual carbon emission of Germany. The recent sequestration potential of the forests is due to a favourable age-class structure in several regions of Germany. In the next centuries, in all probability, reduced sequestration in German forests is likely. It may even be conceivable that the current sink effect in German forests could convert into a carbon source.

Silvicultural options for counteracting this development are restricted. Even with a modification of the rotation time in connection with underplanting, change of tree species and the increased use of slash and wood from thinning operations as an energy source, the mitigation could be expanded to about 2 Mt year^{-1}, or 20% of the recent annual carbon sink.

Enhanced wood utilization has potential for carbon mitigation, through, for example, extending the carbon storage in wood products. With an estimated potential of 340 Mt carbon, product storage currently amounts to roughly 7% of the carbon stored in forests. Due to the utilization of wood and the effects of the substitution of material and energy, 5.1 Mt carbon is mitigated annually. As a result, the carbon emissions of Germany are about 2% lower than without the utilization of wood. To maintain the current role of German forests and forestry in the carbon budget, there is an urgent need to intensify and expand wood utilization. Moreover, this must be considered if the wood supply rises due to intensive afforestation.

Large-scale afforestation of agricultural land is the most favourable possibility to mitigate CO_2 emissions within the forestry sector – potentially between 3.2 and 12.3 Mt year^{-1}. Such large-scale afforestation would appear to be economically feasible, given the current economic environment, although planting will have to be subsidized. This chapter has illustrated that, even in an industrialized country like Germany, forests and forestry might serve to reduce carbon emissions. It would be an economic, ecological and political failure to neglect these extraordinary possibilities.

NOTES

1 It has been estimated that in Germany 28% of the harvested wood is used as construction wood, 19% for the production of furniture and other products and 28% for paper and packing. Moreover, a mean duration of 65 years for construction wood, 15 years for furniture and 1 year for paper and packaging are assumed.

2 On average, 0.28 t of carbon emission is avoided with the use of every cubic metre of construction wood. Such a substitution effect has been considered only for the use of stemwood. Due to the high energy consumption for the production of particle and fibre boards, the energy saving of this type of wood is rather low (Burschel et al., 1993a).

3 Calculating the energy substitution, it has been presumed that the substitution by wood instead of oil combustion avoids, when burning 1 kg wood, the release of 0.26 kg carbon (Burschel et al., 1993a). Carbon is, of course, also released by burning wood. However, at the same time it is sequestrated due to photosynthesis. Consequently, burning wood from sustainably managed forests has a neutral carbon balance.

4   The figures are based on an investigation of Frühwald and Wegener (1993), who assumed an average carbon storage in single and double family dwellings of 7.5 t and in multiple family dwellings 15 t. In every household, a carbon pool of 1 t in wooden products has been considered. The carbon content of the other product groups has been estimated based on statistical data of the German Bureau of Statistics in Wiesbaden (further information is given in Frühwald and Wegener, 1994).

18 Carbon Fixation in Swedish Forests in the Context of Environmental National Accounts

Peter Eliasson

INTRODUCTION

This chapter tackles issues in estimating the net value of forests in their carbon-sink function. The aim of the study is to incorporate the value of actual sequestration of carbon dioxide (CO_2) in forest trees into a satellite account of total net income from Swedish forests. The starting-point is work by Hultkrantz (1992) and Eliasson (1995) concerning estimation of the net national income from natural resources in the Swedish forests. Those calculations included timber, berries, mushrooms and reindeer forage (lichen), as well as environmental stocks, such as soil nutrients, biodiversity and carbon sequestration.

The analysis on carbon sequestration reported in this chapter is based on research that aims to develop resource accounting for the Swedish forests and forestry, forming a satellite account to an adjusted net national product (NNP). The results aim to increase the understanding of important connections between the economy and the environment, with special emphasis given to the forest as a producer of ecological and economic services, including recreation.

The structure of the chapter is as follows. The following section discusses environmental accounting in general and outlines the findings of previous studies, in the framework referred to above. The next section examines in detail how forests in Sweden function as carbon sinks and discusses the sustainability of the carbon sink.

ENVIRONMENTALLY ADJUSTED NATIONAL ACCOUNTS

The current system of national accounts (SNA) integrates information about resource allocation in all nations. The SNA is thereby the basis for the construction, evaluation and comparison of national economic performance in countries throughout the world. Gross national product (GNP) is one of the important economic indicators resulting from the SNA. The GNP has sometimes been interpreted as a measure of welfare, which is valid only if social welfare is understood to be synonymous with the value of all economic production in society. In the perfect economy, this value can be represented, for instance, by the value of consumption of marketed goods and services during a given period of time, provided that certain technical restrictions are met (Dasgupta *et al.*, 1995). The major impetus for environmental adjustments to the SNA is based on the observation that economic growth brings changes in consumption behaviour, which generally has negative impacts on the environment.

The nature of the environmental effects from economic growth depends on the particular dimension of environmental quality or resource use, as well as the prior level of income and other factors. Increasing GNP in a poor country may be associated with worsening environmental conditions, such as reduction in air quality as a result of industrialization. On the other hand, some environmental indicators, such as air and water quality, appear to benefit from economic growth in richer countries (Holtz-Eakin and Selden, 1995). On the other hand, indicators such as municipal waste per capita and CO_2 emissions appear to increase as the economy grows, independently of the level of per capita income (Grossman and Krueger, 1995). True social welfare thus has at least two major components that are poorly represented in the GNP: non-market goods and services and, to some extent, depreciation of the resource stock.

The primary focus of this study is on assessment of timber flow in the SNA. However, the forest ecosystem as a whole produces other important non-market goods and services that contribute to social welfare. Such issues are of special interest in the Nordic countries, where forests cover the majority of the land area. The main purpose of this chapter is to shed light on the economic importance of forests as carbon sinks, thereby indicating their key environmental service role.

The NNP measure within the SNA has been suggested, most recently by Dasgupta *et al.* (1995), to give a theoretical basis for estimating a welfare index. Adjusted for environmental effects, NNP reflects not only the value of the change in resource stocks, but may also serve as a welfare measure, since it is a linear index indicating whether the performance of the economy during a specific period in time improves or reduces the present value of current and future well-being.

RESOURCE ACCOUNTING IN FOREST ECOSYSTEMS

This section presents updates of forest sectoral account adjustments for Sweden, first suggested by Hultkrantz (1992). The aim of this section (following Eliasson, 1995) is to operationalize some of the principles that could be developed for use in modified national accounting to estimate a sustainable national net income based on NNP (for greater detail, see Eliasson, 1995).

Estimation of the value of non-market and environmental goods and services from forestry has been undertaken, using a valuation principle applied in a conservative fashion, based on lower-bound estimates, giving the results in satellite account form for 1987 and 1991. They are shown in Table 18.1. The valuations presented are of an experimental nature and should be interpreted as partial adjustments, with factors such as recreational values or hydrological services omitted, while the components included are now briefly outlined.

Table 18.1. Value of forest resources in Sweden in the years 1987 and 1991.

	1987		1991
	Value (billion SEK)	Value in 1991 (billion SEK)	value (billion SEK)
Timber products	18.05	24.55	20.63
Market value (roadside)	18.60	25.29	21.30
Inputs from other sectors	–3.14	–4.27	–4.30
Increase in growing stock	4.14	5.63	5.67
Silviculture	–1.55	–2.11	–2.04
Other products	1.76	2.39	2.19
Berries	0.50	0.68	0.41
Mushrooms	0.25	0.34	0.30
Game hunting (meat)	0.47	0.64	0.75
Lichens	0.54	0.73	0.73
Changes in environmental stocks	0.27	0.37	0.34
Biodiversity	–1.24	–1.69	–1.46
Carbon sinks	2.25	3.06	2.84
Exchangeable cations in soil	–0.72	–0.98	–0.98
Lichen stock	–0.02	–0.03	–0.06
Total net income	20.1	27.3	23.2

Timber Production

The market value at the roadside of timber in the national accounts is composed of the value of sawn timber, pulpwood, fire wood, Christmas trees and other round timber, as in conventional national accounts. We have added to these values the estimated value of non-market firewood and growing stock increase. Non-market timber volumes are estimates from surveys by Statistics Sweden and the National Board of Forestry (Swedish National Board of Forestry, 1993). The estimation of increases in this growing stock is examined in the following section. Using the average of the unit net conversion values, the total value of the timber volumes is 18.6 and 21.3 billion Swedish crowns (SEK) in the years 1987 and 1991, respectively.

Effects of Air Pollutants on Forest Production

This external cost to the forestry sector is estimated from calculations on the impact from the annual acid deposition (Sverdrup *et al.*, 1994). Liming was assumed to be an adequate measure for mitigating acidification, and acidification is a grave threat to sustainable production. Although there are uncertainties as to whether liming itself has negative impacts on acidified soils, the situation is considered by the Swedish National Board of Forestry to be serious enough to require action. Therefore, it is assumed that the cost due to acidification of Swedish forest soil is at least equal to the cost of preventing the acidification.

Carbon Pool

Approximately 40–60% of the total annual CO_2 discharge in Sweden is assimilated annually in the growing stock of living trees. Other important carbon sinks in forests are vegetation, soil (peatland) and lakes. The forest carbon sink was valued by the effluent tax on CO_2 that is imposed on firms in Sweden.

Berries and Mushrooms

These can be picked free on both public and private lands in Sweden because of the Right of Common Access. Survey data, from the National Forest Inventory, on utilization were combined with recent market price data to estimate the values reported in Table 18.1.

Game Hunting

The total value of hunting in Sweden has been studied in a national survey by Mattsson (1993), where hunters gave their opinions on the value of the game bag. The total value of the meat from hunting was adjusted for price effects. Statistics from the Swedish Hunters Association were also used to adjust the number of animals killed (Table 18.1).

Lichen

About 30% of the forest area in Sweden is used as wintertime pasture for reindeer grazing. The main source of forage for reindeer is ground lichens and those growing in old growth stands of trees. In a similar fashion to timber values, the value of lichen is composed of stocks and flows. The value of grazing (flow) was estimated by calculating the opportunity cost of feeding all reindeer in Sweden. The volumes used for assessing the lichen stock are based on results from a simulation model by Wilhelmsson (1988). The calculation is made assuming that the offset decrease of lichen production follows a linear index: let the value of sustainable lichen production be

$$V_t = \frac{I_t}{r}$$

where I_t is the annual lichen production and r is the interest rate of return chosen (5% in this case, for example). Thus, the change of lichen stock is calculated as

$$\Delta V_t = \frac{\Delta I_t}{r}$$

where Δ denotes the annual change.

Biodiversity

An evaluation of forest diversity, utilizing the so-called permanent inventory method, is used as a basis for the assessment here. The value of the depletion of the stock is assessed by using the cost of maintenance or replacement of habitats required to retain the present level of diversity (Hultkrantz, 1992).

Felling has a negative impact on the majority of red-listed species in Swedish forests (Berg et al., 1995). Part of the cost of maintaining the stock of biodiversity is therefore recognized as preserving certain areas from forest operations. Ecologists in Sweden (Liljelund et al., 1992) have proposed that the minimum level of protection for endangered species should be to exempt at least 15% of the forest-land area from exploitation. Protecting this forest area is thus treated as a minimum effort to maintain biodiversity (provided also that such an area would be allocated effectively).

At present only about 5% of the forest area has been exempted from exploitation. Hence it is assumed that the cost for not protecting the remaining 10% of forest area is equivalent to at least 10% of the total timber value in Sweden. This assessment of minimum maintenance effort is a somewhat questionable aspect of the adjustments to the sectoral accounts. It may be argued, for example, that the value of biodiversity is related directly to average income from forestry per hectare. Further, the possibility cannot be excluded that species in areas that are protected from forestry could be stressed by other land use.

Assessing the value biodiversity, or, rather, the diversity of physical factors creating the necessary conditions for diversity of species and genetic diversity within species, is a very intricate task. In the handbook of *Integrated Environmental and Economic Accounting* (UN, 1993), for example, it is stated that biological assets depend heavily on intact ecosystems and not on individual species, and yet it gives no clear guidance for such assessment.

Reindeer Forage

In Sweden, 30% of the forest area is used as wintertime pasture for reindeer grazing. The main source of forage for reindeer is lichens on the ground, but they also eat lichens growing in old growth stands of trees. The value of this forage has been estimated by calculating the costs of feeding with a simulation model (Wilhelmsson, 1988) to estimate future lichen production.

CARBON-SEQUESTRATION FUNCTION IN THE SECTORAL ACCOUNTS

Increasing concentration of CO_2 in the atmosphere, along with other greenhouse gases, is considered to be the major cause of rising mean annual temperatures and is at least partially caused by present and historical fluxes of carbon associated with land-use change and forestry. In this section, the carbon cycle of the Swedish forest ecosystem as a whole is described and the net sequestration of carbon from this ecosystem is quantified. Finally, a range of values is presented.

The Carbon Cycle

Considerable carbon fluxes occur in the forest ecosystem annually, as represented in Fig. 18.1. Plants and trees consume large amounts of CO_2 during photosynthesis, with the waste product being oxygen. In biological terms, this process is regarded as primary production. Consumption in this context is the reverse process. Insects, mites, worms, bacteria and fungi release the organic carbon back to the atmosphere as they consume organic

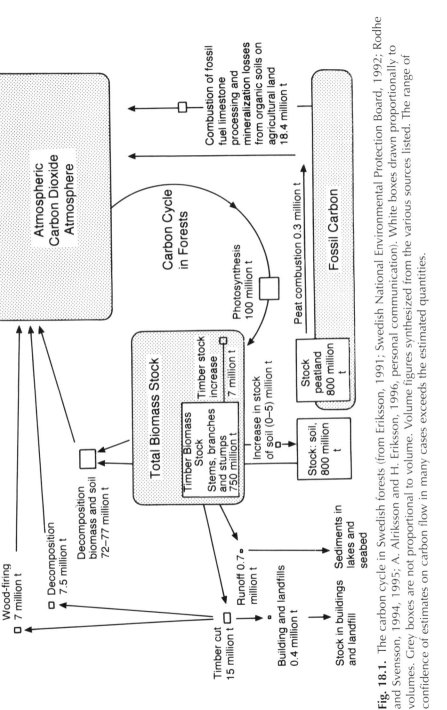

Fig. 18.1. The carbon cycle in Swedish forests (from Eriksson, 1991; Swedish National Environmental Protection Board, 1992; Rodhe and Svensson, 1994, 1995; A. Alriksson and H. Eriksson, 1996, personal communication). White boxes drawn proportionally to volumes. Grey boxes are not proportional to volume. Volume figures synthesized from the various sources listed. The range of confidence of estimates on carbon flow in many cases exceeds the estimated quantities.

matter and oxygen. Carbon dioxide emerges as the metabolic waste product in this decomposition process.

There is an important distinction between the carbon in land biota alluded to above and anthropogenic carbon emissions. The anthropogenic carbon emissions cause a net increase of total carbon, altering the carbon cycle (right box in Fig. 18.1). Combustion and decomposition of biomass, on the other hand, may bring about a faster circulation, but do not contribute to a net increase of carbon as with fossil-fuel combustion (upper left in Fig. 18.1).

The total timber volume in Swedish forests has increased by 60% since forest inventories began in 1923 (Fig. 18.2). The annual increment of growth has exceeded cuts at an average of 15 million m^3 forest throughout the twentieth century. Carbon dioxide flux corresponding to this volume increase has thus been sequestered. From the total annual discharge of CO_2 in Sweden, approximately 40–50% is assimilated in the growing stock of living trees each year.

The Swedish National Board of Forestry (1993) has estimated the flow of carbon in stems, logging residues and stumps, starting out from the approximated average stock increase of 25 million m^3 forest (such an approximation is as good as any, since inventory data contain some uncertainty). In

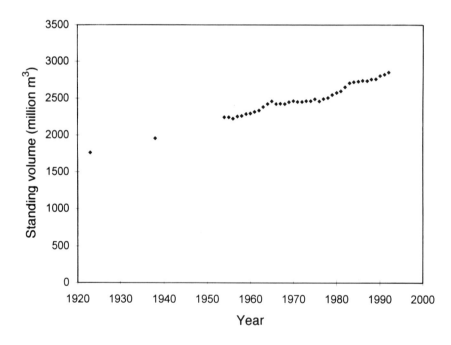

Fig. 18.2. Total standing volume in Sweden's forests, as estimated by the National Forest Inventory.

environmental accounting, there is a need to monitor and report annual fluctuations. Such estimates have been made (Eliasson, 1995), using data from the annual National Forest Inventory in Sweden. Revised estimates of annual stock increment for the years 1987, 1991 and 1993 were determined to be 39, 52 and 32 million m^3 forest, respectively (Tables 18.2 and 18.3). The carbon from total stock increase in the years studied was therefore found to contain 7.6, 10.2 and 6.2 million tonnes of carbon (Table 18.3), although these estimates of annual fluctuations are uncertain.

Table 18.2. Estimates of annual stock increment, harvest and natural thinning in Sweden (from Department of Forest Survey, Swedish University of Agricultural Sciences).

	1987 (million m^3)	1990 (million m^3)	1993 (million m^3)
Stock increment (error: 0.5 ± 5%)	97.8	101.4	98.7
Harvest (error: 2.2 ± 5%)*	65.5	61.9	70.3
Natural thinning (error: 2.2 ± 5%)	4.7	2.4	5.7
Total stock increase	27.5	37.2	22.7

* Cutting data are corrected by a factor of 1.06, as stumps with diameters below 5 cm were not included.

Table 18.3. Estimates of carbon (C) content in forest biomass from annual stock increment. Other carbon sinks, such as vegetation, soil (peatland) and lakes, are omitted from estimates.

	1987 (million m^3)	1990 (million m^3)	1993 (million m^3)
Total biomass in trees (conversion factor from m^3 forest: 1.4)	38.6	52.1	31.8
Oven-dry weight (green weight to oven-dry weight: 0.4)	15.4	20.8	12.7
Carbon weight (oven-dry weight to C: 0.49)	7.6	10.2	6.2
Sequestered carbon dioxide (CO_2)	27.7	37.4	22.8

Economic Value of Carbon-stock Changes

Evaluation of carbon fixation implies that the costs of the damage by carbon emissions are estimable and relevant to present-day fluxes, as discussed in the chapters by Fankhauser and Price (Chapters 5 and 6, this volume), for example. Assuming that the carbon emissions would double from the year 1990 to 2030, following a 'business as usual' scenario (i.e. the rate of increase is unchanged), the economic costs associated with global climatic-change impacts have been variously estimated as 1.0–1.3% of GNP annually. This range of cost assessment refers to conditions in the USA, using the estimates by Cline (1992a) and Nordhaus (1992, 1993a). This would imply a damage cost of 0.02 SEK (based on $US1 = 6 SEK) year^{-1} kg^{-1} CO$_2$ (method 1 in Table 18.4), providing that the valuations above are applicable to Swedish conditions.

Table 18.4. Estimates of the value of CO$_2$ sequestration in 1987, 1990 and 1993 (total emission estimates from Statistics Sweden; see text for other sources).

	1987	1990	1993
Total CO$_2$ emissions Sweden (million t C)	67.4	59.4	60.7
Sequestered carbon (C) in relation to total emissions (%)	41	63	38
Valuation methods			
Method 1: estimated damage costs (million SEK) (0.02 SEK kg^{-1} CO$_2$)	0.55	0.75	0.45
Method 2: emission tax in Sweden (million SEK) (0.08 SEK kg^{-1} CO$_2$)	2.2	3.0	1.8
Method 3: stabilizing the emissions at current level (million SEK) (0.15 SEK kg^{-1} CO$_2$)	4.2	5.6	3.4
Method 4: stabilizing climate (million SEK) (0.32 SEK kg^{-1} CO$_2$)	8.9	12	7.3

Nordhaus has also constructed a model (dynamic integrated model of climate and the economy) (Nordhaus, 1994) for the calculation of costs associated with reaching certain environmental goals. Using such an approach, stabilization of emissions at a level corresponding to 1990 would imply a carbon tax for Sweden of 0.15 SEK kg^{-1} CO$_2$ ($US90 t^{-1}) (method 3 in Table 18.4). Alternatively, to stabilize the climate in order to limit the increase of temperature to 0.2°C per decade would require a tax of 0.32 SEK kg^{-1} CO$_2$ (method 4 in Table 18.4).

A further method of estimating the value of carbon sequestration is to use the shadow value associated with the actual carbon tax introduced in Sweden in 1991 and re-evaluated a number of times since then. The rates are now differentiated so that rates intended for households are 0.32 SEK kg^{-1} CO$_2$, while industry faces a tax rate of 0.08 SEK kg^{-1} CO$_2$. This carbon tax was introduced for political considerations and was preceded only by the assertion that economic instruments in environmental policy were preferable to parliamentary sanctions.

The theoretical basis for using the tax rate in assessing carbon emissions is the assumption that the shadow price is, in effect, a political estimate of willingness to pay (WTP). Since the tax is differentiated, the tax rate directed towards industry has been chosen as the WTP estimate. The rationale for this choice of rate is that industries are more mobile than households: when costs increase, companies are more likely to minimize their costs by relocating their activities abroad, while households are more immobile. The tax on industries is therefore assumed to be the best estimate available for WTP for avoiding CO_2 emissions in Sweden.

CAN CARBON SEQUESTRATION BE SUSTAINED?

As illustrated in the carbon-cycle diagram (Fig. 18.1), CO_2 is consumed as plants and trees produce carbohydrates. The fixed carbon is eventually released again, when insects, mites, worms, bacteria and fungi decompose the carbohydrates. There is a natural progress towards an equilibrium between the organic carbon and CO_2 in a prolonged time perspective. The amount of carbon fixed in a primeval forest ecosystem nearly equals the volumes of carbon released. It is therefore unlikely that a significant long-term increase in forest biomass storage can occur (Eriksson, 1991). Emissions from decomposition will balance out biomass increments at a time when maximum storing capacity of biomass is reached. Given these conditions, annual harvest and natural thinning through tree mortality, would equal the annual increment (total economic production). The possibilities for CO_2 sequestration would then be exhausted.

As human activity adds carbon from fossil-fuel sources into the atmosphere and stimulates the carbon cycle, there may be an equivalent increase in biomass in the long run. If it is important to limit CO_2 concentrations in the atmosphere to avoid the impacts of climate change (as discussed in the chapters by Fankhauser (Chapter 5, this volume), van Kooten and Folmer (Chapter 9, this volume) and others), then effectively there must be a corresponding increase of biomass for all releases of long-term storage from the fossil-fuel carbon reservoir.

There are two major options for increasing biomass in forests. One such option is afforestation, producing not only an accumulation of biomass in timber but also an increasing concentration of carbon in soil on afforested land over time (Alriksson and Olsson, 1995). The other option is, if forests are sufficiently immature, to restrict the volume of annual harvest to less than the growth increment, thereby increasing density. The afforestation option is limited by how much area can be converted to forestry. The density option is limited by biological factors related to climax or feasible density.

Sweden has 23 million hectares of productive forest land, where productive is defined as producing more than 1 m^3 timber ha^{-1} $year^{-1}$. The forest area has not increased more than a few per cent since 1920. Hence, the major volume increase from 1.8 billion m^3 in the early 1920s up to 2.8 billion in the early 1990s (more than 50% increase) results mainly from increased density in young forests and, to a lesser extent, a change of land use. The major part of Sweden is covered by forests (Fig. 18.3); consequently, the

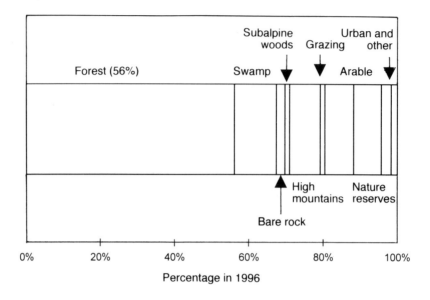

Fig. 18.3. Present land use in Sweden (from Official Statistics of Sweden and National Board of Forestry, 1996).

role of afforestation as a major potential carbon sink is of less importance in Sweden.

As implied in Fig. 18.2, the estimates of volume of sequestered carbon are based on volumes of annual increase of timber, as measured in forest cubic metres, minus decomposition, wood firing, sedimentation, buildings and possible humification or dehumification of soils. The timber stock increase (and with that the accumulation of carbon in timber biomass) has been made possible as a result of extensive cutting in the past (Östlund, 1993), as well as better methods in silviculture.

WHAT IS THE POTENTIAL FOR FOREST CARBON SEQUESTRATION?

Estimating maximum storage capacity of carbon in forests involves several considerations. Firstly, some consideration of the future demand for timber is necessary. Secondly, increased cutting volumes would slow down the annual stock increase. Another problem is how to estimate increment as growth over time, since the tree growth rate in young forests first increases and then decreases again, to a point of steady state, and in the future will be subjected both to elevated levels of CO_2 and to uncertain climatic changes.

A hypothetical summary of how biomass volume changes over time is illustrated in Fig. 18.4. The curve is drawn under the assumption that the increase in biomass has a sigmoid shape (Begon *et al.*, 1990). However, there are no survey data to support the estimates for biomass volumes before the early nineteenth century, with the best guess on prior biomass coming from the Swedish National Environment Protection Board (1992).

Alternatively, a rough approximation of maximum future biomass volumes can be made, using average site indexes. Site indexes give information on average annual growth over one rotation period under ideal conditions. The average site index in Sweden is 5.1 m^3 ha^{-1}, on 23 million ha. The rotation periods range from 80 to 145 years in the northern parts of Sweden to between 70 and 130 years in the south. Using the assumption that an average rotation period is 100 years and based on data from the National Forest Inventory, the theoretical maximum volume of Swedish forests would reach 10,000 million m^3, assuming the absence of major disturbances, such as fires and insects and other pests. A sustainable standing volume could therefore be much lower in reality, even if human management efforts were directed only at storing carbon in timber biomass.

If it is assumed that the upper limit of standing volume could reach 5,000 million m^3 and that the annual cuts would be adjusted to a stock increase amounting to a volume of 25 million m^3 annually, the storage capacity would reach a maximum within 100 years (assuming a linear trend as an approximation of the sigmoid curve in Fig. 18.4). H. Eriksson (1996, personal communication) has estimated that a maximum storage capacity in Sweden will be reached 'within decades'. If the development of events would take us to such a point, the timber volume must be maintained in order not to release its carbon content into the atmosphere, since the alternative of using the forest biomass increase as a carbon sink would then be exhausted. In summary, this discussion illustrates that there may well be an upper limit for carbon storage in forest biomass in the long run.

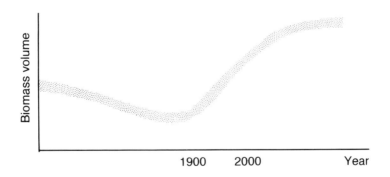

Fig. 18.4. Projections of biomass volume in Swedish forestry beyond 2000.

CONCLUSIONS

This chapter has summarized experimental sectoral environmental accounts of the net value of forests in Sweden, following the work of Hultkrantz (1992) and Eliasson (1995). Of the environmental stocks examined, the only positive change was the value of carbon sequestration, which is also most significant as the major non-timber value in the accounts. In this chapter, the role of the carbon cycle in forests in particular has been scrutinized.

Carbon sequestration, both physically and in terms of net income from the forestry sector, plays a significant role in the Swedish forest-sectoral account. This study shows values ranging from 3 to 50% of net income from forests, depending on how the utility of the carbon sink is evaluated. The lowest value is based on damage costs for a doubling of CO_2 concentrations in the atmosphere from 1990 up to the year 2030. It is argued that the carbon tax on Swedish industry is the best estimate available for the willingness to pay (WTP) for avoiding CO_2 emissions. Such an estimate of WTP amounts to around 12% of net income from forests. Carbon emissions could also be evaluated on estimates of the cost of stabilizing the increase of global mean temperature to 0.2°C per decade or on stabilizing the increase of annual emissions. The choice of such actions would impute a value of carbon sequestration in forests up to 50% of net forest income.

The question of sustainability of the sink has also been raised. On the basis of the present working of the carbon cycle, it is concluded that calculating stock increase alone as the major carbon sink is an oversimplification. Depending on the chosen time perspective, sources of carbon sinks are more or less depletable. Pulp and timber from annual cuts are continuously decomposing, except for that minor portion which is set aside, for example, in buildings and, possibly, in landfills. Carbon from anthropogenic sources is added into the cycle even if it is temporarily sequestered in biomass. There is a limit to the maximum density of biomass in forests. If such an upper limit is reached, there will remain a minor capacity for a carbon sink in peatland, sediments, maybe in stock of new products (houses) and possibly in soil. There is also an option for increasing biomass by afforestation.

The average standing-timber volume in Sweden is increasing by around 25 million m^3 year^{-1}. Such a volume increase could continue for a limited time in the future (from decades up to 100 years). There is an important implication of combustion of fossil fuels, because total carbon is added into the carbon cycle. When carbon is sequestered as biomass, the equilibrium between biomass carbon and atmospheric carbon is shifted towards the biomass. The carbon sequestered must be maintained as biomass by human efforts in order to force the equilibrium towards the biomass. If not, sequestered volumes will eventually be lost into the atmosphere by natural forces.

If increasing the standing inventory of the biomass is used as a strategy to sequester excessive CO_2 originating from the use of fossil fuels, only a temporary respite will be gained. In the long run, new technologies, both for mitigating greenhouse gas-emissions and for adapting to the climatic changes already committed to, must be developed.

19

The Potential Role of Large-scale Forestry in Argentina

Roger A. Sedjo and Eduardo Ley[1]

INTRODUCTION

In recent years, there has been increasing concern over the prospects of a general global warming. In natural systems, the transfer, storage and release of carbon among the ocean, the atmosphere and the terrestrial system take place continually. However, humans are having a significant impact on the carbon cycle through land-use changes and, even more importantly, through fossil-fuel use. The international community has begun to respond to these concerns through activities such as the creation of scientific groups, such as the Intergovernmental Panel on Climate Change (IPCC), whose remit is to examine and summarize the state of the science related with this issue. Additionally, international agreements have been forged, most notably, the Framework Convention on Climate Change (FCCC), which commits most developed countries to stabilizing their emission of greenhouse gas (GHG) by 2000 to their 1990 levels.

This chapter examines the effects on carbon sequestration of alternative afforestation strategies in Argentina. The focus is on subsidies for various types of large-scale plantation forests in different regions of Argentina and on the carbon-sequestration implications of such strategies. Projections of the expected intertemporal carbon sequestration associated with the different strategies are developed. The focus is on alternatives that include the establishment of large-scale, steady-state, short-rotation plantations in the Mesopotamia region and a lower level of longer-rotation plantations in the Patagonia region. The study examines: long-term intertemporal carbon-sequestering implications of these various approaches; the financial costs, including the cost per tonne of carbon sequestered; and the potential of these projects to substantially assist in allowing Argentina to achieve its national sequestration goals.

A critical analysis of the social and economic constraints of the various afforestation scenarios is beyond the scope of the study undertaken. However, the establishment of new plantation areas at the rate of 20,000 ha year^{-1}, as posited in the most extreme scenarios, represents only a continuation of existing trends.[2] Thus it appears to be well within the capacity of the social and economic constraints of the country.

FOREST MANAGEMENT: MITIGATION ACTIVITIES AND SINK ENHANCEMENT

Carbon is sequestered in tree biomass through the process of biological growth. Although there are some differences between species, the amount of carbon is roughly linked to the biological growth of the tree. Carbon is released when the tree decomposes or is burned. Conceptually, for forestry to be effective in mitigating global atmospheric carbon, the global stock of forests must increase. If there is a global increase in the stock of forest, this will also indicate that there is an increase in the stock of carbon embodied in the forest. Generally, carbon in forest soils will increase over time in the area that has maintained forests and will decrease in the early years after clearing.[3] If cleared forest land is rapidly reforested, the soil losses will be modest. For the global forest to increase its carbon stock, the forest stock must increase. Carbon is also held captive in long-lived stocks of forest products, such as those used in construction and furniture. Research has been done to estimate the average life cycle of various types of wood products and hence the extent to which wood products in themselves constitute a significant carbon sink.

Although it is not realistic to expect forests to offset the entire human-generated carbon build-up, forestry is a particularly attractive approach to carbon mitigation, for a number of reasons. Forests have the potential to sequester large amounts of carbon; the technology for establishing large areas of additional forests already exists and has been tried and proved; forests have a number of environmental benefits apart from carbon sequestration; and most studies indicate that the costs of carbon sequestration using forests, at least for low levels of planting in a global context, are relatively modest.

Recent studies have estimated the costs of sequestering carbon via specific forestry activities, usually in developing countries. These include not only forest plantations, but agroforestry activities and forest protection also. Some of these cost point estimates are listed in Table 19.1.

Recently, a number of more sophisticated studies for the US have begun to overcome some of the limitations of the earlier studies.[4] The modifications include: developing a cost function, rather than a simple point estimate;

Table 19.1. Estimated costs* of sequestering carbon through agroforest projects: various regions ($ t⁻¹) (from Dixon et al., 1994a).

			Regime		
			Temperate		
Source	Region agroforestry (1)	Tropical plantation (2)	Plantation (3)	Plantation (4)	Boreal protection (5)
Andrasko et al. (1991)	3–5	3–6	0–2		
Dixon et al. (1994a)	4–16	6–60	2–50	3–27	1–4
Krankina and Dixon (1994)			1–7	1–8	1–3
Houghton et al. (1991)	3–12	4–37			

* These costs are the average costs for the project, but might be viewed as the marginal costs if large-scale sequestration were being undertaken, using a host of projects.

recognizing that land has opportunity costs in the form of some type of use or rental payment; refining the tree-plantation establishment cost estimates by recognizing differences in costs associated with location and site considerations; and utilizing discounting procedures.

The usual approach is to estimate an 'annualized expenditure stream' by amortizing the initial establishment costs, using the appropriate discount rate over the relevant time period, and by combining the amortized value with current expenditures, which include a measure of the opportunity cost of carbon. A more complete review is found in Sedjo et al. (1995).

ARGENTINA: CARBON AND FORESTS[5]

Argentine net carbon emissions were estimated to be about 30 million tonnes of carbon for 1990 (World Resource Institute, 1994). This level was projected to rise by about 1% year⁻¹, or an average of 0.4 million tonnes annually, to 40 million tonnes of carbon by 2020 (SAGyP, 1994). Below, estimates are made of the long-term time profile of carbon sequestered by forestry activities and of the costs associated with the carbon sequestration.

Three major types of forest-management activities are examined, with the use of the creation of 'stylized' forest plantations. The first involves the creation of a 'steady-state' regulated[6] plantation forest in Mesopotamia with 25 age classes and a 25-year rotation. At the end of 25 years, when the forest is fully established, the growth of the forest is just equal to the 'drain', with an annual cut at rotation age, harvesting the oldest age cohort of timber. The second involves a mixed-species plantation, established also on the basis of annual planting, except that harvests are absent and growth rates are slower.

The third stylized forest plantation involves the creation of a plantation forest in Patagonia with a 35-year rotation, but in which only five age classes are created, covering a total span of only 5 years. This might occur if a programme of intensive planting was undertaken for only a 5-year period and gives a lumpy or uneven-aged stand.

In this study, attention is focused on the carbon implications of the forest activities in two future years, 2020 and 2070. Sequestered carbon is viewed as a benefit, since carbon sequestered is a proxy for climate damage foregone. However, our analysis does not attempt to value the benefits provided by carbon sequestration or to discount carbon sequestered at some future time.[7] Costs, however, are discounted. It is assumed, for some of our analysis, that these costs are covered, in part, by a subsidy.

Plantation Forests in Argentina

Most of the plantation forest has been established in eastern Argentina, largely in Mesopotamia, although some plantation forests exist to some extent in most regions. As shown in Table 19.2, large land areas have been judged as suitable for the establishment of plantation forests. Plantations have been established, using exotic, non-native species, particularly pine indigenous to North and Central America and eucalyptus indigenous to Australia. Some poplar species are also planted. Very few native species have been established in plantations, with the exception of araucaria, the plantations of which are limited to specialized site conditions and found almost exclusively in Misiones.

Table 19.2. Land suitable for plantation forests in Argentina (from SAGyP, 1994).

Region	Total area (ha)	Suitable land (ha)	Protected area (ha)
Mesopotamia Media	21,150,000	4,550,000	1,258,743
Buenos Aires	36,150,000	8,825,000	66,706
Patagonia Andina	23,125,000	3,934,000	1,467,111
Total	80,425,000	17,309,000	2,792,560

The Scenarios

In the following, we examine the effects of a subsidy on various types of forest and on their ability to sequester carbon. The alternatives are: industrial plantations in Mesopotamia and Patagonia; mixed-species plantations with no harvests; and plantations in Patagonia.

Industrial plantations
Forest plantations have been seen to offer a substantial potential to sequester carbon. In the following section, we examine the long-term carbon and cost implications of alternative forest plantation and management regimes. These include: (i) the establishment of an even-aged, regulated forest, in which each year is represented by an identical area of forest of that age class; (ii) the establishment of a mixed forest that is not harvested; and (iii) the establishment of a forest, over a 5-year period, with equal areas of forest established in each of 5 consecutive years; this reflects a 5-year grant of a plantation-establishment subsidy.

All of the above result in the creation of new forest stock that sequesters carbon. In (i), the stock is a steady-state, regulated forest, which will maintain a constant size and volume of carbon, as well as timber, if it is harvested only at the given rotation age. In (ii), there is a forest that reaches and is maintained at maturity. In (iii), the establishment conditions did not allow for the creation of a steady-state, regulated forest. Thus, the volume of timber in the inventory and the rotation-harvest level will not remain constant, but will rise and decline predictably.

In addition to the carbon stocks associated with the forest inventory, there is also a carbon stock associated with the inventory or stock of long-lived wood products for (i) and (iii). Thus, if a certain portion of the industrial wood harvested goes in the long-lived wood products, such as building materials or furniture, a stock of products is created which has carbon-sequestration implications of its own.

Carbon implications of an even-aged, steady-state forest
Below, we examine the carbon-sequestration implications of the continuation of that planting rate for a period of 25 years, until the year 2020, at which time a 'stylized' fully regulated, even-aged, steady-state forest has been created. Figure 19.1(a) presents a stylized assessment of the results of a continuation of the current plantation-establishment activity in Mesopotamia for 25 years until the year 2020. In the figure, we assume that 20,000 ha of new plantation is established each year for 25 years, for a total of 500,000 ha in the year 2020. At that point, the first harvest is made (thus there is a rotation of 25 years). It is assumed that each harvest is followed by the replanting of the area just harvested. In this context, the result of this management scheme will be what foresters call an even-aged, regulated forest, where annual harvest equals annual growth. Since the forest stock is stable, the carbon sequestered within the forest will be essentially stable.

The net growth will now be harvested annually and wood products produced. Assuming that one-third of the harvested volume goes into long-lived wood products with a straight-line decay function, a stock of carbon will accumulate in long-lived wood products (Fig. 19.1b). Given the decay function, this stock will move to a steady-state maximum in the year 2070,

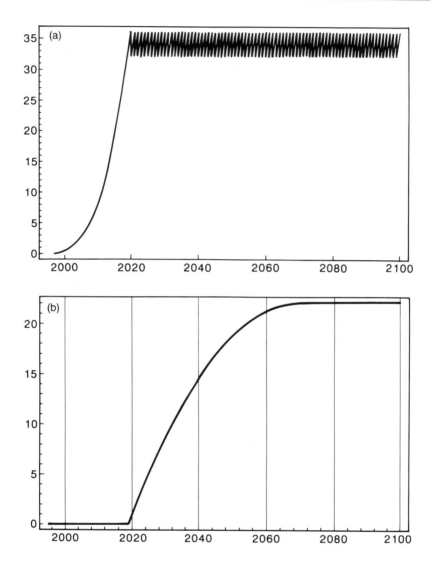

Fig. 19.1. (a) Forest carbon stock in industrial wood plantations in Mesopotamia, Argentina. (b) Carbon in wood products in Mesopotamia, Argentina.

after which time it will stabilize, as will its carbon stock. Thus, in the long term, two steady-state carbon stocks are created (Fig. 19.2). This first is the carbon captured in the forest, which reaches its steady-state maximum in 2020 and maintains that carbon stock indefinitely thereafter. The second is the carbon in the steady-state stock of long-lived wood products, which reaches its steady state in the year 2070. After 2070, the stock of carbon in the forest and products system remains stable.

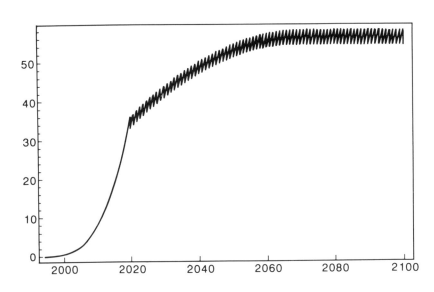

Fig. 19.2. Total carbon in forest stock and wood products in Mesopotamia, Argentina.

Carbon-cost implications of a plantation-subsidy policy in Mesopotamia
This section examines the cost implications of a continuation of the stylized analysis of the current rate of plantation establishment, as discussed above. For each year in which 20,000 ha of new plantation is established, that forest can be expected to generate an average of 104,000 t of carbon annually in the industrial wood (20 m^3 ha^{-1} year^{-1} × 0.26 t m^{-3} × 20,000 ha) for 25 years or accumulating about 2.6 million tonnes of carbon over the 25-year growth period before harvest. If current plantation-establishment rates continue for the 25 years until 2020, the cumulative total for the 25-year period would be 500,000 ha of new forest area capturing 27.5 million tonnes of carbon in the industrial wood on the stand (note that what has been created is an even-aged forest, with all but one of the age cohorts less than 25 years old) and roughly 0.4 million tonnes more in the branches, roots, litter and soil carbon, for a total of about 36 million tonnes of carbon. This represents the carbon capture in the steady-state, even-aged forest established over the period 1995–2020. If we now assume that the new planting would cease after 2020, but that the newly established forest would be annually harvested, in accordance with the 25-year rotations and replanted, the forest would create a steady-state sink of about 36 million tonnes of carbon by 2020, i.e. the forest would be stabilized at a size that would maintain the carbon stock of 36 million tonnes continuously into the future. Note that the carbon sequestration in this forest is roughly equal to the annual projected net carbon emissions of Argentina in 2020 (SAGyP, 1994).

If the establishment costs have a present value of $US1,000 ha^{-1} in the initial year, the total cost for the 20,000 ha established each year would be $US20 million. Using a 10% discount rate and assuming that 20,000 ha would be replanted indefinitely, the present value (PV) of the 500,000 ha of plantations in year 1995 would be about $US200 million. Thus, for $US200 million a total of 42 million tonnes of carbon would be sequestered by the year 2020 indefinitely (for ever). The average cost per tonne of carbon sequestered for ever would be about $US5.56.

However, as noted above, such a forest could produce roughly 10 million m^3 of industrial wood annually from incremental growth (thus not reducing its total stock of timber or carbon) and thereby create an additional carbon sink. Assuming that one-third of the industrial wood harvested from the Argentina forest was used in the long-lived wood products, 866,667 t of additional carbon would be sequestered annually (i.e. one-third of 2.6 million tonnes), once harvesting from the new plantations had begun (in 2020). Assuming a straight-line decay function, whereby all of the carbon in the solid wood would be released after 50 years (the wood would all be destroyed), a new wood-products carbon sink would have been created, which would reach a steady-state carbon stock of about 21.5 million tonnes after 50 years. By the year 2070, this steady-state forest and the steady-state wood-products stock would reach their joint steady-state maximum, in which the system would be sequestrating 58.12 million tonnes of carbon indefinitely. This would have been 'purchased' for a present value of $US200 million in 1995 or for an average total cost of $US3.44 per tonne of sequestered carbon.

It will be noted that the calculation of the cost of the carbon includes the full costs of the plantation establishment but makes no provision for the value of the wood products produced. Thus, the full costs are borne by the carbon benefits. In fact, of course, net costs associated with the carbon benefits should be calculated by adjusting the gross costs for the non-carbon benefits of the industrial wood produced.

To do this, let us assume that the initial expenditure of $US1,000 ha^{-1} can be divided into two components: the cost incurred by the private forest owner and the cost incurred by the subsidy. Assuming that the $US1,000 cost is approximately equally divided between the two (as is currently the situation in Argentina) and treating the subsidy as promoting carbon sequestration in the pursuit of its international obligations, the marginal cost of sequestering carbon is reduced by half, since the private expenditures of $US500 ha^{-1} were provided only to produce industrial wood. In this case, we argue that the incremental cost necessary to establish the steady-state forest that permanently sequesters 36 million tonnes of carbon in 2020 does so for a total present value of $US100 million or only $US2.78 per tonne of carbon sequestered. When the stock of wood products reaches a steady state in 2070, the total stock of sequestered carbon – 58.12 million tonnes – has been achieved at an average net cost of only about $US1.72 per tonne.

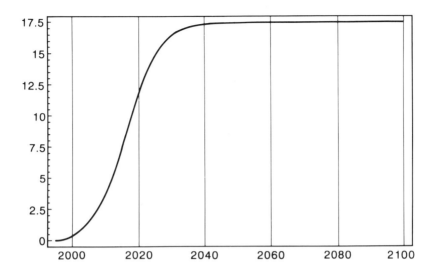

Fig. 19.3. Carbon from mixed-species plantations: Mesopotamia, Argentina, under no harvest regime and indefinite growth.

Mixed-species plantations
Data on mixed-species costs and growth in Argentina are not well developed. As noted, these activities are still in the experimental stage. Nevertheless, it is probable that the costs will be somewhat higher and the biomass growth somewhat more modest. A prototype projection is provided in Fig. 19.3, using the same area of forest, i.e. planting 20,000 ha year^{-1}, but only for 5 years, with less rapid growth and eliminating the harvest. In this scenario, we reduce the average growth rate to 8 m^3 ha^{-1} year^{-1}, with the forest reaching maturity (approximately zero net growth) at 50 years.

Our scenario assumes that 20,000 ha are planted each year for 5 years, at a cost of $US1,000 ha^{-1}, beginning in 1995. In the year 2020, there would be 18 million m^3 of industrial wood equivalent, or 6.55 million tonnes of carbon, when adjustment is made for branches, roots, etc. By the year 2070, the forest would approximate maturity, i.e. no net growth, with a stock of carbon of about 14 million tonnes. The present value of the cost would be roughly $US90 million, for an average cost of about $US6.50 per tonne of carbon sequestered.

In the second mixed-species scenario, we assume an average growth rate of 5 m^3 ha^{-1} year^{-1} in Patagonia and allow the plantation to grow to biological maturity in 70 years, after which it maintains its volume indefinitely (Fig. 19.4). Using the above parameters, the planting of a mixed-species forest in Patagonia would be generated only as a result of a subsidy focused on the planting of mixed species. Our scenario assumes that 20,000 ha are planted each year for 5 years at a cost of $US1,000 ha^{-1}, beginning in 1995.

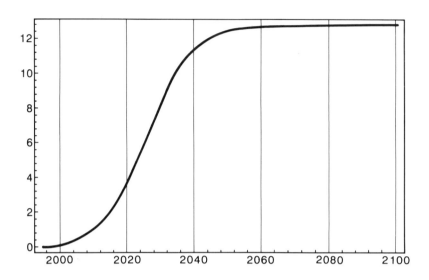

Fig. 19.4. Mixed-species plantation in Patagonia, Argentina, under no harvest regime.

In the year 2020, there would be 11.25 million m³ of industrial wood, or 4.1 million tonnes of carbon, when adjustment is made for branches, roots, etc. By the year 2070, the forest would approximate maturity, i.e. no net growth, with a stock of carbon of about 12.7 million tonnes. The present value of the cost would be roughly $US90 million, for an average cost of about $US7.8 per tonne of carbon sequestered.

Carbon and cost implications of a plantation subsidy over a 5-year period
It may be the case that a large subsidy might be available for a relatively short period of time, e.g. if an international organization, such as the Global Environmental Facility (GEF), undertakes the subsidy of plantation establishment for purposes of sequestering carbon. Suppose that a programme promotes tree planting for a limited period – perhaps 5 years – and that a subsidy of $US20 million year^{-1} is provided for 5 years to establish pine plantations in less attractive tree-growing areas, as in Patagonia. Assume that the subsidy required was $US1,000 ha^{-1} for successful plantation establishment. If it were established over the period 1995–2000, with a growth average of 13 m³ ha^{-1} year^{-1} for 25 years, by 2020 there would be 29.25 million m³ or 7.6 million tonnes of carbon in the industrial wood or 10.6 million tonnes in the forest system.

As shown in Fig. 19.5(a), if the forest with only five age classes were harvested with a 35-year rotation and immediately replanted, the highly skewed age distribution would result in sharp fluctuations in the intertemporal levels of carbon storage. The carbon sequestered would rise over the

35-year growth period, but decline dramatically over the 5-year harvest period, only to rise again upon replanting and to decline once again at the subsequent harvest. However, as in our earlier example, when a substantial portion of the industrial wood goes into long-lived wood products, a second stock of captive carbon will be created. The skewed harvest cycle will skew the age of the stock of wood products, as shown in Fig. 19.5(b). Note that, although the stock of captive carbon exhibits fluctuations, there is a 'floor'

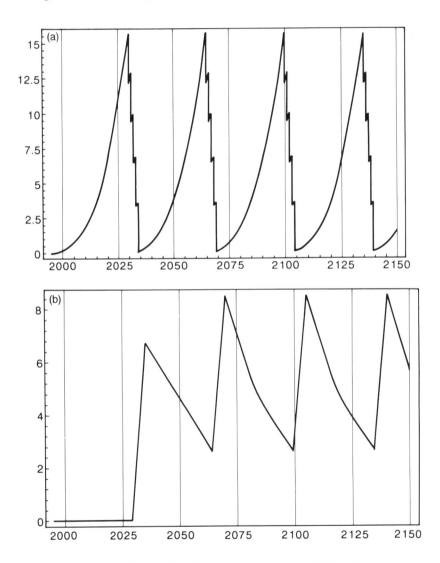

Fig. 19.5. (a) Forest carbon stock in Patagonia, Argentina. (b) Wood products carbon in Patagonia, Argentina.

level below which it does not fall. The total stocks of sequestered carbon are represented in Fig. 19.6, which combines both the carbon sequestered in the forest and that sequestered in the stock of wood products.

Thus, the average costs of carbon sequestration would depend upon the point in time that was chosen to estimate costs. Suppose that the carbon credit were based upon an estimate of the long-term average annual carbon held in sequestration. Thirty-five years after the beginning of the 5-year period of forest establishment, the forest would have sequestered about 156 t ha^{-1} or 15.6 million tonnes on the 100,000 ha, given that four of the age cohorts are slightly less than 35 years old. This volume is roughly consistent with an annual average volume of about 7.8 million tonnes, which is an approximate mean average on the site for all years. The PV of the establishment costs would be about $US70 million. Thus, the average cost per tonne of carbon based on the average carbon captured on the site is about $US8.97.

If 50% of the wood harvested made its way into long-lived products, there would be 67 t of carbon captured per hectare of harvest, or about 6.7 million tonnes. After 35 years, this is reduced to about 2.01 million tonnes, after which new harvests occur and the total rises to 8.52 million tonnes. Thereafter, the sequestered stock of carbon cycles, first declining over 35 years to 2.01 million tonnes, only to rise abruptly to 8.52 million tonnes, after which it then declines back to 1.74 million tonnes, and so forth. If we use an average of 5.27 million tonnes, the average through time of both sinks becomes 13.07 million tonnes, at an average cost of $US5.36 per tonne of carbon sequestered.

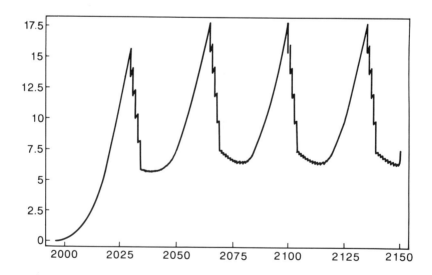

Fig. 19.6. Total carbon in forests in Patagonia, Argentina.

In isolation, one might question which level of sequestered carbon is appropriate for analysis, although even in isolation the way to treat the intertemporal volatility is not clear. However, if there are a number of similar projects under way in various countries, one might expect that, when Argentina's carbon stock is low, some other country may have a high carbon stock.

SUMMARY OF CARBON SEQUESTERED AND COSTS

Table 19.3 summarizes the findings of the scenarios discussed above. By far the greatest single new carbon sink would be created by establishing the steady-state forest through continuation of the current planting level of 20,000 ha annually in new industrial plantations until the year 2020. Additionally, this is also the low-cost scenario: because it involves planting in a high-growth area; because many of the costs are borne by the industrial timber grower; and because the maintenance costs are borne by the industrial grower. Movement to other regions or mixed species suffers from both lower growth and the need to subsidize all of the costs. Mixed species on high-quality lands in Mesopotamia also involve land opportunity costs, which are not captured in the numerical estimates. The analysis demonstrates that afforestation can play a significant role in GHG mitigation in Argentina.

Table 19.3. Summary of carbon sequestered and cost scenarios.

	Carbon 2020 (million tonnes)	Cost per tonne ($US)	Carbon 2070 (million tonnes)	Cost per tonne ($US)
Mesopotamia				
Industrial plantation (25 years, 20,000 ha year^{-1}, harvesting)	36	5.56 (2.78)*		
Industrial plantation and wood stocks (25 years, 20,000 ha year^{-1}, harvesting)			58	3.44 (1.72)
Mixed-species plantation (25 years, 20,000 ha year^{-1}, no harvesting)	6.55	13.74[†]	14	6.50[†]
Patagonia				
Industrial (5 years, 20,000 ha year^{-1}, harvesting)	10.6	7.6	13.07[‡]	5.36
Mixed species (5 years, 20,000 ha year^{-1}, no harvesting)	4.1	21.95	12.70	7.87

* Half of the costs are attributed to industrial wood production.
[†] Does not include the opportunity costs of higher-quality lands.
[‡] Long-term average.

NOTES

1. This chapter draws on a larger report, 'Argentina: Carbon and Forests,' 4 December 1995, prepared by the authors for the Global Environmental Facility (GEF) and the World Bank (WB). Although that study and this chapter benefit from comments from the staff of the GEF and WB, this chapter is the work of the authors and does not necessarily represent the view of the GEF or the WB. Any errors that remain are the responsibility of the authors.
2. A number of countries, including Chile, Brazil and New Zealand, have experienced forest-plantation establishment well in excess of 20,000 ha annually, often in excess of 100,000 ha, without experiencing any apparent serious social or economic difficulties.
3. It is recognized that both grassland soils and forest soils are capable of sequestrating large amounts of carbon. This study assumes that the soil carbon sequestered by grass and forests will be approximately the same over the long term. The actual empirical situation is scientifically an open question. To the extent that this assumption is incorrect, the estimates of the net carbon sequestered through plantation forests are biased.
4. Examples of more sophisticated approaches include Moulton and Richards (1990) and Adams *et al.* (1993).
5. Much of the discussion and results of this section are drawn from the report prepared for the GEF and the WB.
6. The forester's 'regulated' forest refers to a forest with the distribution of timber by age, such that each year is equally represented from year 1 to the nth year, that is, the number of years in the rotation. The harvest at rotation age will then generate an even-flow harvest, where growth equals harvest. Thus, where land productivity is identical across the forest, the forest area would be divided equally into the same number of areas as there were years in the rotation. Each forest area would then have a different age of timber from year 1 to the rotation age. In these circumstances, the harvest of the oldest cohort at rotation age would equal the net annual growth of the timber in the forest, thus providing a 'steady-state' harvest.
7. The issue of whether sequestered carbon should be discounted is more complicated than it might at first appear, due to the surrogate nature of carbon. Only if there were a one-to-one mapping of sequestered carbon to damage avoided would discounting be appropriate. For a full discussion, see van Kooten *et al.* (1997).

Forestry and Agroforestry Land-use Systems for Carbon Mitigation: an Analysis in Chiapas, Mexico

20

Ben H.J. De Jong, Lorena Soto-Pinto,
Guillermo Montoya-Gómez, Kristen Nelson,
John Taylor and Richard Tipper[1]

INTRODUCTION

The potential role of forests in carbon sequestration has been evaluated by a number of authors (Marland, 1988; Sedjo and Solomon, 1989; Andrasko *et al.*, 1991; Houghton *et al.*, 1991; see also Ciesla, Chapter 3, this volume). They suggest that forest conservation, establishment and management, as well as agroforestry, could contribute to global carbon mitigation. Forestry and agroforestry compensate for greenhouse-gas emissions in two ways: (i) creating new sinks for carbon dioxide (CO_2) by increasing the mass of woody material within growing trees and in harvested timber converted to durable products; and (ii) protecting natural forests and soils as carbon stores.

Strategies for carbon sequestration or conservation in the forestry and agroforestry sectors include: the establishment of permanent agroforestry plots to substitute slash-and-burn agriculture; and the conservation of standing old-growth forests as carbon sinks. In addition, carbon sequestration can be enhanced through increased harvesting efficiency in forests and utilizing a higher percentage of total biomass; improving forest productivity on existing forest lands through management and genetic manipulation; establishing plantations on surplus crop land and pastures; restoring degraded forest ecosystems through natural regeneration and enrichment planting; establishing plantations and agroforestry projects with fast-growing and high-biomass species on short rotations for biomass and timber; and increasing soil carbon by leaving dead wood, litter and slash from harvest (Andrasko, 1990).

Estimates of the global biological and economic potential of forest-management practices for controlling atmospheric CO_2 concentrations are

© CAB INTERNATIONAL 1997. *Climate-change Mitigation and European Land-use Policies* (eds W.N. Adger, D. Pettenella and M. Whitby)

highly speculative. Dixon *et al.* (1993a) consider that economically viable forestry and agroforestry management practices could roughly conserve and sequester 1 billion tonnes of carbon (t C) per year globally. They estimate the marginal cost of implementing these options at \$US10 for each tonne. Trexler (1991) identifies a range of forestry policy options for US forests, which could make emission savings of 0.75–1.15 Gt C year^{-1}, at a marginal cost of \$US30–50 t^{-1}. A number of private projects, sponsored by power generators, have claimed sequestration costs of less than \$US5 t^{-1} (FACE, 1994), with some less than \$US1 t^{-1} (Swisher, 1991).

These carbon-storage technologies can potentially be achieved at relatively low costs in most tropical countries, because the two principal inputs, land and labour, can, in some localities, be relatively inexpensive. With a greater understanding of the local social context of an (agro)forestry project, it appears possible to develop projects of CO_2 mitigation and simultaneously help finance promising sustainable development programmes (Swisher and Masters, 1992). Accordingly, Gay *et al.* (1995) state that efforts should only be made to implement options that are inexpensive or economically profitable; consistent with social and cultural factors; provide tangible benefits at the local level; and provide environmental benefits in addition to greenhouse-gas mitigation. Typically, these requirements need the identification of options that can be implemented on different scales, for various orientations of output (commercial, subsistence), and with a range of beneficiaries (rural communities, private sector, governments).

Mexico ranks among the top ten countries in carbon releases from forest clearing during recent decades (Detwiler and Hall, 1988). Carbon-sequestration scenarios developed by Masera and colleagues (1995) suggest that Mexico can significantly reduce net national carbon emissions to the atmosphere, through forest management of native temperate and tropical forests and reforestation of degraded areas. Masera *et al.* (1995) outline seven forest-mitigation options for Mexico: four options related to increasing forested areas (restoration, pulpwood and energy plantations and agroforestry systems) and three related to the conservation of existing forests and resources (conservation of natural protected areas, management of native forests and dissemination of improved wood-burning cooking stoves). Apart from the fact that planting trees to sequester carbon is a prominent strategy for mitigating greenhouse-gas emissions (Adams *et al.*, 1993; De Jong *et al.*, 1995), it can also create a low-cost source of biomass and an alternative source of fuel and wood-based material for construction and other uses. The previously mentioned studies results suggest that a modest carbon-sequestration programme can achieve reasonable goals, without major negative impacts on agriculture and commodity production.

Based on the 1992 land reform, Mexico has sought to implement a progressive forest policy, particularly in tropical areas, which contain the most important forest extensions of the country. This policy has been underlined by a number of activities: the development of a national Tropical Forestry Action Plan (SARH, 1994); the establishment of the headquarters of the Forest Stewardship Council in Oaxaca, southern Mexico; the elimination of subsidies and credit for extensive cattle ranching in the tropical states; and government support to various community-controlled forest-management projects (examples are Plan Piloto Forestal Quintana Roo, Plan Piloto Forestal Marques de Comillas and Selva Lacandona).

The 1992 change in the Mexican Land Tenure Law (Article 27) gives legal title to the rural communities for the land they manage as an *ejido* or community. *Ejido* is a term used for a productive grouping of people with land, given in common ownership after the 1917 revolution. The 1992 changes in the Land Tenure Law allow rural farmers legal status to establish joint ventures with investors, so that capital can be invested in alternative land-use systems. To ensure mutually profitable and just business partnerships, farmers' organizations will need to play an important role in the negotiations between the farmers they represent and interested investors.

Mexico has also played an important role in the development of joint initiatives for greenhouse-gas emission reductions, particularly in conjunction with the US Initiative on Joint Implementation (USIJI). Joint implementation (JI) refers to the idea that greenhouse-gas emissions reductions or offsets in one country may counterbalance emissions generated in other countries, in order to achieve reductions agreed to under the United Nations Framework Convention on Climate Change. Legal provisions to allow co-funding of emission-mitigation projects in Mexico are in place. However, discussions relating to the distribution of the 'carbon credits' from JI projects are still ongoing (SEMARNAP, 1996).

The Commission for Environment Cooperation, formed by the environmental side-agreement to the North American Free Trade Agreement (NAFTA), is supporting greenhouse-gas mitigation projects in Mexico. Current free-trade negotiations between Mexico and the European Union could possibly enhance the Mexican JI potential. Certain countries, such as Germany, Austria and the Netherlands, which are having difficulties approaching their emission targets, appear to be potential partners for a JI programme with Mexico.

The Mexican government is very keen to identify potential projects that are acceptable for rural communities and ecologically and economically viable, in order to incorporate them into a JI programme. To this end, a case-study has been carried out to identify farmer-preferred and ecologically viable forestry and agroforestry systems and their *ex ante* carbon-sequestration potential and to assess the economic potential of carbon offsets of such systems.

Using participatory methods, current agroforestry systems have been analysed and potential system improvements and new alternative systems proposed, discussed and evaluated, based on social, economic, technical and carbon-sequestration criteria. Selected systems were used for carbon-offset calculations. In this chapter, total accumulated carbon for each system and region is presented. The costs of carbon sequestration for each system are estimated, based on the discounted direct costs of improving the current systems or establishing the new systems and the discounted opportunity costs during the first rotation for those systems where land use is diverted from agriculture or animal husbandry to forestry. Costs are presented on a per-hectare basis. In addition, we discuss the importance of community organization for an agroforestry project.

We point out the advantages and disadvantages of available carbon-sequestration models tested in this study for the calculation of the potential carbon mitigation of the selected systems. Combining systems that provide commercial products and ecological services, such as carbon mitigation, could generate the necessary capital for the long-term investment in farm-forestry projects which has always formed a constraint on such activities.

THE STUDY AREAS

The state of Chiapas is located in southern Mexico. Tropical rain forests, pine–oak forests and montane rain forests are among the important vegetation types. The majority of the rural communities consist of subsistence or semisubsistence farmers, who rely heavily on forest resources for fuel wood, construction materials and food supplements.

The study was carried out in two ecological zones, one inhabited by Maya-Tojolabal (Tojolabal region) and the other by Maya-Tzeltal (Tzeltal region) indigenous groups. These particular zones were selected by the farmers' organization Union de Credito Pajal Ya Kac'tic (Pajal), based on the biological potential of the regions for agroforestry and the interest demonstrated by their members in these zones. Five communities from each region were invited to participate in the feasibility study. All but one maintained their interest throughout the project. All the participating communities retain legal land-tenancy titles as *ejidos* (for further details, see De Jong *et al.*, 1995; Montoya *et al.*, 1995).

The climate of the Tojolabal region is subtropical subhumid (García, 1973), with a mean annual temperature of 18°C and 1,030 mm of annual rainfall. The communities lie at an average altitude of 1,600 m above sea-level. Within these communities are well-preserved extensions of pine–oak forest. The five Tojolabal communities – Juznajab, Yaluma, Lomantan, Bajucu and Palma Real – are situated in the municipalities of Comitán and Las Margaritas. Each community is an *ejido*. The total area of the five communities is 9,281 ha, with 5,336 habitants, of which 439 are members of Pajal.

The climate of the Tzeltal region is tropical subhumid (García, 1973) with a mean annual temperature of 24°C and 1,800 mm of annual rainfall.

The communities lie at an average altitude of 800 m above sea-level. There remain in the communities only fragments of primary evergreen rain forest and large extensions of shaded coffee. The five Tzeltal communities – Chapullil, Segundo Cololteel, Alan Cantajal, Muquenal and Jol-Cacualha – are situated in the municipality of Chilón. Participants have shown great interest in improving their land management with agroforestry systems. The total area of the five communities is 2,387 ha, with 907 habitants, of which 170 are members of Pajal.

METHODS

The suitability of agroforestry for carbon sequestration requires insights into the institutions involved at the implementation scale. The study therefore uses participatory methods to identify constraints to sustainability in the case-study areas. To begin the project, a multidisciplinary team of scientists and farmers was established. The Pajal members of each community appointed two delegates to represent them during the study. These two representatives gathered information and designed the community agroforestry/forestry options together with the members in the communities, while receiving technical assistance from the scientists of El Colegio de la Frontera Sur (ECOSUR) and Pajal.

The design and evaluation of the options were carried out during two workshops, in which all of the representatives and scientific advisers participated. During the first workshop, the agroforest-management alternatives were discussed. This was also the first time that the concept of carbon sequestration was presented as a potential component of a forestry project. During the sessions, ECOSUR scientists and Pajal technicians explained the technical and social implications of agroforestry alternatives, while the community representatives supplied detailed descriptions of the land-use systems in their communities, the common tree species of the regions and how they manage these species. They explained their major land-use constraints and which of the farm-forestry and agroforestry alternatives they considered attractive.

Between the first and second workshop, the Pajal representatives collected data on: (i) community members' interest in a farm-forestry and agroforestry project; (ii) how much land they would have available for each activity; and (iii) the system(s) they would prefer, including species selection and tree distribution within the system. Data were also collected on the current social organizations, the land and tree tenure, benefit distributions, current community norms for forest use and possible target groups.

During the second workshop, the final community proposals were developed and evaluated, incorporating the results of the first workshop, the visits and data collected by the representatives. The following variables were included in the system analysis: (i) area available for farm forestry; (ii) potential species or species combinations; (iii) the current land-use system in which they would plant the trees; (iv) planting design; and (v) how to obtain the planting material.

The simulation model CO_2FIX described by Mohren and Klein-Goldewijk (1990) was used for the calculations of the carbon fluxes for each system. The model describes the carbon cycle from annual growth and loss rates of the main biomass compartments of the tree component of the systems, in combination with accumulation and turnover of soil organic matter. In this simulation programme, a carbon accounting procedure is used to drive carbon accumulation in the entire biomass through proportionality coefficients derived from biomass measurements. Adjustments can be made for the rotation age, thinning procedures, product use and other silvicultural factors, etc., according to the species and silvicultural system (Nabuurs and Mohren, 1993). Existing local field data and published data on growth rates, decomposition rates and amounts of biomass in the various land-use types were used in the model. Calculations of the growth rates are based on the site-quality indices developed for the regions. Mean annual increment (MAI) and total height of mature trees were used as the parameters to define the quality index. The MAI in the Tojolabal region was calculated, taking the average of the yearly increase of *Pinus* trees under different growing conditions. In the Tzeltal region, where trees in general do not present yearly growth rings, the MAI was determined by using the average diameter at breast height of the ten largest trees in forest fallows of known age and dividing this average by the age of the fallow. In Table 20.1, we present the parameters used for this simulation (for a detailed description and application of the model, see Mohren and Klein-Goldewijk, 1990; Nabuurs and Mohren, 1995). The net accumulated amount of carbon over the first six cycles for the Tzeltal region (25 years for each cycle) and five cycles for the Tojolabal region (30 years for each cycle) was estimated. Total net carbon accumulation was thus calculated for a period of 150 years for each region and system. To account for local variation in soil fertility and varying management practices, we included a 20% variation in estimated annual increment (m^3 ha^{-1} $year^{-1}$). To use the model, it was assumed that: (i) the incorporation of the tree component in the system does not affect the total carbon dynamics of the rest of the system; and (ii) the site quality remains unaltered during the cycles.

The costs of carbon sequestration (t^{-1} C ha^{-1} sequestered) for each system were calculated by summing the costs of: (i) the establishment and maintenance of the tree component within the system; and (ii) opportunity costs of lost benefits from the alternative system. The cost factors include labour, equipment, site preparation, site protection and planting stock for the first rotation of the trees. Costs for additional rotations are expected to be covered from the sale of the tree products of the first rotation. Labour costs are based on local minimum wages. The costs are represented on a

Table 20.1. Parameters for the CO$_2$FIX simulation model to estimate carbon fluxes in potential agroforestry systems in Chiapas, Mexico.

Parameters	Tzeltal			Tojolabal		
Rotations (years)	6 × 25			5 × 30		
Initial humus content of the soil (t C ha^{-1})	75			75		
Basic wood density (kg m^{-3})	500			450		
Carbon content (% of dry weight)	50			50		
	Years after planting			Years after planting		
	0–10	10–20	20–25	0–10	10–20	20–30
Dry-weight increment relative to stem increment during one rotation						
Foliage	0.7	0.4	0.4	0.8	0.6	0.2
Branches	0.6	0.4	0.4	0.8	0.5	0.2
Roots	0.7	0.4	0.4	0.9	0.6	0.3
Turnover rates						
Foliage	0.5			0.3		
Branches	0.05			0.05		
Roots	0.07			0.07		
Humification factor	0.1			0.05		
Litter residence time (years)	13					
Stable-humus residence time (years)	100			200		
Product allocation for thinnings						
First thinning after 8 years		60% fuel wood				
		40% dead wood				
Second thinning after 16 years		40% fuel wood				
		40% fence poles				
		20% dead wood				
Expected lifetime of products (years)						
Fuel wood		1				
Fence poles		5				
Dead wood		10				
Timber		35				

per-hectare basis, as are the carbon-sequestration potentials. Interest rates used to calculate costs to the present date were 5 and 10%. Dividing the total costs by the net carbon-sequestration potential during the 150-year cycle determined the final cost of 1 t C sequestered in each selected system.

RESULTS

Land Available for a Carbon-sequestration Project

As a result of the workshops and project presentations in the communities, a total of 171.5 ha are made available for the first year of a carbon-sequestration project: coffee plantations (46.5 ha), fallow (32 ha), maize fields (8 ha), forest (5 ha) and pasture (2 ha) for the Tzeltal region, and fallow (35.5 ha), forest (25 ha), maize fields (12.5 ha) and pasture (5 ha) for the Tojolabal region. The communities possess a considerable amount of communal forest areas; however, management decisions for these areas have to be taken by the whole community, which requires general community assemblies. At this stage of the project, the members of the Pajal are undecided concerning the possible management alternatives for the communal forests and therefore prefer to start a project on an individual basis. Therefore, we have omitted the analysis of the carbon-sequestration potential of forest management. However, it is expected that, once the project has started, the rest of the communities will gain interest and will discuss the possibilities of sustainable management of the communal forests.

Selected Agroforestry Systems

As a result of the workshops and field data collection, in which alternatives of forestry and agroforestry practices were discussed with the community members, a series of systems were evaluated. A total of five systems were considered viable, with local adjustments to species selection, planting arrangements and rotation. Table 20.2 shows the systems for each region that were considered technically viable, socially acceptable and economically feasible, including the planting arrangement of the tree component, the species to be used in the systems, the most important silvicultural treatments after planting and before harvesting and the expected wood production of each system, according to three production levels: level I for poor sites, level II for medium-fertile sites and level III for fertile sites. All systems require regular weed control and replacement of dead seedlings in the first year.

Carbon-sequestration Potential of the Selected Systems

The selected systems are used for carbon-offset calculations. Figure 20.1 illustrates the current annual increments used for each system and region during one rotation. The estimations are based on analysis of field data on MAI and maximum heights measured in each region. The carbon fluxes for each system and region are calculated for a 150-year period to overcome the

Table 20.2. Forestry and agroforestry systems considered viable in Chiapas, Mexico.

Tree and treatment	Planting distance (m)	Production level (m³ ha⁻¹ year⁻¹)		
		I	II	III
Tzeltal				
Live fence *Cedrela odorata*	3	4.8	6.0	7.2
Coffee with *Cedrela odorata* or *Cordia alliodora* as shade trees	10 × 10	6.0	7.5	9.0
Taungya with *Cedrela odorata*: thinnings after 8 and 16 years (25% of total stand)	10 × 3	11.9	14.9	17.9
Enriched fallow with *Cedrela odorata*, *Cordia alliodora* or *Calophyllum brasiliense*: thinnings after 8 and 16 years (25% of total stand)	10 × 2	11.9	14.9	17.9
Tojolabal				
Live fence *Pinus oocarpa*, *P. michoacana* or *Cypressus* sp.	3	3.2	4.0	4.8
Plantation of *Pinus oocarpa*, *P. michoacana* or *Cypressus* sp.	2 × 3	8.0	10.0	12.0
Taungya with *Pinus oocarpa*, *P. michoacana* or *Cypressus* sp.: thinnings after 8 and 16 years (25% of total stand)	4 × 4	8.0	10.0	12.0
Enriched fallow with *Pinus oocarpa*, *P. michoacana* or *Cypressus* sp: thinnings after 8 and 16 years (25% of total stand)	7 × 2	8.0	10.0	12.0

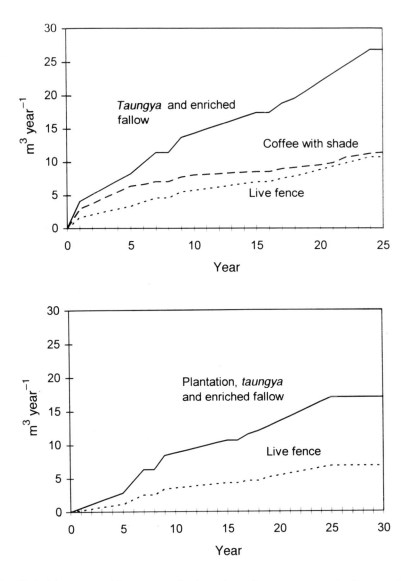

Fig. 20.1. Mean annual increment of selected agroforestry systems for first rotation in (a) Tzeltal and (b) Tojolabal regions.

tendency to overestimate carbon stocks in short-rotation forest systems (Nabuurs and Mohren, 1995).

Table 20.3 presents the total accumulated carbon during the 150-year cycle and the average of the net annual carbon flux within each system during the first rotation, separated for the three production levels. Carbon-sequestration potential varies highly between the systems and regions

Table 20.3. Net accumulated carbon (t C ha^{-1}), including product decomposition in 150 year forest rotation in Chiapas, Mexico.

System	Production level					
	I		II		III	
	Net carbon accumulation (t C ha^{-1})	Net annual accumulation (t C ha^{-1} year^{-1})	Net carbon accumulation (t C ha^{-1})	Net annual accumulation (t C ha^{-1} year^{-1})	Net carbon accumulation (t C ha^{-1})	Net annual accumulation (t C ha^{-1} year^{-1})
Tzeltal						
Live fence	65.6	(1.66)	92.3	(2.19)	118.9	(2.72)
Coffee with shade trees	84.5	(2.11)	115.9	(2.76)	147.2	(3.41)
Taungya	214.6	(4.71)	276.8	(6.00)	338.9	(7.23)
Enriched fallow	214.6	(4.71)	276.8	(6.00)	338.9	(7.23)
Tojolabal						
Live fence	26.7	(0.96)	39.1	(1.25)	51.5	(1.54)
Plantation	92.5	(2.58)	121.4	(3.27)	150.4	(3.97)
Taungya	94.5	(2.61)	123.9	(3.31)	153.3	(4.01)
Enriched fallow	94.5	(2.61)	123.9	(3.31)	153.3	(4.01)

(26.7–338.9 t C ha^{-1}), with the lowest potential for live fences in the Tojolabal region and the highest for the plantation systems (*taungya* and enriched fallow) of the Tzeltal region. Differences of the systems within a region are due to differences in added tree densities, while the high potential of the systems in the Tzeltal region, compared with the Tojolabal region, is due to the differences in current annual increments of the species to be included in the systems.

Economic Costs of Carbon Sequestration

The estimated costs of carbon sequestration for each system are based on the discounted direct costs of implementing the systems, plus the discounted opportunity costs during the first rotation for those systems where land use is diverted from agriculture or animal husbandry to forestry. It is expected that the farmer will be able to cover the costs of subsequent rotations from the income of timber production. In Table 20.4, we present the costs of carbon sequestration for each system, using the expected carbon sequestration for production level II. Costs are calculated on a per-hectare basis and include costs for establishment, maintenance, silvicultural treatments and harvesting for year 0 to 25 for the Tzeltal region and for year 0 to 30 for the Tojolabal region. Costs are presented separately for the discount rates of 5% and 10%, which are used to calculate the costs to the present date. Total cost t^{-1} C ha^{-1} varies between $US1.84 and $US3.98 for the systems selected for the Tzeltal region and between $US1.47 and $US11.15 for the systems of the Tojolabal region. The differences in costs within the same region are due to differences in costs of establishment and opportunity costs, while the differences between the two regions are due to differences in carbon-sequestration potential.

SUMMARY AND CONCLUSIONS

In Mexican rural communities, the different models of social organization affect the outcome of a farm forestry project definition, as illustrated in this study through the experiences of a farmers' productive organization, with individual members from various communities. The structure of decision-making, however, implies that members lack decisive power in relation to community matters. The advantage of working with such a farmers' organization is that a project can be extended rapidly to various communities, after a successful pilot stage. The disadvantage is that the project can only work on individually based systems. Many farmers are interested in starting small-scale forestry or agroforestry activities on their own land, but consensus for managing communal areas of natural forest is much harder to achieve, since

Table 20.4. Estimated costs of carbon sequestration for selected forestry systems in Chiapas, Mexico, based on the total carbon accumulation for each system (production level II), calculated with 5 and 10% discount rates.

	Live fence		Coffee		Taungya		Enriched fallow	
Tzeltal region								
Discount rates (%)	5	10	5	10	5	10	5	10
Direct costs ($US ha⁻¹)	214	170	358	274	454	317	608	420
Opportunity costs ($US ha⁻¹)					493	329	493	329
Total costs ($US ha⁻¹)	214	170	358	274	947	646	1,101	749
Total carbon accumulation (t C ha⁻¹)	92.3	92.3	115.9	115.9	276.8	276.8	276.8	276.8
Costs ($US t⁻¹ C)	2.32	1.84	3.09	2.36	3.42	2.33	3.98	2.71

	Live fence		Plantation		Taungya		Enriched fallow	
Tojolabal region								
Discount rates (%)	5	10	5	10	5	10	5	10
Direct costs ($US ha⁻¹)	98	58	1,176	966	386	248	952	816
Opportunity costs ($US ha⁻¹)	0	0	177	113	177	113	177	113
Total costs ($US ha⁻¹)	98	58	1,353	1,079	563	361	1,130	929
Total carbon accumulation (t C ha⁻¹)	39.1	39.1	121.4	121.4	123.9	123.9	123.9	123.9
Costs ($US t⁻¹ C)	2.50	1.47	11.15	8.89	4.54	2.91	9.12	7.50

these have to pass community-level decision-making structures. Social and political divisions often require some degree of resolution before control of *ad hoc* or legal exploitation of the common resource can be achieved.

In the agroforestry projects outlined, individuals plant trees in their coffee plots, fallow areas, pastures and maize fields, but the participants work cooperatively to organize seed-beds, nurseries and the marketing of wood products. This use of the reciprocal work-system '*ayuda-por-ayuda*' takes advantage of economies of scale and strengthens the new project by embedding it in the traditional labour-exchange system.

The current project will probably benefit the male members of Pajal and possibly their immediate families. To promote a community-level forestry project, however, there would need to be a greater distribution of benefits. Women's agroforestry projects for home gardens could increase the possibilities of benefits for children and female-headed households. Community-administered work teams for tree planting, management and harvesting on communal lands could distribute salaries to families without land and a percentage of profits for all community members. In all cases, the use of traditional labour-exchange systems, decision-making for communal-land management and equitable distribution of benefits will be essential ingredients for a successful carbon-sequestration programme through community forestry.

To estimate the effect of an agroforestry project on the regional balance of carbon fluxes, the simulation models tested have certain shortcomings. The landscape-based models to estimate carbon fluxes through different land-use scenarios lack a dynamic approach to calculate the fluxes through existing systems (e.g. Makundi *et al.*, 1995), while stand-based carbon-flux models (e.g. CO_2FIX) lack a quantitative landscape component. Cairns *et al.* (1995) have shown that significant differences in carbon-flux calculations can occur, if the carbon dynamics in, for example, secondary and primary forests are included. Models that calculate carbon fluxes under different land-use scenarios and which consider the growth potential of all components within the landscape are required to monitor carbon fluxes on a landscape basis of community-forestry projects.

The carbon-sequestration potential of different agroforestry systems depends highly on the quantity of trees to be managed in the system. However, low-tree-density systems, such as live fences and commercial trees, combined with (perennial) crops that require shade (such as coffee and cacao), are relatively inexpensive alternatives that can contribute substantially to the carbon budget on a regional basis, since the costs of establishment and maintenance are low and opportunity costs are negligible.

Although excluded from our analysis, the improvement of seminatural communal-forest management is likely to yield additional significant net economic benefits for the communities because of: (i) the relatively low initial costs of a management project, since the system is already established;

(ii) the fact that standing timber is a capital asset, which can help finance development; (iii) the lower opportunity costs (not replacing agricultural systems); and (iv) the higher conservation values (biodiversity, watershed protection, soil conservation). However, since the forests already have relatively high levels of carbon stored in the system, the forest-management options would represent a low carbon-sequestration potential but a high potential for carbon sinks. However, discussions about the economic value of conserving carbon sinks are still unresolved.

The costs and benefits associated with bringing a particular forested area under sustainable management would certainly depend upon the particular circumstances at each location. Where forests in good condition exist and are currently subject to low levels of exploitation, there may be immediate short-term benefits, in terms of increased annual harvests and increased productivity, when improved forestry practices are to be implemented. However, where forests are severely damaged, the costs of reparative actions may outweigh the benefits for several years.

As forestry and agroforestry concern long-term investments, with final products to be obtained after more than 20 years, mechanisms are required that guarantee the financial aspects of these projects. Combining systems that provide commercial products, such as wood for timber, and long-term ecological services, such as carbon sequestration, can generate the necessary capital for the investments in the systems.

The long-term characteristics of these types of project also require stable organizational structures that can support decision-making, sound financial management and good planning, including consulting local people and selecting appropriate species and management regimes for particular sites and social conditions. Whether the proposed project involves sustainable management of a large area of communal forest or establishment of small-farm woodlands, the quantity, quality, timing and distribution of both inputs and outputs must be carefully considered (De Jong *et al.*, 1995). As such, communities, groups and individual farmers have to take responsibility for the projects, not only in planning, but also for implementation, management and monitoring of progress, which implies a structuring of the project around existing social organization patterns.

NOTE

1 The authors thank participating institutions and organizations in their research, especially Unión de Crédito Pajal Ya Kac'tic for workshop facilities, and access to the communities, and El Colegio de la Frontera Sur (ECOSUR) for assistance. They also gratefully acknowledge the assistance of the following for the collection of data: Jerónima Alvaro, Martha Gómez, Manuela Moreno Moreno, Manuela Moreno Jiménez, Juana Moreno, Alejandra López, Aljandra López,

María de Tránsito, Angelica García, Teresa Pérez, Regina Pérez, y Pascuala López, Manuel Moreno Hernández, Juan Moreno Gómez, Gilberto Alvaro Jiménez, Mariano Moreno Jiménez, Mariano Moreno Moreno, Jerónimo Gómez Pérez y Nicolas Hernández Pérez, Julian Pérez López, Pablo Santís García, Julio Jiménez Román, Antolino Pérez Hernández, Ventura Aguilar Santís, Abelardo García Pérez, Ramón Vázquez Vázquez, José R. Vázquez Vázquez, Caralampio Gullén López and Fernandeo López Aguilar. The study was financed by the National Institute of Ecology of the Secretary of Environment, Natural Resources, and Fisheries, European Community (B-73014/92/412/9) and project R6320 of the UK Department for International Development, Forestry Research Programme.

Institutional Premises for the Fulfilment of Carbon-credit Requirements by Russia

Andrei A. Gusev and Nina L. Korobova

INTRODUCTION

Direct reduction of carbon dioxide (CO_2) emissions from the energy industry and from other fossil-fuel sources can be very costly. As has been demonstrated in many chapters in this volume, it is possible, however, to redistribute investment to offset emission reduction through sequestration in the biosphere. Institutional arrangements to facilitate carbon credits for this purpose are currently being developed within the framework of the Russian–American pilot project, RUSAFOR. Proposals for the improvement of current Russian legislation and institutional strengthening in the forestry area have been made, both at the federal and at the regional level. These suggested changes, which are analysed in this chapter, cover the issues of economic, legislative and administrative support for carbon credits.

GENERAL INSTITUTIONAL PREMISES AT THE FEDERAL LEVEL

Russia ratified the United Nations (UN) Framework Convention on Climate Change (FCCC) in October 1994, with compliance with its targets dependent upon the elaboration and realization of a national economic strategy. One of the main concepts, concerning the structural reorganization of the economy adopted by the Russian government, is the adoption of energy and resource saving. Fifty per cent of total CO_2 emissions are from the energy-generation sector. Hence, the potential role of power saving is significant.

However, it is also reasonable to consider reafforestation and carbon-credit options, since forests can provide greenhouse-gas emission reductions at relatively low cost, both through old-growth preservation and as offsets through sink enhancement.

A carbon-credit system entails the allowable increase of CO_2 emission quotas for one country to be compensated by concurrent offsets in another. The country providing the offsets is compensated by the emitting country. Planting trees to sequester CO_2 could provide an offset to emissions from fossil fuels. The cost effectiveness of different options, including the use of tropical forests as a compliance option in greenhouse-gas emissions control, has been considered by Dudek and Leblanc (1992). Emissions trading and forestry offsets between different states and countries were shown in that study to be the most likely options in FCCC compliance. As has been shown by other scientists and by environmental economists, direct reduction of emissions is often more costly, and so reafforestation and old forest preservation can be an attractive option under present conditions (see Crabtree, Chapter 14, and Sedjo and Ley, Chapter 19, this volume for example).

Russian forest ecosystems occupy 884 million ha, accounting for more 20% of the world forest resources (Anuchin et al., 1985). The Russian forest area per capita is 4.75 ha per person, compared with 3.43 for Scandinavian countries, 2.71 in North America and 0.25 for Europe without Scandinavia (Federal Forestry Service, 1994). Forty per cent of total forests are young and middle-aged (up to 60 years old) forest stands which are the most efficient in removing CO_2 from the atmosphere (Karaban et al., 1993). An important factor is thus the natural extension of forest areas and young and middle-aged forests' shares of growth. The potential ability of those territories to mitigate global climate change is probably one of the most valuable (although as yet intangible) parts of the Russian stock of natural capital with an average growth rate equivalent to 3–7 t of CO_2.

The northern Arctic territories in Russia are predominantly covered by boreal forests and wetlands. The Russian share of overall wetland territory in the Northern hemisphere is about 60%. The world's high-latitude wetlands are often considered to be more efficient carbon sinks than tropical forests (Danilov-Daniljan et al., 1994). Boreal forests usually grow much more slowly than tropical ones and the annual incremental rate of sequestration of the latter is higher. The total amount of carbon fixed by the two types of forests at maturity are almost the same (Woods Hole Research Center, 1994).

The potential carbon-sequestration capacity of the northern territories could be substantially increased if those territories obtain the status of specially protected areas, along with additional funding for protection and conservation. The total area of specially protected territories in Russia (natural preserves and national parks) is 40 million ha (Federal Forestry Service, 1994), which is about 5–6% of the undisturbed territories. In the current difficult economic situation, annual expenditures for maintaining specially protected areas are as low as $US0.3 ha^{-1} (Ministry of Environment Protection and Natural Resources, 1995).

According to existing evaluations, under proper funding the wood-covered area could be increased by at least 10%, providing an additional annual sequestration of 140–160 Mt of carbon. Unit costs per tonne of carbon fixed depend upon the type of project (afforestation, reafforestation or agroforestry, for example). Credible comparative assessments of carbon-credit benefits in Russia and abroad are not available, but the widely held view is that the unit cost per tonne of carbon fixed could be less than in other countries. Analysis of the options for conserving and sequestering carbon by Krankina *et al.* (1996), for example, shows that approximately 200 Mt of carbon could be sequestered from various sources, as shown in Table 21.1, based on an assessment of available land area.

Table 21.1. Estimates of forest-management options for conserving and sequestering carbon (C) in the Russian forest sector (from Krankina *et al.*, 1996).

Management option	Available land area (million ha)	Additional carbon storage (t C ha^{-1})	Carbon-sequestration potential (Mt C)
Plantation on agricultural land	19.5	36.0	70
Plantation on other land	4.1	18.0	7
Increasing stand productivity through silviculture	96.8	3.6	35
Reduction of fires	13.3	36.0	48
Harvest reduction and increased length of rotation	96.8	4.3	42
Total			202

An analysis provided by the Federal Agency for Hydrometeorology and Ecological Monitoring (FAHEM) shows that Russian forests remove 30% of total CO_2 emissions in Russia. But the FAHEM analysis is based upon an estimate of the Russian territory covered by forests as 772 million ha (Metalnikov, 1994), although some specialists use the 884 million ha cited earlier (Dixon *et al.* 1996). The FAHEM has also presented CO_2 removal forecasts, as shown in Fig. 21.1. The estimates in Fig. 21.1 take into account both forest area extension and the increased concentration of CO_2 in the atmosphere, as well as global climate change.

The extent of carbon sequestration is, however, disputed: some authors (Danilov-Daniljan *et al.*, 1994) provide estimates of annual sequestration by Russian forests of approximately 500–600 t of atmospheric carbon. An approximate estimate of total annual CO_2 emission was 650–700 Mt of carbon in 1990 (FAHEM evaluations), which would mean that current carbon sequestration is not much less than total carbon emissions.

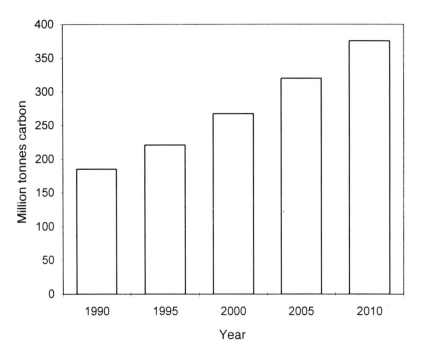

Fig. 21.1. Projections of CO_2 sequestration in Russian forests (Mt carbon) (from Metalnikov,1994).

The present economic situation, however, impedes the implementation of carbon-sink and storage projects, in comparison with carbon emission-reduction projects that are closely connected with energy saving. The major governmental objectives in a transition period in Russia include macro-economic and financial stabilization, privatization and restructuring of the national economy. Environmental problems thus have a very low priority. In 1994, overall Russian expenditures for forest and protected-area maintenance amounted to 7% of total environmental expenditure in the country and 0.1% of gross domestic product (GDP) (Ministry of Environment Protection and Natural Resources, 1995).

Taking the current economic situation into account, tangible funding for carbon-storage-orientated projects from internal sources, public or private, is not expected. So the only possible chance for major finance for funding is through joint implementation or carbon credit; with Russia providing forest resources, with consequent legislative and institutional support, for foreign investors funding corresponding projects and providing technology and know-how transfers.

In accordance with the forecast of the Ministry of Economy, Russian Federation crude-oil recovery and gas-condensate recovery in 2000 will be

30% lower than production levels in 1990 and coal recovery 15% lower (Metalnikov, 1994). In the present economic situation, even partial fulfilment of an energy-saving strategy would reduce the CO_2 emission level in 2000 compared with 1990, to achieve Russian commitments under the Climate Convention. Russia could therefore potentially provide the carbon credit, and the implementation of a reafforestation and afforestation project could attract foreign investments under a favourable investment climate. Russia has a federal political structure, with three levels of management: federal, regional and local. All the problems, in the context of the carbon credits, concerning natural-resource use should be regulated first at the federal level, as 80% of all natural resources are federal property, while for forests this share is 90%.

CARBON CREDITS AS A STRATEGY TO REDUCE EMISSIONS

As the provision of carbon credits is a small part of the complex overall problem of greenhouse-gas emission reductions, with different economic sectors involved, they need to be considered within the framework of the federal programme on greenhouse-gas emission reductions. This is currently being elaborated under the auspices of the Federal Agency of Hydrometeorology, despite most environmental matters being dealt with under the Ministry of Environment Protection and Natural Resources (MEPNR).

Since global climate change is presently perceived as an international problem, with international agreements to limit emissions, the policy issue is of federal importance and receives federal status for resolution in many countries. The status of a federal programme would also mean its inclusion in the list of federal programmes financed from the federal budget and having governmental guarantees. Although current practice shows that budget financing usually does not exceed 20% of programme costs, this support and budget input is important for foreign investors. This explains the emphasis in the first institutional scheme on the whole cycle of programme elaboration, including the consideration, adoption and creation of the agency responsible for programme implementation (Vinson *et al.*, 1996).

Although measures mitigating global climate change include more than energy saving and conversion to other fuels, it is suggested that all of this activity should be a subprogramme within the framework of the federal 'Fuel and Energy' programme. Energy policy is of vital importance in terms of greenhouse-gas emissions and the main decisions as to its scope have already been made. By such means, it may be easier to add some institutional and legislative elements to facilitate and accelerate the process of global climate-change mitigation and adaptation management.

Problems of funding for the most important state long-term goal-orientated programme are specified and regulated by a Government Order 'Concerning good deliveries for state needs' (Jurisdicheskaya Literatura, 1995, N594). A package of goal-orientated programmes for budget finance is also being considered by the Federal Government. Those programmes which are of great national economic, environmental, innovative and international significance are given priority. Hence energy saving and environment protection are among state priorities. A Presidential Decree 'Concerning main directions of energy policy and restructuring of fuel and energy complex' (7 May 1995, N472) stipulates the elaboration of a state programme 'Fuel and Power' for 1996–2000.

It is considered that all direct and offset CO_2 emission-reduction activities should be included in this programme, in accordance with the Act 'Concerning energy saving' (20 March 1996), since these emissions are supposed to be included in the remit of the Federal Agency for the realization of a power-saving strategy. An earmarked fund for energy saving could be created under the Agency in these circumstances. The main purpose of the fund would be to finance the federal programmes for energy-saving activities. The main means of financing such a programme would be fuel and power taxes. In our opinion, this Act could be expanded to include a provision that the orientated programmes must have state customers, who would be obliged to realize the programme, in accordance with the Russian Government Order N594.

This Federal Agency could be a customer, and budgetary finances for the national programme for energy-saving activity could be directed to this earmarked fund, if the programme were adopted by the government. One of the functions of the fund could be to fund the activities to offset CO_2 emissions, in conformity with activities of special regional associations and funds. Both foreign investments and earmarked funds could be the sources of federal-programme funding, according to Government Order N594. A schematic representation of the institutional and economic programmes that have been suggested to assist Russia's policies and programmes to meet international requirements for direct and offset CO_2 emission reduction is presented in Fig. 21.2.

In Fig. 21.2, the various processes involved in setting up a system of CO_2 offsets are indicated by numbers. At 1, the MEPNR, together with FAHEM, prepare proposals for a national programme of activities for direct and offset reduction of CO_2 emission within energy-saving policy of the national programme on fuel and power. Then, at 2, the coordination of the development of the Orientated Scientific and Technical Programme of the Russian Hydrometeorological Committee forms monitoring organizations regarding anthropological CO_2 emission.

The MEPNR and FAHEM send these proposals to the Ministry of Economy for consideration, together with the Ministry of Finance, recommending

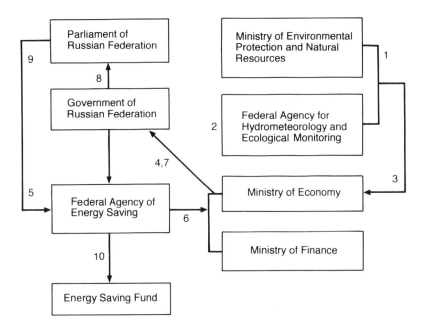

Fig. 21.2. Institutional support for compliance with direct and offset CO_2 emission-reduction requirements in Russia.

them for inclusion in the national programme of the package of targeted programmes of the Ministry of Economy, at 3. The Ministry of Economy, together with the Ministry of Finance, consider the proposals and send them, at 4, to the government for approval.

The package of the orientated programmes is considered by the government (5), a decision on programme elaboration is made and a state customer is selected. The state customer, in this case the Federal Agency of Energy Saving (FAES) prepares a draft programme and coordinates it with the ministries involved, prepares a business plan and a statement about budget and earmarked funding and sends all the documents, together with those on the investment climate for foreign investors, to the Ministry of Economy (6).

After approval of the programme, the Ministry of Economy, together with the Ministry of Finance, sends the final version to the government (7). Then an expert council under the government assesses the programme and it is recommended to be included in the federal budget (8). The budget must be coordinated and approved by Parliament, and the state customer is appointed officially (9). Finally (10), a fund for energy saving is formed to provide finance for the orientated programme.

LEGISLATIVE AND ECONOMIC PREREQUISITES FOR CARBON CREDIT AT THE REGIONAL LEVEL

Internal and external rules for exploitation of forest areas are specified in 'Foundations of Forestry Legislation in the Russian Federation' and the Law 'On Environmental Protection'. Russian law, especially laws regarding real property rights, are currently in a state of flux. There are regional differences in legislation regulating the ownership and use of land and forests (Petrov, 1994). A clarification of Russian legislation on land ownership, use rights and contracts will facilitate the establishment of joint-implementation forestry projects, such as carbon-credit projects.

In the current legal climate, obtaining approval at all levels of government and with all government agencies involved would help provide legal security for a project and the formal agreements concerning the project. For example, for the Vologda reafforestation project described below, it was recommended that the status of the site should be changed to a special ecologically protected status, under the jurisdiction of the Ministry of Environmental Protection, in addition to placing it under the supervision of the National Park Administration (Dudek and Leblanc, 1995).

The regime of sites allocated for carbon credit must therefore correspond to the status of special protected territories. To meet these requirements, special regulations specifying the characteristics of forest land used for carbon credit need to be developed and approved by the Ministry of Environmental Protection and Natural Resources, which forms a constraint.

The institutional mechanism for carbon credit at a regional level should be based on a model, funded, carbon-credit association. The main functions of the association are to organize and manage reafforestation activities and also to be responsible for the rational spending of financial resources. The earmarked fund should be formed under the association. It must have the status of an environmental organization. In accordance with the Federal Law on Environmental Protection, all environmental funds are free of taxation. Because the reafforestation fund is not included in the list of environmental funds, it is necessary to include an article about such funds in an amendment of this law. In our case, contributions of every enterprise in the region to this fund would be tax-free and donations could be one of the potential financial sources for the fund.

The main financial sources for the fund would be foreign investments for CO_2 credit. If the FAES is created, some share of the finances of the Agency fund would be transferred to the regional fund. The main purpose of the fund is to finance reafforestation activities to offset CO_2 emissions. In the region, the following measures can be used as economic support for CO_2 credit. In accordance with Order N67 (5 February 1992), payments for economic damage due to irrational forest use (such as clear cutting) could be introduced and partly transferred to the fund, in accordance with a resolution of the regional authority. In accordance with the Law 'Concerning Amendments and Additions to the Taxation System of Russia', the local authorities can render tax privileges within the local budget to support carbon-credit activities, as well as institutional and economic aid for CO_2 credit on a regional level.

Within the proposed scheme, the purchaser of the carbon credits pays for the carbon credits produced and monitors the scheme at the project site, as should be stated in the basic project agreement. The seller delivers the carbon credits through reafforestation; secures land for the project site and appropriate land-use rights, by concluding agreements with users or relevant government authorities; performs reafforestation and land maintenance; monitors the carbon-credit benefits; and administers the project or employs others to do this.

THE CARBON CREDIT ASSOCIATION: A PILOT SCHEME

This scheme was proposed as part of the working documents, and the main conclusions on the institutional set-up were derived from it in pilot projects in Vologda and Saratov regions. The experience and the main conclusions were included in the report *Reforestation in Russia: Building Institutional Capacity for Joint Implementation Through the Vologda Demonstration Project*, which was prepared by a group of Russian experts under the leadership of Daniel Dudek and Alice Leblanc (1995). On 26 October 1996, the Russian government adopted the federal programme 'Mitigation of Dangerous Climate Change and Its Negative Consequences', where extra budget sources of financing are mainly from joint project implementation. A special Interagency Commission for Global Climate Change, with representatives of different ministries involved, is supposed to promote these projects and act as the intermediary for contacts between foreign investors.

The Vologda region is located approximately 480 km north of Moscow, has an area of 150,000 km^2 and 1.4 million inhabitants. Forests cover about 70% of the land area and most of them are state property (Krjukov, 1987). The forests are dominated by spruce, pine, aspen and birch. The Vologda region was chosen because of its relative autonomy and the stability of the regional government, the willingness of the local government to implement the project and its proximity to Moscow and as a typical forest area.

In the Vologda region, four different sites were examined, some of them being rejected due to unfavourable environmental conditions. To meet environmental cost-effective criteria, a site without forests and not used for agriculture was chosen. Carbon dioxide sequestration and emission-monitoring methods were used and adjusted. Institutional and legislative support problems for carbon credit were considered, together with the monitoring of carbon benefits, land-cost estimation and foreign investment and insurance.

Estimated carbon benefits from above-ground and root biomass from the project are 1.4 Mg carbon ha^{-1} year^{-1} for the first 10 years. The cost per megagram of carbon is estimated to be approximately $US11.5 for the first 10 years. For years 11–60, the average carbon benefits are estimated to be 4,000 Mg carbon ha^{-1} year^{-1} (Dudek and Leblanc, 1995).

CONCLUSIONS

To meet the UN Convention on Global Climate Change requirements, Russia has begun to introduce novel institutional and legislative measures, on both federal and regional levels. Carbon-emission targets will best be reached both by implementing energy-saving programmes and by forestry activities. Russia is rich in forests, some of which fulfil the criteria of being both environmentally stable and effective in carbon fixation and hence could be used for carbon-credit provision, thus minimizing overall CO_2 direct emission-reduction costs for the country. If international funding can be mobilized for such a programme, this will have the additional benefit of providing foreign investment for the Russian forest sector.

Issues and Implications for Agriculture and Forestry: a Focus on Policy Instruments

Paola Gatto and Maurizio Merlo[1]

THE RESEARCH AND POLICY ISSUES AT STAKE

This chapter reflects discussion at the Workshop on 'Instruments for Global Warming Mitigation: the Role of Agriculture and Forestry', held at Padova University from which this volume stems. We review research and policy issues arising from the discussions and papers presented on land-use-related greenhouse-gas (GHG) fluxes and sinks, as well as options for offsetting emissions and increasing uptake. The framework in which mitigation and adaptation policies should be implemented is outlined; mandatory and voluntary policy instruments are considered, both at national and international levels, with regard to land use. Advantages and constraints of the various instruments are considered, emphasizing their effectiveness and the transaction costs, in the light of the existing European Union (EU) Common Agricultural Policy (CAP) and environmental policies. Finally, we suggest that the gradual development and continuity of existing EU policy instruments should be made in the context of multiobjective 'no-regret' policies.

After the publication of the two comprehensive Intergovernmental Panel on Climate Change (IPCC) assessments (Houghton *et al.*, 1990, 1996), the relationship between human activities and climate change is now well proven. However, although scientific knowledge in this field has increased enormously, the global-warming picture is still complex, multifaceted and occasionally contradictory. Indeed, the anthropogenic influence on GHG cycles and the quantities involved in the various cycles are still far from being well understood. The existing evidence can be, and sometimes has been, used inappropriately. For instance, poor basic data seem to have justified, and even stimulated, economic analysis that does not adequately represent the underlying physical process (see critique by Price, Chapter 6, this volume). Oversimplifications and approximations have sometimes been misleading, although the policy implications of this debate are likely to be very significant, leading *The Economist* to comment on the final wording of the IPCC 1996 assessment as 'strong stuff for an international committee' (*The Economist*, 23 March 1996).

The chapters in this volume, written in the light of the latest IPCC assessments, highlight a number of research needs related to data and models, as well as to better estimates of the costs and benefits of GHG variations.

1. Empirical data on GHG fluxes are often missing or are very variable, depending on models, assumptions and approximations. Scandinavian countries seem more advanced in this field, reflecting long histories of monitoring and forestry research[2] (see Seppälä and Pingoud, Chapter 16, and Eliasson, Chapter 18, this volume). This undoubtedly reflects the important economic role of forestry and the need to legitimize silvicultural and management techniques in these regions.

2. Uncertainties remain on GHG cycles, especially on the extent and the dynamics of carbon sinks in soils and in oceans, in which time-lags and buffering effects substantially affect climate mitigation but are usually overlooked.

3. Many different carbon, nitrogen, chlorofluorocarbon (CFC) and sulphur emissions interact in affecting climate change. In particular, a sulphur dioxide (SO_2) cooling effect may be offsetting atmospheric warming by other GHGs. Despite the potential for these emissions to affect the distribution of vegetation and opportunities for adaptation, this aspect seems to have been underresearched up to now (see Zecca and Brusa, Chapter 4, this volume).

4. Particular emphasis has been given so far to global estimates of the economic costs of impacts. The positive consequences of carbon concentration increase in the atmosphere (in particular, the carbon dioxide (CO_2) fertilization effect on vegetation growth) may have been downplayed. Similarly, little attention has been paid by economists and policy analysis to the adaptability of living systems, although this issue is now considered by the IPCC itself.

5. In conclusion, only 'best-guess estimates' are at present available for assessing the potential costs and benefits of climate change. Even established institutions seem far from reaching firm conclusions. In this context, the discounting issue, with its ethical implications, is a key empirical issue highlighted in this volume (see Fankhauser, Chapter 5, and Price, Chapter 6, this volume).

Land-use Implications

According to current estimates of GHG emissions, about one-quarter of human-induced GHG emissions are related to land use, while the majority are linked to consumption of fossil fuels (Adger and Brown, 1994). Land use concerns both forestry and agriculture on a global and a regional scale. Usually, forestry is seen as an opportunity for enhancing carbon retention, while agriculture is blamed for causing net GHG emissions. As shown in this volume, the issue is complex.

Forestry-related fluxes can be seen, globally, as maintaining or expanding sinks through forest management or afforestation (see Ciesla, Chapter 3, this volume) and reducing sources of GHGs. Forestry cannot be seen as a panacea for mitigation, only as part of an overall strategy. Afforestation, when carried out at the expense of natural ecosystems (e.g. peatland, pastures, Mediterranean *maquis*), often results in a net flux of carbon into the atmosphere, as shown, for example, for Britain in this century by Adger *et al.* (1992). In European countries, afforestation faces the problems of competing with agriculture. Incentives to private landowners, as currently designed, may not take account of the diverse social benefit of afforestation. In general, profitable forestry is only evident on low-quality land. Incentives for afforestation of high-value land should take account of carbon sequestration and other environmental benefits (see Crabtree, Chapter 14, this volume). The different impact of plantation forestry (poplars, eucalyptus, Sitka spruce), on the one hand, and nature-orientated forestry, on the other, is not always acknowledged. From this point of view, regulations, such as 2080/92 within the EU, calling for carbon storage, are sometimes contradictory. Other constraints are linked to property-rights issues, which may prevent afforestation (see Lippert and Rittershofer, Chapter 15, this volume).

Important carbon sinks can be created through large-scale forestry projects outside Europe, in areas such as Central and South America. Here, land availability at apparently low opportunity cost can be a key factor in promoting afforestation. Timber plantations would seem to be an efficient means for achieving carbon storage in these circumstances (see Sedjo and Ley, Chapter 19, this volume). However, whenever public financial sources are involved, a clear analysis of the overall environmental impacts, costs and benefits is required, as highlighted for Argentina in Chapter 19. Where afforestation is recorded by international statistics, particularly in tropical and subtropical regions, this does not necessarily constitute an increased forest area but often means substitution of natural forests (more or less degraded) with tree plantations. What can be gained in short-term carbon storage by wood may be lost through reduction in soil organic matter. In addition, loss of biodiversity and other negative environmental impacts must be accounted for.

In contrast to forestry, reduction in emissions from agriculture involves greater uncertainty and variability concerning estimates of fluxes, due in part to greater interaction between diverse gases and the relevance of anthropogenic factors involved. Human intervention and its environmental impact are clearly more evident in agriculture than in forestry. Chemical input, technology and products are more varied, which in turn increases complexity in modelling emissions. The two main agricultural production methods affecting GHG releases are arable crops and animal husbandry. Methodologies are currently being developed for modelling the arable sector and assessing the effects of agricultural policies on carbon fluxes. For example, about one-third of targeted reduction in UK emissions can be offset through application of existing agri-environmental measures (see Armstrong Brown and colleagues, Chapter 10, this volume), though with higher than previously estimated emissions of nitrous oxide. Technical feasibility often contrasts with economic rationality and costs, however, as demonstrated in Germany by Löthe and colleagues (Chapter 12, this volume).

The contribution of the livestock sector to GHG emissions from agriculture has been highlighted as critical in several chapters. In line with the second IPCC assessment, extensive rearing systems tend to involve higher emissions per unit of product than intensive systems. This can be seen in terms of both direct emissions (using more concentrate feeds reduces methane (CH_4) release) and opportunity costs, since rural areas currently used for grazing or for growing arable crops for stock feed could be used to produce a range of sources of renewable energy, including energy crops, forest residue, straw and farm waste (see Subak, Chapter 11, this volume).

In summary, the major releases altering global ecological systems come from the use of fossil fuels. In fact, considering that major emissions usually described as land-use-related – e.g. CH_4 from animal husbandry and landfills – are actually more linked to the production and consumption process, the effective role of land cover can be reduced from one-fourth to about one-fifth of the total emissions. Soil uptake and emission and the possible missing sinks related to forestry can play a substantial role in global climate-mitigation strategies. These, however, must not be exaggerated. The 500 million ha to be afforested to offset CO_2 emissions (sometimes proposed) represent an extreme hypothesis which is certainly unfeasible and would undoubtedly lead to negative and unpredictable consequences (FAO, 1993a).

POLICY INSTRUMENTS AND MEASURES FOR CLIMATE-CHANGE MITIGATION

The following elements are central in considering policy instruments and are integrated into the subsequent discussion.

1. The inertia of policy-makers with regard to the agri-environmental arena (local, regional, national and European).

2. The present agri-environmental policy instruments and measures.
3. The pervasive uncertainty surrounding both the mitigation and the impacts of climate change.

A possible framework for the classification of policy instruments and measures for climate-change mitigation that are applicable to land-use GHG sources and sinks is presented in Fig. 22.1. A basic distinction is made between mandatory and voluntary policy instruments, including economic–financial and market instruments. Persuasion can be considered a complementary instrument accompanying and supporting both mandatory and voluntary ones. The different instruments can be, and often are, employed simultaneously and have international, regional and local dimensions. Each of the major approaches – juridical, market-based instruments and taxes, and persuasion – are now considered in turn.

Juridical Policy Instruments

Conventions and international agreements

Juridical policy instruments relevant for global-warming mitigation include conventions and international agreements, assuming obligatory compliance by signatory countries. The major example in this context is the Framework Convention on Climate Change (FCCC). Of course, international agreements must be implemented at a national, regional and local level. Enforcement remains the responsibility of individual countries and is one of the weakest points of this instrument (Hodge, 1995; see also Adger and colleagues, Chapter 1, this volume). The difficulties in implementation of an international agreement on climate are at least comparable to those for oil production and for milk quotas, which deal directly with tangible goods, rather than, in the climate-change case, polluting emissions, which are by-products of a plethora of economic activities. Monitoring of the FCCC in the land-use sector may be relatively easier over time, due to increasingly sophisticated remote sensing.

Definition and assignment of property rights

The definition and assignment of property rights, as traditionally conceived and often consolidated into constitutions, represent an intrinsic part of any mechanism for controlling land-use change in the European context. In Alpine countries, for example, deforestation and pasture ploughing have been forbidden at least since the last century. In addition, landowners were obliged to follow specific codes of practices. These measures were originally aimed at watershed management and at other environmental benefits, including local climate regulation, and were based on an existing framework of property rights, supported by a strong social consensus since the Middle

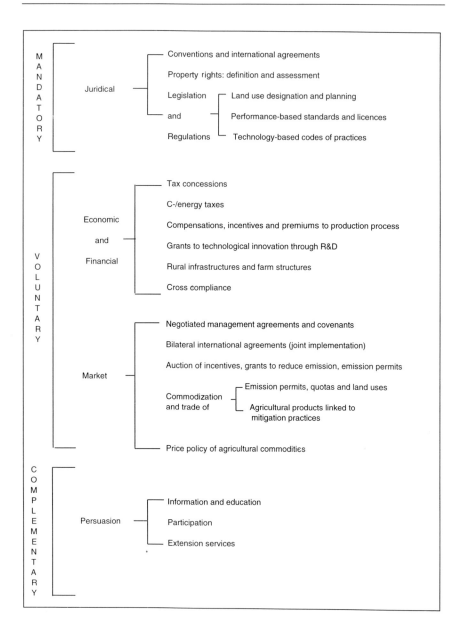

Fig. 22.1. A framework for the classification of policy instruments and measures for climate-change mitigation in agriculture and forestry.

Ages. In the present context, there are clear potentials for carbon fixation, biodiversity and the provision of other environmental benefits.

Legislation and regulations
Legislation and regulations in the European land-use sector include designation for favourable environmental impacts. Such regulations should be integrated with structures of property rights and with physical planning by local authorities. The role of local authorities, particularly the local communes or municipalities, in regulating land-use changes in many EU countries is considerable. Crop control is also applied by the EU on an extensive scale through remote sensing. Standards and licences based on performance of different land uses can be the object of regulation at both a national and an international level. Codes of practice, including equipment and technologies complete and fully implement legislation and regulations. Agricultural measures in this context can include, for instance, limitations on the use of chemical inputs, controlling manure spreading, fostering minimum tillage, conservation of permanent pastures and meadows, and forbidding straw burning. Naturally orientated forestry – increasing growing stocks, enhancing certain species and providing for forest soil conservation – can also be regulated and included into forestry plans, thereby contributing to reducing GHG emissions. Wetland and peatland conservation can also be considered, through designation under the Habitats Directive within the EU or under the Ramsar Convention, for example.

All the above regulatory measures are, to a large extent, already part of EU member states' existing agricultural, forest and environmental policies. Extension of these to include climate change is therefore certainly possible and in line with established policies. However, extending mandatory measures may be unpalatable to farmers if not supported by compensations or other positive instruments. In fact, drawbacks and failures with the use of mandatory instruments are linked to their 'command and control' nature. The imposition of land-use regulation for the single objective of carbon sequestration is likely to be socially unacceptable if it results in disaffection and the abandonment of agriculture and forestry activities in rural areas, as shown in Mediterranean and mountainous areas. A further shortcoming of the regulatory approach is the high transaction cost for policy enforcement and control. Explicit administrative costs can be particularly relevant in countries lacking *ad hoc* administrations. Finally, mandatory instruments at the EU level cannot easily be considered, because of their link with established assignments of property rights, which are the prerogative of member states' constitutional and legislative framework, local habits and traditions.

Economic and Financial Policy Instruments

Economic and financial instruments proposed in this area follow quite clearly a Pigouvian rationale, aimed at internalizing externalities. The Keynesian side of these instruments, often applied for creating jobs and activating the economy, must not be neglected. This generation of instruments is, to a large extent, more 'democratic' and based on the 'carrot', rather than on the 'stick' as in mandatory instruments. They present the common characteristic of being positive policy instruments aimed at convincing people that they should implement certain measures in exchange for financial subsidies or other economic advantages.

Tax concessions
Tax concessions are perhaps the oldest financial instrument applied to land uses. For example, in Italy and France, tax reliefs are well-established measures to secure afforestation and other land uses producing positive environmental impacts. Similar measures have been applied at times in the UK. Of course, the main objectives in these countries have been timber production and environmental benefits in general. Carbon storage could now be another good reason for revisitation and confirmation of these measures. Advantages claimed are ease of administration. Disadvantages include the regressive nature of such measures and their reduced impact when marginal tax rates are adjusted for other macroeconomic reasons.

Taxes on energy or its carbon content
Taxes on energy or its carbon content are perhaps the most commonly envisaged instruments for controlling emissions. In fact, they follow the much advocated and little applied 'polluter pays' principle. Such measures are supported, to some extent, by the EU (European Commission, 1992) and by the second IPCC assessment. It has been demonstrated, however, that unilateral taxes can be very costly for industries competing on international markets (see Komen and Peerlings, Chapter 13, this volume). Exemptions for certain sectors of the economy were therefore often considered by governments considering these unilateral taxes. The application of carbon or energy taxes to agriculture and forestry is certainly possible, however, taking into account that in these sectors fossil-fuel consumption is rather unresponsive to price changes. Carbon or energy taxes could also be applied, at least in principle, to land-use changes producing GHGs. It is often suggested also, at EU level (European Commission, 1994a), that revenues from environmental taxation should be profitably employed in other sectors of the economy (mainly in measures to reduce unemployment), so to achieve double dividends. However, there is not complete agreement as to whether positive impacts on both

employment and the environment can be effectively accomplished (for example, Carraro *et al.*, 1995; Goulder, 1995b; Bohm, 1997).

Compensation, incentives and premiums
Compensation may be provided for farmers and foresters, to make up for higher costs and lower revenues and to obtain environmental goods and services, which could feasibly include carbon sequestration. This is also a long-established instrument for promoting afforestation and landscape quality and, in general, securing the environmental goods and services provided by agriculture and forestry. Again, extension to carbon storage is perfectly in line with existing policies. Incentives and premiums for low-impact production processes are, to a large extent, annual compensations for loss of income, augmented by a benefit aimed at stimulating landowners' participation. When they are given once and for all as a lump sum, discounting the flow of annual incentives, they can be defined as grants.

Grants for technological innovation
Grants for technological innovation, both through research and development and fostering the adoption of new technologies, are another policy instrument that could be employed to favour environmentally friendly production processes (see Dosi and Moretto, Chapter 7, this volume). Such policies have always been integral to agriculture and forestry, through, for example, public investment in research and development, education and extension. Such investment could be feasibly directed toward new technologies, for example, in the field of energy saving and the development of long-lived wood products. This represents a so-called 'technological escape' to overcome present and future environmental problems, including GHG emissions (De Benedictis, 1997). Considering the timing of innovation *vis-à-vis* the abandonment of polluting technology, benefits become uncertain if investment grants cannot be suitably timed.

Rural infrastructures and farm structures
When rationally designed, rural infrastructures and farm structures can reduce fuel consumption and GHG emissions. These would include rationalization of land-holding and related mechanization patterns and consolidation, drainage and irrigation systems, all of which have been subsidized in the past. Rationalization could have a positive impact on energy use in agriculture, but the impact of biomass changes associated with the loss of hedgerows needs to be considered. Up to now, structural policy has been used to achieve environmental policy only to a very limited extent and has often been demonstrated to produce a net negative impact on the environment. There are, however,

opportunities to condition structural and infrastructural rural policies to environmentally friendly land uses. The positive impact is not negligible; after all, it is estimated that more than 10% of farmland is given to structures and infrastructures.

Cross-compliance
Finally, cross-compliance, which is popular at present and much advocated, is to a certain extent an indirect financial instrument. The concession of existing payments, such as income support, could become conditional upon the adoption of environmentally friendly techniques, including those aimed at an improved carbon balance. In other words, cross-compliance could be considered strongly persuasive, in line with voluntary financial instruments.

The above policy instruments can be considered in the context of past and present CAP and EU environmental policies. The liaison with and the potential for a climate-change mitigation policy are evident. European Union compensations, incentives, premiums and grants could play a crucial role, particularly since the 1992 CAP reform and in its evolution over the coming years. Structural measures and now the Structural Funds policies of the EU, applied in rural areas, could incorporate climate-change-aimed measures. Meanwhile, energy taxes should be agreed at an EU level, being up to now the individual countries' prerogative and hence only sparsely implemented within the EU.

Measures accompanying the CAP reform – Regulations 2078/92 and 2080/92 – represent the starting-point of a new environmentally orientated CAP. It is self-evident that the positive impact on climate change of a reduction in chemical inputs, the maintenance of the landscape, the introduction of organic agriculture, the upkeep of abandoned woodlands and the set-aside of farmland may lead to a reduction of carbon emissions, enhancement of carbon retention by soils and overall increase of carbon sinks. Other measures, such as those aimed at reducing livestock densities, may not have clear-cut implications for climate change. The objective of global-warming prevention and carbon storage is clearly stated as one of the four objectives of Regulation 2080/92, which remains, however, quite ambiguous. Plantation forestry overlaps and is confused with naturally orientated forestry. In Article 3, where levels of compensation are considered, reference is made to coniferous and broadleaf species, including such trees as Sitka spruce, eucalyptus and poplars. Clear boundaries between species, silvicultural techniques and forest management should be drawn in order to make clear the different impacts. Perhaps incentives should be differentiated with reference, first of all, to forest typologies rather than to species.

Coming to the nature of the various financial instruments, it is sometimes difficult to clarify the rationale behind them: in particular, the difference between compensation aimed at re-establishing the status quo and incentives and tax concessions aimed at encouraging otherwise neglected practices. In the measures accompanying the 1992 CAP reform, payments are provided for farmers who undertake certain practices or forestry investments. It is clearly stated that these payments have to be proportional to any increase in costs or loss of income. However, application of financial instruments according to a flat rate inevitably gives some farmers compensation higher than the foregone income. Revenue for producing environmental goods contains a certain amount of producers' surplus or soil rent, at least for the non-marginal farmers. Therefore, it has been argued that codification and standard rates constitute the main shortcomings of financial instruments (Colman *et al.*, 1992; Bishop and Phillips, 1993).

Financial instruments have also been criticized for undermining farmers' ethical commitment to stewardship (Colman, 1994). Therefore an inherent risk in financial instruments is paying for something farmers would do anyway, with lower payments or none at all. An effective solution is to adopt differentiated standard payments, referred to sites and specific practices, as in many Zonal Programmes of EU Regulations 2078/92 and 2080/92 (Povellato, 1996). Another substantial drawback of voluntary financial–economic instruments, however, in common with mandatory approaches, is given by transaction costs, the so-called 'omitted variables' (Whitby, 1995).

Market-led Instruments

The market is considered in this section as a possible source of integration with or alternative to economic–financial instruments. This approach, widely discussed in recent years, advocates a Coasian rationale. In particular, it should overcome the objection often made about financial instruments of being indiscriminate and unethical. A wide range of market-led instruments may be identified.

Management agreements and covenants

Management agreements provide payments subject to negotiation between farmers and the responsible public authority. With respect to standard payments, they should avoid possible excess payments, resulting in farmers' rent. The main drawback is represented by the time-consuming negotiation process, meaning high transaction costs on both sides. However, as far as carbon storage is concerned, payment parameters should be more easily definable than in the case of other environmental goods and services. At least in theory, the agreed compensation should approximate the marginal cost incurred by the farmers, plus the profit necessary to stimulate the agreement. A more extended view of management agreements, requiring contract registration, is given by the so-called covenants – the *servitutes praediorum* of Roman law.

These are legally binding land-management agreements. If permanent, they are attached to the land. From the community's point of view, they represent a stronger commitment and a guarantee of environmentally friendly farming and forestry, especially by extending beyond the end of present finite agreements.

Joint implementation: bilateral agreements or multilateral funding
Bilateral agreements or multilateral funding for offsetting carbon emissions can also be negotiated at an international level among sovereign countries – the so-called joint implementation. Under these agreements, highly emitting countries or firms buy themselves further emission credits by investing in carbon-offsetting projects (see De Jong and colleagues, Chapter 20, this volume). This ratio could be the basis for international cooperation and the underlying compensation among countries and should cut the global costs of reducing emissions. Certain countries have, in fact, comparative advantages – that is, lower costs for mitigation policies. This may be the case for Russian endowment of forested areas and land available for further expansion (see Gusev and Korobova, Chapter 21, this volume).

Auctions
Auctions of permits to emit pollutants are now very sparsely applied in Europe. Such auctions work through competitive bidding, directly submitted by those wishing to participate in an environmental programme, including certain standards, environmental restrictions and land uses. Auctions can feasibly be used to sell and buy carbon-emission permits. In the case of agri-environmental programmes, bids represent the amount of payment for which farmers are willing to accept the restrictions imposed by the programme. There is a proved cost-effectiveness in switching from fixed-rate compensations to auction schemes, and this is enhanced by information asymmetry (van der Hamsvoort and Latacz-Lohmann, 1996). Negotiating costs are not so important here, although other costs, such as those related to the preparation of bids, must be considered. In addition, compliance to the schemes must be checked and monitored.

Commoditization and trade
Commoditization of and trade in emission permits, quotas and land uses, that is, positive and negative externalities affecting climate-change mitigation, represent a rather sophisticated market instrument. This approach foresees the creation of specific markets, according to a pure Coasian rationale. For instance, GHG emission permits and quota or land uses, such as forestry, undertaken according to environmentally friendly codes of practice could be the object of an international multilateral market. Of course, the market organization must be based on an international convention defining and allocating quotas. Exercises like this are now the subject of investigation

(UNCTAD, 1992), although far from being implemented. The main premises required by Coase's theorem for such schemes to be feasible are often lacking, however. Definition, assignment and allocation of property rights on the atmosphere are far from being clear. In addition, quantification of GHG fluxes ante- and post-market transaction is difficult. Moreover, transaction costs in general make it difficult to market environmental goods or 'bads', regardless of whether costs are met by public bodies or by the contracting parties. Finally, there are equity implications in the allocation of emission permits between industrialized countries and less developed ones. In conclusion, although these instruments are certainly important in climate-change policies, the role of a pure Coasian approach must not be exaggerated (Bromley, 1991; Zamagni, 1994).

Agricultural products
Agricultural products whose image and prices are affected by environmentally friendly practices can represent a further market opportunity for achieving climate-change mitigation. It has to be made clear to consumers that certain traditional quality products embody environmental quality in general, including a potential to prevent climate change. Of course, product quality, techniques and practices stated in protocols should be well known and trusted by the consumers. As a consequence, market rules could be modified, giving rise to differentiation and monopolistic competition, allowing farmers to receive remuneration not only for the quality of the product, but also for the related agricultural practices and the favourable climate-change impact. This market rationale has been traditionally used for promoting agricultural products, such as wine and cheese in Latin countries. Now, it is proposed by the EU for many other products of controlled origin and quality – for instance, EU Regulation 2092/91 on organic farming and Regulations 2081/92 and 2082/92 on origin and specificity labels. In this context, attention must be paid to the policy of forest-stewardship certification, which should guarantee the provision of environmental benefits, including carbon retention and sink enhancement, produced through sustainable forestry (Hansen, 1997). Given the growth of 'green consumerism' and the sophistication and willingness to pay of European consumers, these policies seem to have a future and therefore an impact on the conservation of rural amenities, including carbon fixation.

Price policy for agricultural commodities
The price policy with regard to agricultural commodities, including protection tariffs, is clearly a market instrument that can also encourage mitigation behaviour. Needless to say, a voluminous literature is available on this issue (Brouwer and van Berkum, 1996; Whitby, 1997). For instance, the last 30 years of CAP, characterized by the high prices of cereals, milk and livestock, have put forestry, pastures and the related carbon storage at a disadvantage.

Persuasion Policy Instruments

Persuasion instruments include complementary instruments, such as information, participation and education, and advice and extension services. They represent an important support to juridical, economic–financial and market instruments. Their help is essential in pursuing the social acceptability of climate mitigation (see Storey and McKenzie-Hedger, Chapter 2, and De Jong and colleagues, Chapter 20, this volume). An argument often advanced is that a key factor in the success of land-use policy is the 'bottom-up' approach. The consensus of local populations, when forestry and agroforestry programmes are carried out, also guarantees the acceptance and success of afforestation programmes with carbon-sequestration purposes. Involvement of rural communities retaining common ownership of land is particularly important.

CLIMATE-CHANGE POLICY PERSPECTIVES

Further increases in GHG rates are apparently inevitable. Development currently under way in large areas of the world will potentially involve large GHG emissions, if, as appears to be the case, present technologies continue to be adopted (Kats, 1992). For example, China's reservation with regard to the IPCC documents or certain assumptions by the cohesion states discussed in the introductory chapter by Adger and colleagues (Chapter 1, this volume) are 'hard facts' that must be taken into account. Things seem to be exacerbated by the 'flywheel effects' or positive feedbacks associated with global warming. Therefore, no justification can be admitted for delaying action, as agreed throughout this volume.

'No-regret' policies have emerged as the primary way ahead, for reasons of political feasibility as well as negative real cost. They should, however, be aimed at various environmental and economic objectives, as far as they are compatible with each other. In general terms, the sustainability of agriculture or forestry practices is correlated with an improved GHG balance. This case has been widely supported within this volume, where sound multipurpose agriculture and forest policies have been advocated, independent of predictable effects on GHG emissions and other pollutants (for example, see Ciesla, Chapter 3, this volume). The provision of a wider mix of environmental benefits would also give grounds for high-incentive forestry schemes, otherwise not justifiable in terms of carbon fixation alone. The above orientation is supported by international organizations, such as the Food and Agriculture Organization (FAO) (Ciesla, 1995). In wider terms, the IPCC assessment itself states that policy reforms which help improve the efficiency of an economy are to be welcome, 'whatever the consequences for climate change' (IPCC, 1996b).

It is significant that adaptation policies have gained momentum in the climate-change debate. After all, climate changes will occur slowly in time and homeostasis of both ecosystems and economic systems will allow adaptation. The models presented in this volume shows that adaptation to climate change by farmers is not only possible, but can be accomplished without a reduction in global output (see Sohngen and colleagues, Chapter 8, this volume). However, suitable methods for measuring adaptation costs still need refining and appropriate cost–benefit analysis frameworks need to be set up to compare performances of different policies.

To what extent are existing agriculture and forestry policies within Europe in line with mitigation policies? The question has been posed by various chapters, with some incisive criticisms of the present environment policy. For example, it has been stressed that, at the moment, government intervention in agricultural markets and government regulations on land use constitute rigidity with regard to not only implementation of mitigation strategies but also adaptation to climate change (see van Kooten and Folmer, Chapter 9, this volume).

In reviewing the policies of individual EU countries, various chapters suggest a clear need for explicit mitigation policies. This does not mean, however, separate new policies, but rather an adaptation of existing policies. It seems possible that climate-change issues could be integrated within existing agri-environmental policies and mechanisms. This seems particularly feasible in the light of EU CAP reform and further development. In addition, it is clear that such a complex phenomenon as climate change cannot be faced with single-aimed policy instruments. A set of integrated actions in different sectors of production and consumption is needed. However, existing policies and official policy positions, even within the various EU countries, are very different (see Adger and colleagues, Chapter 1, this volume). Application to carbon storage could be considered and integrated into the overall package of compensatory payments to environmentally orientated farmers, as seem to be the perspectives of post-reform CAP development (Ferro, 1997). Provision of appropriately designed compensations, incentives and grants, together with information, policy announcements and institutional transparency, can facilitate the procedure.

Due to the global character of GHGs, international, if not global, agreements are required, in addition to CAP and EU environmental policies. International institutions thus become the most crucial, and a largely unresolved, issue. If environmental policies are difficult to apply within individual countries, owing to specific administrations and enforcement means, it has been shown to be even more difficult to apply them internationally. The issue of cooperation on reducing GHG emissions is particularly crucial when redistribution between rich and poor countries is considered, an issue often discussed but never really taken into account by either economists or policy-makers.[3]

Redistribution between generations is even more difficult. While a market of present environmental goods and services can at least be conceived, it is almost impossible for future environmental goods and 'bads' (Zamagni, 1994), thereby invoking the unresolved issue of discounting. Fankhauser (Chapter 5, this volume) distinguishes a 'prescriptive' from a 'descriptive' rate. Price (Chapter 6, this volume) goes even further, highlighting intergenerational unfairness in discounting mitigation strategies. There are, however, no firm conclusions on the most appropriate rate for examining preferences in the climate-change area (Turner *et al.*, 1994). It can only be noted that in Alpine areas village communities and even private landowners have often adopted as a rule a zero discount rate, having in mind as a paramount value the community's sustainability. However, whenever such communities are in danger (wars, famines or plagues), overexploitation of forest resources has been sanctioned. The validity of this attitude has been acknowledged by Pareto himself, having in mind village communities' common property, when he stated that 'since it has undergone the trial of free competition for centuries, collective land tenure responds better (than other forms of land tenure) to certain specific needs' (Pareto, 1896).

Price's radical statement in Chapter 6 (this volume) that 'we have two viable mitigation strategies: to replicate the geological time-scale processes which fixed this carbon, or else to desist from releasing' is no doubt a gloomy perspective. Consideration of GHGs and the related climate change must be undertaken in the wider context of geological and economic history. To avoid a parochial view of present-day concerns, a reconsideration of Lovelock's (1995) Gaia hypothesis produces interesting insights. Global perspectives on the ultimate fate of the world under climate change, and environmental change generally, may be summarized in four scenarios. The first two certainly do not appear rosy, while the second two are more optimistic and in line with a Gaian perspective and observed economic history.

1. Collapse of the world's ecosystem because of the impossibility of realizing draconian changes in the model of development and consumption. Even if changes were accepted in developed countries, they appear to be politically and morally impossible in countries currently undertaking rapid industrialization based on fossil fuels, which make up two-thirds of the world's population. This is the scenario of humans as parasites of the world's natural resources, who destroy themselves, together with their host.

2. The realization (or imposition) of draconian changes by the world's ecosystem itself, which is unable to support the present model of development, once it has gone beyond its carrying capacity. According to this scenario, the ecosystem reacts automatically to unplanned shocks through self-defence mechanisms (Gaian 'regeneration'). This would mean drastic changes in modes of life and development and adaptation to ecological drives – in other words, a new Middle Ages, whose connotations are yet to be defined.

3. Technological innovations of such significance that the current technologies which form the major sources of GHGs would become redundant, with the adoption of new technologies reducing emissions and enhancing sinks. It is the 'technological escape',[4] based on research and development and pushed automatically by market forces. A recent contribution (UNESCO/ICEC, 1996) appears to point out that this is the road to follow and it has already been taken up by some multinational groups, who are exhibiting greater foresight than agriforestry policy-makers, who do not always transmit coherent signals to the market.

4. A further scenario emphasizes the capacity of the world's ecosystem to regenerate itself through reactions that have so far not been predicted, but are not such as to pose a threat to models of development. Humans would be able to adapt to the needs of the ecosystem in a dialectical relationship with the planet on which they live. This scenario, compatible with the 'technological escape', does not appear to be mere science fiction, in the light of the uncertainties mentioned several times in this volume, such as those associated with the missing global carbon sinks.

In line with the optimistic technological, political and ecological viewpoints, the history of economic development over the past 200 years can also be shown to give some qualified cause for optimism. Pessimists, from Ricardo to Malthus and, to a certain extent, the Club of Rome (1972), have all been belied by the course of events.

The above arguments do not imply that climate change is not a significant issue. It is clear that interventions in present energy consumption, as well as in land use, are necessary. The major message emerging from the volume is that 'no-regret' policies should be applied in a multipurpose context, aimed at general improvements of the environment. This can be achieved by a plurality of policy instruments: regulatory, economic–financial and market-based. Overall support is required in terms of persuasion and consensus. In fact, various chapters in this volume have questioned or criticized those mitigation measures that are solely aimed at carbon sequestration and forest investments based on timber production alone, without other positive effects on the environment (see Ciesla, Chapter 3, Crabtree, Chapter 14, and Eliasson, Chapter 18).

In conclusion, the most desirable options to be pursued by the EU can be summarized as 'no-regret' mitigation policies, aimed at a multiplicity of environmental objectives, for increasing the multiple benefits associated with the use of natural resources. Present experience also suggests that a multiplicity of policy instruments should be employed, on the basis of existing policies and institutions. Evolutionary approaches are therefore the most practicable and, in the European context, will inevitably be based on present CAP reform and structural policies. The final aim and result should be the creation of 'multiple dividends', in terms of environmental quality, employment and rural development, with all their associated cultural and social values.

NOTES

1 Paola Gatto is responsible for the introductory section, Maurizio Merlo for the final one. The section on policy instruments has been jointly designed.
2 This undoubtedly reflects the important economic role of forestry and the need to legitimize silviculture and management techniques in these regions.
3 See on this subject the argument about 'historical responsibility' of industrialized countries in causing the present GHG concentrations and the recent attempt in estimating indices of responsibilities (Adger and Brown, 1994).
4 In line with writers such as Verne and Asimov.

References

Adams, R.M., Chang, C.C., McCarl, B.A. and Callaway, J.M. (1992) The role of agriculture in climate change: a preliminary evaluation of emission-control strategies. In: Reilly, J.M. and Anderson, M. (eds) *Economic Issues in Global Climate Change: Agriculture, Forestry and Natural Resources.* Westview Press, Boulder, Colorado, pp. 273–287.

Adams, R.M., Adams, D.M., Callaway, J.M., Chang, C.C. and McCarl, B.A. (1993) Sequestering carbon on agricultural land: social cost and impact on timber markets. *Contemporary Policy Issues* 11, 76–87.

Adams, R.M., McCarl, B.A., Segerson, K., Rosenzweig, C., Bryant, K.J., Dixon, B.L., Conner, R., Evenson, R.E. and Ojima, D. (1995) *The Economic Effects of Climate Change on US Agriculture.* Report prepared for the Electric Power Research Institute, Palo Alto, California.

Adger, W.N. and Brown, K. (1994) *Land Use and the Causes of Global Warming.* John Wiley, Chichester.

Adger, W.N., Brown, K., Shiel, R.S. and Whitby, M.C. (1992) Carbon dynamics of land use in Great Britain. *Journal of Environmental Management* 36, 117–133.

Agro Business Consultants (1994) *Agricultural Budgeting and Costing Book No. 38.* Agro Business Consultants, Melton Mowbray, UK.

Ahrens, H. and Lippert, C. (1994) Tinbergen-Regel und Agrapolitik. *Gesellschaftliche Forderungen an die Landwirtschaft. Schriften der Gesellschaft für Wirtschafts- und Sozialwissenschaften des Landbaues* 30, 151–160.

ALB (1989) *Emissionen von Ammoniak.* Bundesamt für Ernährung und Forstwirtschaft, Frankfurt-am-Main.

Allison, S.M., Proe, M.F. and Matthews, K.B. (1994) The prediction and distribution of general yield classes of Sitka spruce in Scotland by empirical analysis of site factors and a geographic information system. *Canadian Journal of Forestry Research* 24, 2166–2171.

Alriksson, A. and Olsson, M.T. (1995) Soil changes in different age classes of Norway spruce *Picea abies* L. Karst. on afforestated farmland. *Plant and Soil* 169, 103–110.

Anderson, D. (1991) *The Forestry Industry and the Greenhouse Effect.* Forestry Commission and the Scottish Forestry Trust, Edinburgh.

Anderson, D. and Williams, R. (1993) *The Cost Effectiveness of GEF Projects.* Working Paper No. 6, Global Environment Facility, Washington, DC.

Andrasko, K. (1990) Forestry. In: Lashof, D. and Tirpak, D. (eds) *Policy Options to Stabilize Global Climate: Report to Congress.* US Environmental Protection Agency, Washington, DC, pp. 175–237.

Andrasko, K., Heaton, K. and Winnett, S. (1991) Evaluating the costs and efficiency of options to manage global forests: a cost curve approach of site, national and global analyses. In: Howlett, D. and Sargent, C. (eds) *Proceedings of the Technical Workshop to Explore Options for Global Forestry Management, April 1991, Bangkok, Thailand.* IIED, London, pp. 216–233.

Anuchin, N.P. (1985) *Forestry Encyclopedia.* State Committee for Forestry, Moscow, USSR.

Anz, C. (1993) Community afforestation policy. In: Volz, K.-R. and Weber, N. (eds) *Afforestation of Agricultural Land.* Commission of the European Communities, Brussels, pp. 9–10.

Apps, M.J., Kurz, W.A., Luxmoore, R.J., Nilsson, L.O., Sedjo, R.A., Schmidt, R., Simpson, L.G. and Vinson, T.S. (1993) Boreal forests and tundra. *Water, Air and Soil Pollution* 70, 39–54.

Armstrong Brown, S., Rounsevell, M.D.A., Annan, J.D., Phillips, V.P. and Audsley, E. (1996) Agricultural policy impacts on UK nitrous oxide fluxes. *Proceedings of the Workshop on Mineral Emissions from Agriculture.* January, Oslo, Norway.

Arrow, K.J., Cline, W.R., Mäler, K.-G., Munasinghe, M. and Stiglitz, J.E. (1996) Intertemporal equity, discounting, and economic efficiency. In: Bruce, J., Lee, H. and Haites, E. (eds) *Climate Change 1995: Economic and Social Dimensions of Climate Change.* Cambridge University Press, Cambridge, pp. 125–144.

Arthur, L.M. and Abizadeh, F. (1988) Potential effects of climate change on agriculture in the prairie region of Canada. *Western Journal of Agricultural Economics* 13, 215–224.

Assmann, E. and Franz, F. (1963) Vorläufige Fichtenertragstafel für Bayern. In: Hilfstafeln für die Forsteinrichtung (ed.) *Bayerisches Staatsministerium für Ernährung.* Landwirtschaft und Forsten, Munich, p. 334.

Audsley, E. (1987) *The Effects of Field Burning Restrictions, Straw Incorporation and Alternative Uses as Fuel, Fibre or Feed on the Amount of Field Burning and the Supply and Demand for Straw and the Economic Consequences on the Farm of Baling and Briquetting.* Report No. 52, AFRC Institute of Engineering Research, Silsoe, UK.

Audsley, E. (1993) Labour, machinery and cropping planning. *Proceedings of the XXV CIOSTA-CIGR V Congress,* Wageningen, The Netherlands, pp. 83–88.

Ayres, R. and Walter, J. (1991) The greenhouse effect: damages, costs and abatement. *Environmental and Resource Economics* 1, 237–270.

Barrow, P., Hinsley, A.P. and Price, C. (1986) The effect of afforestation on hydroelectricity generation: a quantitative assessment. *Land Use Policy* 3, 141–151.

Bateman, I.J., Diamond, E., Langford, I.H. and Jones, A. (1996) Household willingness to pay and farmers' willingness to accept compensation for establishing a recreational woodland. *Journal of Environmental Planning and Management* 39, 21–44.

Begon, M., Harper, J.L. and Townsend, C.R. (1990) *Ecology: Individuals, Populations and Communities*. Blackwell, Cambridge, Massachusetts.

Berg, Å., Ehnström, B., Gustafsson, L., Hallingbäck, T., Jonsell, M. and Weslien, J. (1995) Threat levels and threats in two red listed species in Swedish forests. *Conservation Biology* 9, 1629–1633.

Bergman, L. (1991) General equilibrium effects of environmental policy: a CGE-modeling approach. *Environmental and Resource Economics* 1, 43–61.

Bernthal, F.M. (ed.) (1990) *Formulation of Response Strategies*. Intergovernmental Panel on Climate Change, Working Group III, WMO and UNEP, Geneva.

Beuermann, C. and Jäger, J. (1996) Climate change politics in Germany: how long will the double dividend last? In: O'Riordan, T. and Jäger, J. (eds) *Politics of Climate Change: A European Perspective*. Routledge, London, pp. 186–227.

Bishop K.D. and Phillips A.C. (1993) Seven steps to market: the development of the market-led approach to countryside conservation and recreation. *Journal of Rural Studies* 9, 315–338.

BML (1990) *Bundeswaldinventur, Band I: Inventurbericht und Übersichtstabellen für das Bundesgebiet nach dem Gebiestsstand vor dem 01.10.1990 einschließlich Berline (West). Band II: Grundtabellen für das Bundesgebiet nach dem Gebiestsstand vor dem 01.10.1990 einschließlich Berlin (West)*. Bundesministerium für Ernährung, Landwirtschaft und Forsten, Bonn.

Bodansky, D. (1993) The United Nations Framework Convention on Climate Change: a commentary. *Yale Journal of International Law* 18, 451–558.

Bohm, P. (1997) Environmental taxation and the environmental dividend: fact or fallacy. In O'Riordan, T. (ed.) *Ecotaxation*. Earthscan, London, pp. 106–124.

Bonan, G.B., Pollard, D. and Thomson, S.L. (1992) Effects of boreal forest vegetation on global climate. *Nature* 359, 716–718.

Bonny, S. (1993) Is agriculture using more energy? A French case study. *Agricultural Systems* 43, 51–66.

Borman, F.J. and Likens, G.E. (1979) *Patterns and Process in a Forested Ecosystem*. Springer, New York.

Böswald, K. (1995) Wald und Forstwirtschaft im regionalen Kohlenstoffhaushalt Bayerns. *Allgemeine Forst Zeitschrift* 6, 291–295.

Böswald, K. (1996) Zur Bedeutung des Waldes und der Forstwirtschaft im Kohlenstoffhaushalt – Eine Analyse am Beispiel des Bundeslandes Bayern. Schriftenreihe der Forstwissenschaftlichen Fakultät der Universität Munchen und der Bayerischen Landesanstalt für Wald und Forstwirtschaft, Band 159.

Böswald, K. and Wierling, R. (1995) *Wald und Forstwirtschaft Niedersachsens im Kohlenstoffhaushalt*. Gutachten fur das Niedersächsische Ministerium für Ernährung, Landwirtschaft und Forsten. Unpublished.

Bouwman, A.F., van den Born, G.J. and Swart, R.J. (1992) *Land-Use Related Sources of CO_2, CH_4, and N_2O*. Report 222901004, National Institute of Public Health and Environmental Protection, Bilthoven, the Netherlands.

Bovenberg, A.L. (1993) Policy instruments for curbing CO_2 emissions: the case of the Netherlands. *Environmental and Resource Economics* 3, 233–244.

Bovenberg, A.L. and de Mooij, R.A. (1993) *Environmental Policy in a Small Open Economy with Distortionary Labour Taxes: A General Equilibrium Analysis*. Research Centre for Economic Policy Research, Erasmus University, Rotterdam.

Bovenberg, A.L. and de Mooij, R.A. (1994) Environmental levies and distortionary taxation. *American Economic Review* 84, 1085–1089.

Bovenberg, A.L. and Goulder, L.H. (1994) *Optimal Environmental Taxation in the Presence of Other Taxes: An Applied General Equilibrium Analysis.* Working Paper, Stanford University, Stanford, California.

Bovenberg, A.L. and van der Ploeg, F. (1994a) *Consequences of Environmental Tax Reform for Involuntary Unemployment and Welfare.* Working Paper 56, Centre for Economic Studies, University of Munich, Munich.

Bovenberg, A.L. and van der Ploeg, F. (1994b) Environmental policy, public finance and the labour market in a second best world. *Journal of Public Economics* 55, 349–390.

Bovenberg, A.L. and van der Ploeg, F. (1994c) Environmental policy, public goods and the marginal cost of public funds. *Economic Journal* 104, 444–454.

Brabänder, H.D., Haber, W., Köhne, N., Sturies, H.J. and Thoroe, C. (1992) Konzeption zur verstärkten Aufforstung landwirtschaftlicher Flächen. *Agra Europe* 51, 1–7.

Bradley, R.I. (1993) *The Century Model of Soil Organic Matter. SSLRC Research Report for MAFF Project on the Impact of Climate Change on Soil Workability.* Soil Survey and Land Research Centre, Silsoe.

Bramryd, T. (1982) Fluxes and accumulation of organic carbon in urban ecosystems on a global scale. In: Bornkam, R., Lee, J.A. and Seaward, M.R.D. (eds) *Urban Ecology.* Blackwell Scientific Publications, Oxford, pp. 3–12.

Brazee, R. and Mendelsohn, R. (1990) A dynamic model of timber markets. *Forest Science* 36, 255–264.

Breuss, F. and Steininger, K. (1995) *Reducing the Greenhouse Effect in Austria: A General Equilibrium Evaluation of CO_2 Policy Options.* IEF Working Paper 7, Research Institute for European Affairs, Vienna.

Bromley, D. (1991) *Environment and Economy: Property Rights and Public Policy.* Blackwell, Cambridge, Massachusetts.

Bromley, D. and Hodge, I. (1990) Private property rights and presumptive policy entitlements: reconsidering the premises of rural policy. *European Review of Agricultural Economics* 17, 197–214.

Broome, J. (1992) *Counting the Cost of Global Warming.* White Horse Press, Cambridge.

Brouwer, F.M. and van Berkum, S. (1996) *CAP and Environment in the European Union.* Agricultural Economics Research Institute, The Hague.

Brown, K. and Adger, W.N. (1993) Estimating national greenhouse gas emissions under the Climate Change Convention. *Global Environmental Change* 3, 149–158.

Brown, K. and Pearce, D. (1994) The economic value of non-market benefits of tropical forests: carbon storage. In: Weiss, J. (ed.) *The Economics of Project Appraisal and the Environment.* Edward Elgar, Aldershot, pp. 102–123.

Brown, L.R. (1995) *Who Will Feed China? Wake-Up Call for a Small Planet.* W.W. Norton, New York.

Brown, S. (1996) Present and potential roles of forests in the global climate change debate. *Unasylva* 185(47), 3–8.

Brown, S., Cannell, M.G.R., Kauppi, P.E. and Sathaye, J. (1996) Management of forests for mitigation of greenhouse gas emissions. In: Watson, R.T., Zinyowera,

M.C. and Moss, R.H. (eds) *Climate Change 1995: Impacts, Adaptations and Mitigation of Climate Change: Scientific–Technical Analyses.* Cambridge University Press, Cambridge, pp. 773–797.

Bundesministerium fuer Ernaehrung, Landwirtschaft und Forsten (ed.) (1995) *Agrarbericht der Bundesregierung.* Bonn.

Burschel, P., Kürsten, E. and Larson, B.C. (1993a) *Die Rolle von Wald und Forstwirtschaft im Kohlenstoffhaushalt – eine Betrachtung für die Bundesrepublik Deutschland.* Schriftenreihe der Forstwissenschaftlichen Fakultät der Universität Munchen und der Bayerischen Forstlichen Versuchs- und Forschungsanstalt Munchen, Munich.

Burschel, P., Weber, M., Böswald, K. and Felbermeier, B. (1993b) *Sitzung der Enquete-Komission des Deutschen Bundestages zum Schutz der Erdatmosphäre am 15.03.01993.* Stellungnahme zum Fragenkatalog: Potentiale der Kohlenstoffixierung durch Ausweitung der Waldflächen. Unpublished.

Burton, M. and Young, T. (1996) The impact of BSE on the demand for beef and other meats in Great Britain. *Applied Economics* 28, 687–693.

Cairns, M.A., Baker, J.R., Shea, R.W. and Haggerty, P.K. (1995) Carbon dynamics of Mexican tropical evergreen forests: influence of forestry mitigation options and refinement of carbon-flux estimates. *Interciencia* 20(6), 401–408.

Cannell, M.G.R. (1995) *Forest and the Global Carbon Cycle in the Past, Present and Future.* Research Report 2, European Forest Institute, Joensuu, Finland.

Cannell, M.G.R. and Dewar, R.C. (1995) The carbon sink provided by plantation forests and their products in Britain. *Forestry* 68, 35–48.

Cannell, M.G.R. and Milne, R. (1995) Carbon pools and sequestration in forest ecosystems in Britain. *Forestry* 68, 361–378.

Cannell, M.G.R., Dewar, R.C. and Pyatt, D.G. (1993) Conifer plantations on drained peatlands in Britain: a net gain or loss of carbon? *Forestry* 66, 353–369.

Capros, P., Georgakopoulos, P., Zografakis, S., Proost, S., Van Regemorter, D., Conrad, C., Schmidt, T., Smeers, Y. and Michiels, E. (1995) *Double Dividend Analysis: First Results of a General Equilibrium Model (GEM-E^3) Linking the EU-12 Countries.* Nota Di Lavoro 26.95, Fondazione Eni Enrico Mattei, Milan.

Carraro, C., Galeotti, M. and Gallo, M. (1995) *Environmental Taxation and Unemployment: Some Evidence on the 'Double Dividend Hypothesis' in Europe.* Nota Di Lavoro 34.95, Fondazione Eni Enrico Mattei, Milan.

Charlson, R.J. and Wigley, T.M.L. (1994) Sulfate aerosols and climate change. *Scientific American* 270(2), 48–54.

Charlson, R.J., Schwartz, S.E., Hales, J.M., Cess, R.D., Coakley, J.A.J., Hansen, J.E. and Hoffmann, D.J. (1992) Climate forcing by anthropogenic aerosols. *Science* 255, 423–430.

Chichilnisky, G., Gornitw, V., Heal, G., Ring, D. and Rosenzweig, C. (1996) *Building Linkages Among Climate, Impacts and Economics: A New Approach to Integrated Assessment.* Working Paper, New York Global Systems Initiative, New York.

Cicerone, R.J. and Oremland, R.S. (1988) Biogeochemical aspects of atmospheric methane. *Global Biogeochemical Cycles* 2, 299–327.

Ciesla, W.M. (1995) *Climate Change, Forests and Forest Management: An Overview.* FAO Forestry Paper 126, FAO, Rome.

Cleveland, C.J. (1996) Resource degradation, technical change, and the productivity of energy use in US agriculture. *Ecological Economics* 13(3), 185–202.

Cline, W.R. (1992a) *The Economics of Global Warming*. Institute for International Economics, Washington, DC.
Cline, W.R. (1992b) *Optimal Carbon Emissions over Time: Experiments with the Nordhaus DICE Model*. Mimeo, Institute for International Economics, Washington, DC.
Cline, W.R. (1993) Give greenhouse abatement a chance. *Finance and Development* 30(1), 3–5.
Club of Rome (1972) *I limiti dello sviluppo*. Edizioni Scientifiche e Tecniche Mondadori, Milan.
Coase, R.H. (1960) The problem of social costs. *Journal of Law and Economics* 3, 1–44.
Cole, V., Cerri, C., Minami, K., Mosier, A., Rosenberg, N. and Sauerbeck, D. (1996) Agricultural options for mitigation of greenhouse gas emissions. In: Watson, R.T., Zinyowera, M.C. and Moss, R.H. (eds) *Climate Change 1995: Impacts, Adaptations and Mitigation of Climate Change: Scientific–Technical Analyses*. Cambridge University Press, Cambridge, pp. 747–771.
Colman, D. (1994) Ethics and externalities: agricultural stewardship and other behaviour. *Journal of Agricultural Economics* 45, 299–311.
Colman, D., Crabtree B., Froud J. and O'Carrol, L. (1992) *Comparative Effectiveness of Conservation Mechanisms*. Department of Agricultural Economics, University of Manchester, Manchester.
Conrad, K. and Schröder, M. (1991) The control of CO_2 emissions and its economic impact: an AGE model for a German state. *Environmental and Resource Economics* 1, 289–312.
Conrad, K. and Schröder, M. (1993) Choosing environmental policy instruments using general equilibrium models. *Journal of Policy Modeling* 15, 521–543.
Cox, D.R. and Miller, H.D. (1965) *The Theory of Stochastic Process*. Chapman and Hall, London.
Coxworth, E., Hultgreen, G. and Leduc, P. (1994) *Net Carbon Balance Effects of Low Disturbance Seeding Systems on Fuel, Fertilizer, Herbicide and Machinery in Western Canadian Agriculture*. Prairie Agricultural Machinery Institute, Humboldt, Saskatchewan.
Crabtree, J.R. (1995) Agricultural reform and farm forestry. In: OECD (ed.) *Forestry, Agriculture and the Environment*. OECD, Paris, pp. 55–79.
Crabtree, J.R. (1996) *Evaluation of the Farm Woodland Premium Scheme*. Report to Scottish Office, Edinburgh.
Cramer, W.P. and Leemans, R. (1993) Assessing impacts of climate change on vegetation using climate classification systems. In: Solomon, A.M. and Shugart, H.H. (eds) *Vegetation Dynamics and Global Change*. Chapman and Hall, London, pp. 190–217.
Crutzen, P.J. and Andreae, M.O. (1990) Biomass burning in the tropics: impacts on atmospheric chemistry and biogeochemical cycles. *Science* 250, 1669–1678.
Crutzen, P.J., Aselman, I. and Seiler, W. (1986) Methane production by domestic animals, wild animals, other herbivorous fauna and humans. *Tellus* B38, 271–284.
Curtin, D., Selles, F., Campbell, C.A. and Biederbeck, V.O. (1994) *Canadian Prairie Agriculture as a Source and Sink of the Greenhouse Gases, Carbon Dioxide and Nitrous Oxide*. Research Branch Publication Number 379M0094, Agriculture Canada, Swift Current.

Czerkwaski, A.W. (1969) Methane production in ruminants and its significance. *World Review of Nutrition and Dietetics* 11, 240–282.

Danilov-Daniljan, V.I., Gorshkov, V.G., Arsky, J.M. and Losev, K.S. (1994) The environment between the past and the future: the world and Russia. *Environment and Economic Analysis.* Trial, Russia.

D'Arge, R.C., Schulze, W.D. and Brookshire, D.S. (1982) Carbon dioxide and intergenerational choice. *American Economic Review Papers and Proceedings* 72, 251–256.

Darwin, R., Tsigas, M., Lewandrowski, J. and Raneses, A. (1995) *World Agriculture and Climate Change: Economic Adaptations.* Agricultural Economic Report No. 703, Natural Resources and Environment Division, Economic Research Service, US Department of Agriculture, Washington, DC.

Dasgupta, P., Kriström, B. and Mäler, K.-G. (1995) Current issues in resource accounting. In: Johansson, P.-O., Kriström, B. and Mäler. K.-G. (eds) *Current Issues in Environmental Economics.* Manchester University Press, Manchester, pp. 117–153.

De Alessi, L. (1990) Development of the property rights approach. *Journal of Institutional and Theoretical Economics* 146, 6–23.

De Benedictis M. (1997) From protection to market liberalisation. Italian viewpoints of CAP developments – rural and environmental issues at stake in CAP's future. In: Ferro O. (ed.) *What Future for the CAP? Perspectives and Expectations for the Common Agricultural Policy of the European Union.* Wissenschaftverlag Vauk, Kiel. (in press).

de Haan, M. and Keuning, S.J. (1995) *Taking the Environment into Account: The Netherlands NAMEA's for 1989, 1990 and 1991.* Occasional Paper NA-074, Netherlands Central Bureau of Statistics, The Hague.

De Jong, B.H.J., Montoya-Gómez, G., Nelson, K., Soto-Pinto, L., Taylor, J. and Tipper, R. (1995) Community forest management and carbon sequestration: a feasibility study from Chiapas, Mexico. *Interciencia* 20(6), 409–416.

Delcourt, G. and van Kooten, G.C. (1995) How resilient is grain production to climatic change? Sustainable agriculture in a dryland cropping region of western Canada. *Journal of Sustainable Agriculture* 5, 37–57.

de Melo, J. and Tarr, D. (1992) *A General Equilibrium Analysis of US Foreign Trade Policy.* MIT Press, Cambridge, Massachusetts.

Demeyer, D.I. and van Nevel, C.J. (1975) Methanogenesis, an integrated part of carbohydrate fermentation and its control. In: McDonald, I.W. and Warner, A.C.I. (eds) *Digestion and Metabolism in the Ruminant.* University of New England Publishing Unit, Armidale, pp. 366–382.

Demmel, S. and Alefeld, G. (1993) CO2-Reduktion und Wirtschaftlichkeit – ein Gegensatz? Vergleich verschiedener Systeme für die Strom – und Wärmeeerzeugung. *Energiewirtschaftliche Tagesfragen* 43, 112–117.

Detwiler, R.P. and Hall, C.A.S. (1988) Tropical forests and the global carbon cycle. *Science* 239, 42–47.

Deutscher Bundestag (1993) *Schutz der grünen Erde. Klimaschutz durch umweltgerechte Landwirtschaft und Erhalt der Wälder.* Enquete-Komission 'Schutz der Erdatmosphäre' des Deutschen Bundestages, Economia Verlag, Bonn.

Dewar, R.C. (1991) Analytical model of carbon storage in the trees, soils and wood products of managed forests. *Tree Physiology* 8, 239–258.

Dewar, R.C. and Cannell, M.G.R. (1992) Carbon sequestration in the tree, products and soils of forest plantations: an analysis using UK examples. *Tree Physiology* 11, 49–71.

de Wit, G. (1995) Belastingverschuiving en werkgelegenheid. *Economische Statistische Berichten* 3996, 134–137.

Dixit, A. (1993) *The Art of Smooth Pasting*. Harwood Academic, Chur, Switzerland.

Dixon, R.K., Andrasko, K.J., Sussman, F.G., Lavinson, M.A., Trexler, M.C. and Vinson, T.S. (1993a) Forest sector carbon offset projects: near-term opportunities to mitigate greenhouse gas emissions. *Water, Air and Soil Pollution* 70, 561–577.

Dixon, R.K., Winjum, J.K. and Schroeder, P.E. (1993b) Conservation and sequestration of carbon: the potential of forest and agroforest management practices. *Global Environmental Change* 3, 159–173.

Dixon, R.K., Winjum, J., Andrasko, K., Lee, J. and Stirred, P. (1994a) Integrated systems: assessment of promising agroforest and alternative land-use practices to enhance carbon conservation and sequestration. *Climatic Change* 27, 71–92.

Dixon, R.K., Brown, S., Houghton, R.A., Solomon, A.M., Trexler, M.C. and Wisniewski, J. (1994b) Carbon pools and flux of global forest ecosystems. *Science* 263, 185–190.

Dixon, R.K., Krankina, O.N. and Kobak, K.I. (1996) Global climate change adaptation: examples from Russian boreal forests. In: Smith, J.B. and Guill, S. (eds) *Adapting to Climate Change: Assessments and Issues*. Springer-Verlag, New York, pp. 359–373.

DLO (Dienst Landbouwkundig Onderzoek) (1993) *Meetmehoden NH3-emissie uit stallen*. DLO, Wageningen.

Dosi, C. and Moretto, M. (1996a) *Environmental Innovation and Public Subsidies under Asymmetry of Information and Network Externalities*. Nota di Lavoro, Fondazione Eni Enrico Mattei, Milan.

Dosi, C. and Moretto, M. (1996b) Toward green technologies: switching rules and environmental policies. *Rivista Internazionale di Scienze Economiche e Commerciali* 43, 13–30.

Dosi, C. and Moretto, M. (1997) Pollution accumulation and firm incentives to promote irreversible technological change under uncertain private benefits. *Environmental and Resource Economics* 7 (in press).

Dowlatabadi, H. and Morgan, M.G. (1993) A model framework for integrated studies of the climate problem. *Energy Policy* 21, 209–221.

Downing, T.E., Olsthoorn, A.A. and Tol, R.S.J. (eds) (1996) *Climate Change and Extreme Events: Altered Socio-economic Impacts and Policy Responses*. Institute for Environmental Studies, Free University, Amsterdam.

Dudek, D.J. and Leblanc, A. (1995) *Reforestation in Russia: Building Institutional Capacity for Joint Implementation Through the Vologda Demonstration Project*. Summary Final Report, Environmental Defence Fund, Washington, DC.

Dykstra, D.P. (1994) *Wood Residues from Timber Harvesting and Primary Processing: A Global Assessment for Tropical Forests*. Draft Working Paper, FAO, Rome.

Dykstra, D.P. and Heinrich, R. (1992) Sustaining tropical forests through environmentally sound harvesting practices. *Unasylva* 169(43), 9–15.

Edmonds, J. (1992) Why understanding the natural sinks and sources of CO_2 is important: a policy analysis perspective. *Water, Air and Soil Pollution* 64, 11–21.

Eggleston, H.S. and Williams, M.L. (1989) *UK Emissions of Carbon Dioxide and Methane, 1960–1987.* Warren Spring Laboratory Report to Department of Environment, London.

Eichner, M.J. (1990) Nitrous oxide emissions from fertilised soils: summary of available data. *Journal of Environmental Quality* 19, 272–280.

Ekins, P. (1995) Rethinking the costs related to global warming: a survey of the issues. *Environmental and Resource Economics* 5, 1–47.

Eliasson, P. (1995) *Miljöjusterade nationalräkenskaper för den svenska skogen åren 1987 och 1991.* Report 108, Department of Economics, Swedish University of Agricultural Sciences, Umeå, Sweden.

Emanuel, W., Shugart, H. and Stevenson, M. (1985) Climate change and the broad-scale distribution of terrestrial ecosystem complexes. *Climatic Change* 7, 29–43.

Energie Beheer Nederland (1995) *Energiebulletin 3300.* Heerlen, the Netherlands.

England, R.A. (1984) *Reducing the Nitrogen Input to Cereals: 1. Effect of Nitrogen Price. 2. Price Effects at Farm Level.* National Institute of Agricultural Engineering, Wrest Park, Silsoe.

Enquete Commission (1992) *Klamaänderung gefährdet globale Entwicklung – Zunkunft sichern – Jetzt handeln.* Erster Bericht der Enquete Kommission des 'Schutz der Erdatmosphäre' des 12 Deutschen Bundestages, Deutscher Bundestag, Bonn.

Enquete Commission (1995) *Protecting our Green Earth: How to Manage Global Warming through Environmentally Sound Farming and Preservation of the World's Forests.* Report by 12th German Bundestag, Bonn.

Eriksson, H. (1991) Sources and sinks of carbon dioxide in Sweden. *Ambio* 20, 146–150.

European Commission (1987) *Farm Structure: 1985 Survey, Main Results.* EC, Brussels, Belgium.

European Commission (1992) The climate challenge – economic aspects of the Community's strategy for limiting CO_2 emissions. *European Economy* No. 51, May.

European Commission (1994a) Taxation, employment and environment: fiscal reform for reducing unemployment. *European Economy* No. 56, Part C, Analytical Study No. 3, March.

European Commission (1994b) EC Agricultural Policy for the 21st century. In: *European Economy: Reports and Studies.* Report No. 4, Directorate General for Economic and Financial Affairs. European Commission, pp. 70–72.

European Commission (1995) *Agriculture: Statistical Yearbook.* EC, Brussels, Belgium.

Eurostat (1995) *Europe's Environment: Statistical Compendium for the Dobris Assessment.* Official Publications of the EC, Luxemburg.

FACE (1994) *Forests Absorbing Carbon Dioxide Emission.* Annual Report 1993, FACE, Arnhem, the Netherlands.

Fankhauser, S. (1994) The social costs of greenhouse gas emissions: an expected value approach. *Energy Journal* 15, 157–184.

Fankhauser, S. (1995a) *Valuing Climate Change: The Economics of the Greenhouse.* Earthscan, London.

Fankhauser, S. (1995b) Protection vs. retreat. the economic costs of sea level rise. *Environment and Planning A* 27, 299–319.

Fankhauser, S. and Tol, R.S.J. (1996) Recent advancements in the economic assessment of climate change costs. *Energy Policy* 24, 665–673.

FAO (1990) *Climate Change and Agriculture, Forestry and Fisheries.* Position paper for the Second World Climate Conference, Geneva, Switzerland, FAO, Rome.
FAO (1992) *FAO Yearbook of Forest Products.* FAO Forestry Series No. 125 /FAO Statistics Series No. 103, FAO, Rome.
FAO (1993a) *The Challenge of Sustainable Forest Management – What Future for the World's Forests?* FAO, Rome.
FAO (1993b) *Forest Resources Assessment 1990 – Tropical Countries.* FAO Forestry Paper 112, FAO, Rome.
FAO (1994) *Readings in Sustainable Forest Management.* FAO Forestry Paper 122, FAO, Rome.
FAO (1995) *State of the World's Forests.* FAO, Rome, 48 pp.
FAO (1996) *Production Yearbook.* FAO, Rome.
Federal Forestry Service (1994) *Russian Forests.* Russian Research and Development Centre for Forest Resources, Moscow.
Federal Ministry of Environment, Germany (1994) *First Report of the Government of the Federal Republic of Germany Pursuant to the United Nations Framework Convention on Climate Change.* Federal Ministry of Environment, Bonn.
Ferro, O. (ed.) (1997) *What Future for the CAP? Perspectives and Expectations for the Common Agricultural Policy of the European Union.* Wissenschaftverlag Vauk, Kiel (in press).
Fischer, A.C. and Hanemann, W.M. (1990) Information and dynamics of environmental protection: the concept of the critical period. *Scandinavian Journal of Economics* 92, 399–414.
Foerster, W. and Böswald, K. (1994) Methodik und Zuwachsergebnisse einer Zuwachsberechnung für Bayerns Wälder. *Forstwissenschaftliches Centralblatt* 113, 142–151.
Foerster, W., Böswald, K. and Kennel, E. (1993) Überraschend hoher Zuwachs in Bayern – Vergleich der Inventurergebnisse von 1971 und 1987. *Allgemeine Forst Zeitschrift* 23, 1178–1180.
Forestry Commission (1995) *Annual Report and Accounts.* Forestry Commission, Edinburgh.
Franz, F. and Kennel, E. (1973) *Bayerische Waldinventur 1970/71, Inventurabschnitt I: Großrauminventur, Basistabellen.* Forstliche Forschungsberichte 12, Munich.
Froud, J. (1994) The impact of ESAs on lowland farming. *Land Use Policy* 11, 107–118.
Frühwald, A. and Wegener, G. (1993) Energiekreislauf Holz – Ein Vorbild für die Zukunft. *Holz-Zentralblatt* 119, 1949–1952.
Frühwald, A. and Wegener, G. (1994) *Bewertung von Holz im Vergleich mit a Werkstoffen unter dem Aspekt der CO_2-Bilanz.* Forschungsbericht, bewilligt und koodiniert uber die deutsche Gesellschaft für Holzforschung, Munich, and Forstabsatzfond, Bonn.
Furubotn, E. and Pejovich, S. (1972) Property rights and economic theory: a survey of recent literature. *Journal of Economic Literature* 10, 1137–1162.
Galinski, W. and Kuppers, M. (1994) Polish forest ecosystems: the influence of changes in the economic system on the carbon balance. *Climatic Change* 27, 103–109.
García, E. (1973) *Modoficaciones al Sistema de Climatica de Koppen.* Instituto de Geografia, UNAM, Mexico DF, Mexico.

Gaskins, D.W. and Weyant, J.P. (1993) Model comparisons of the costs of reducing CO_2 emissions. *American Economic Review Papers and Proceedings* 83, 318–330.

Gay, C., Imaz, M., Wiel, S., Wisniewski, J., Ruíz-Suarez, L. and Goldberg, B. (1995) Workshop summary: Regional Workshop on Greenhouse Gas Mitigation Strategies for Latin American Countries, Cancún, México, 10–13 July 1995. *Interciencia* 20, 3–15.

Giampietro, M., Cerretelli, G. and Pimentel, D. (1992) Assessment of different agricultural production practices. *Ambio* 21, 451–459.

Gibbs, M.J. and Leng, R.A. (1993) Methane emissions from livestock. In: van Amstel, A.R. (ed.) *Methane and Nitrous Oxide: Methods in National Emission Inventories and Options for Control.* International IPCC Workshop Proceedings, National Institute for Public Health and Environmental Protection, the Netherlands, pp. 73–79.

Gibbs, M.J. and Woodbury, J.W. (1993) Methane emissions from livestock manure. In: van Amstel, A.R. (ed.) *Methane and Nitrous Oxide: Methods in National Emission Inventories and Options for Control.* International IPCC Workshop Proceedings. National Institute for Public Health and Environmental Protection, the Netherlands, pp. 81–91.

Goulder, L.H. (1995a) Effects of carbon taxes in an economy with prior tax distortions: an intertemporal general equilibrium analysis. *Journal of Environmental Economics and Management* 29, 271–297.

Goulder, L.H. (1995b) Environmental taxation and the 'double dividend': a reader's guide. *International Tax and Public Finance* 2, 157–183.

Grabherr, G., Gottfried, M. and Pauli, H. (1994) Climate effects on mountain plants. *Nature* 369, 448.

Grainger, A. (1990) Modelling the impact of alternative afforestation strategies to reduce carbon dioxide emissions. *Proceedings – Tropical Forestry Response Options to Climate Change.* US EPA, São Paulo, Brazil, pp. 93–104.

Gregory, R. (1989) Political rationality or 'incrementalism'? *Policy and Politics* 17, 139–153.

Grossman, G.M. and Krueger, A.B. (1995) Economic growth and the environment. *Quarterly Journal of Economics* 110, 353–377.

Grubb, M. (1993) The costs of climate change: critical elements. In: Kaya, Y., Nakicenovic, N., Nordhaus, W.D. and Toth, F. (eds) *Costs, Impacts and Benefits of CO_2 Mitigation.* IIASA Collaborative Paper Series, CP93-2, Laxenburg, Austria.

Grubb, M. (1995a) European climate change policy in a global context. In: Bergeson, H.O. and Parmann, G. (eds) *Green Globe Yearbook 1995.* Oxford University Press, Oxford, pp. 41–50.

Grubb, M. (1995b) *Renewable Energy Strategies for Europe,* Vol. 1. Royal Institute of International Affairs and Earthscan, London.

Haigh, N. (1996) Climate change policies and politics in the European Community. In: O'Riordan, T. and Jäger, J. (eds) *Politics of Climate Change: A European Perspective.* Routledge, London, pp. 155–185.

Hanley, N., Simpson, I., Parsisson, D., Macmillan, D., Bullock, C. and Crabtree, R. (1996) *Valuation of the Conservation Benefits of Environmentally Sensitive Areas.* Report to the Scottish Office Agriculture, Environment and Fisheries Department, Edinburgh.

Hansen, E. (1997) Forest certification and its role in marketing strategy. *Forest Product Journal* 47(3), 16–22.

Harmon, M.E., Ferrel, W.K. and Franklin, J.F. (1990) Effects on carbon storage of the conversion of old-growth forests to young forests. *Science* 247, 699–702.

Harold, C. and Runge, C.F. (1993) GATT and the environment: policy research needs. *American Journal of Agricultural Economics* 75, 789–793.

Harrison, J.M. (1985) *Brownian Motion and Stochastic Flow Systems*. John Wiley, New York.

Health, L.S., Kauppi, P.E., Burschel, P., Gregor, H.-D., Guderian, R., Kohlmaier, G.H., Lorenz, S., Overdieck, D., Scholz, F., Thomasius, H. and Weber, M. (1993) Contribution of temperate forests to the world's carbon budget. *Water, Air and Soil Pollution* 70, 55–69.

Henrichsmeyer, W. (1988) Agrarwirtschaft: räumliche Verteilung. *Handwoeterbuch der Wirtschaftswissenschaft*, Vol. 1, Stuttgart, pp. 169–185.

Henrichsmeyer, W. and Witzke, H. (1994) *Agrarpolitik Bd 2*. Bewertung und Willensbildung, Stuttgart.

Heyer, J. (1994) Methane. In: Deutscher Bundestag (ed.) *Studienprogramm Landwirtschaft* Band 1, Teilband 1. Economia Verlag, Bonn.

Hodge, I. (1995) *Environmental Economics*. Macmillan, London.

Hoel, M. (1996) Should a carbon tax be differentiated across sectors? *Journal of Public Economics* 59, 17–32.

Hoen, H.F. and Solberg, B. (1995) CO_2 *Taxing, Timber Rotations, and Market Implications*. Draft, European Forestry Institute, Jonesuu, Finland.

Holdrige, L. (1947) Determination of world plant formation from simple climatic data. *Science* 105, 367–368.

Holling, C.S., Gunderson, L. and Peterson, G. (1993) *Comparing Ecological and Social Systems*. Beijer Paper No. 36, Beijer Institute, Stockholm.

Holtz-Eakin, D. and Selden, T.M. (1995) Stoking the fires? CO_2 emissions and economic growth. *Journal of Public Economics* 57, 85–101.

Hope, C., Anderson, J. and Wenman, P. (1993) Policy analysis of the greenhouse effect: an application of the PAGE model. *Energy Policy* 21, 327–338.

Horlacher, D. and Marschner, H. (1990) Schätzrahmen zur Beurteilung von Ammoniakverlusten nach Ausbringung von Rinderflüssigmist. *Zeitschrift für Pflanzenernährung und Bodenkunde* 153, 107–115.

Houghton, J.T., Jenkins, G.J. and Ephraums, J.J. (eds) (1990) *Climate Change: The IPCC Scientific Assessment*. Cambridge University Press, Cambridge.

Houghton, J.T., Meiro Filho, L.G., Callander, B.A., Harris, N., Kattenberg, A. and Maskell, K. (eds) (1996) *Climate Change 1995: The Science of Climate Change*. Cambridge University Press, Cambridge.

Houghton, R.A. (1990) Projections of future deforestation and reforestation in the tropics. *Proceedings – Tropical Forestry Response Options to Climate Change*. US EPA, São Paulo, Brazil, pp. 87–92.

Houghton, R.A. (1995) Changes in the storage of terrestrial carbon since 1850. In: Lal, R., Kimble, J., Levine, E. and Stewart, B.A. (eds) *Soils and Global Change*. CRC Lewis Publishers, Boca Raton, Florida, pp. 45–65.

Houghton, R.A. and Skole, D.L. (1990) Carbon. In: Turner, B.L., Clark, W.C., Kates, R.W., Richards, J.F., Mathews, J.T. and Meyer, W.B. (eds) *The Earth as*

Transformed by Human Action. Cambridge University Press, Cambridge, pp. 393–408.

Houghton, R.A., Unruh, J. and Lefebvre, P.A. (1991) Current land use in the tropics and its potential for sequestering carbon. In: Howlett, D. and Sargent, C. (eds) *Proceedings of the Technical Workshop to Explore Options for Global Forestry Management. April 1991, Bangkok, Thailand.* IIED, London, pp. 297–310.

Hourcade, J.C., Halsnaes, K., Jaccard, M., Montgomery, D., Richels, R., Robinson, J. and Shukla, P.R. (1996) A review of mitigation cost studies. In: Bruce, J., Lee, H. and Haites, E. (eds) *Climate Change 1995: Economic and Social Dimensions of Climate Change.* Cambridge University Press, Cambridge, pp. 297–366.

House of Lords (1995) *Report from the Select Committee on Sustainable Development.* HMSO, London.

Howes, R. (1995) *Greenhouse Gas Emissions from the Agricultural Sector.* Report to WWF International, IIED, London.

Hultkrantz, L. (1992) National account of timber and forest environmental resources in Sweden. *Environmental and Resource Economics* 2, 283–305.

IPCC (1994) *Climate Change 1994 – Radiative Forcing of Climate Change and an Evaluation of IPCC IS92 Emission Scenarios.* Intergovernmental Panel of Climate Change, Cambridge University Press, Cambridge.

IPCC (1996a) *Climate Change 1995. Impact, Adaptation and Mitigation of Climate Change – Scientific–Technical Analysis.* Contribution of Working Group II to the Second Assessment Report of the Intergovernmental Panel on Climate Change, Cambridge University Press, Cambridge.

IPCC (1996b) *The IPCC Scientific Assessment. The Summary for Policymakers.* IPCC, Geneva.

IPCC/OECD (1994) *Greenhouse Gas Inventory Reference Manual.* IPCC Draft Guidelines for National Greenhouse Gas Inventories, Paris.

Janssens, S. (1990) *Animal Feed: Supply and Demand of Feedingstuffs in the European Community.* Office for Official Publications of the European Communities, Luxemburg.

Jarvis, S.C. and Pain, B.F. (1993) Gaseous emissions from an intensive dairy farming system. *Proceedings of the IPCC AFOS Workshop*, Canberra. IPCC, Geneva, pp. 55–59.

Jarvis, S.C. and Pain, B.F. (1994) Greenhouse gas emissions from intensive livestock systems: their estimation and technologies for reduction. *Climatic Change* 27, 27–38.

Jenkinson, D.S. and Rayner, J.H. (1977) The turnover of soil organic matter in some of the Rothamsted classical experiments. *Soil Science* 123, 298–305.

Jepma, C.J., Asaduzzaman, M., Mintzer, I., Maya, R.S. and Al-Moneff, M. (1996) A generic assessment of response options. In: Bruce, J.P., Lee, H. and Haites, E.F. (eds) *Climate Change 1995: Economic and Social Dimensions of Climate Change.* Cambridge University Press, Cambridge, pp. 229–262.

Jones, P.D. and Raper, S.C.B. (1990) *Global Average Surface Temperature Data.* Climatic Research Unit, University of East Anglia, Norwich.

Jorgenson, D.W. and Wilcoxen, P.J. (1993) The economic impact of the clean air act amendments of 1990. *Energy Journal* 14(1), 159–182.

Josling, T. (1994) The reformed CAP and the industrial world. *European Review of Agricultural Economics* 21, 513–527.

Juridicheskaya Literatura (1994–1995) *Collection of Legislation of Russian Federation.* Moscow.

Kaiser, H.M. and Drennen, T.E. (eds) (1993) *Agricultural Dimensions of Global Climate Change.* St Lucie Press, Delray Beach, Florida.

Kaiser, H.M., Riha, S.J., Wilks, D.S. and Sampath, R. (1993) Adaptation to global climate change at the farm level. In: Kaiser, H.M. and Drennen, T.E. (eds) *Agricultural Dimensions of Global Climate Change.* St Lucie Press, Delray Beach, Florida, pp. 136–152.

Karaban, R.T., Kokorin, A.O., Nazarov, I.M. and Shvidenko, A.Z. (1993) CO_2 absorption by Russian forests. *Meteorology and Hydrology* N1, 5–14.

Karjalainen, T. and Kellomäki, S. (1993) Carbon storage in forest ecosystems in Finland. In: Kanninen, M. (ed.) *Carbon Balance of World's Forested Ecosystems: Towards a Global Estimate: Proceedings of the IPCC AFOS Workshop.* Publication of the Academy of Finland 3/93, Helsinki, pp. 40–51.

Karjalainen, T., Kellomäki, S. and Pussinen, A. (1995) Carbon balance in the forest sector in Finland during 1990–2039. *Climatic Change* 30, 451–478.

Karl, T.R., Jones, P.D., Knight, R.W., Kukla, G., Plummer, N., Razuvaiev, V., Gallo, K.P., Lindeseay, J., Charlson, R.J. and Peterson, T.C. (1993) A new perspective on recent global warming: asymmetric trends of daily maximum and minimum temperature. *Bulletin of the American Meteorological Society* 74, 1007–1023.

Kats, G. (1992) Achieving sustainability in energy use in developing countries. In: Holmberg, J. (ed.) *Policies for a Small Planet.* Earthscan, London, pp. 258–288.

Kauppi, P.E. and Tomppo, E. (1993) Impact of forests on net national emissions of carbon dioxide in Europe. *Water, Air and Soil Pollution* 70, 187–196.

Kauppi, P.E., Mielikainen, K. and Kuusela, K. (1992) Biomass and carbon budget of European forests: 1971 to 1990. *Science* 256, 70–74.

Kauppi, P., Tomppo, E. and Ferm, A. (1995) C and N storage in living trees within Finland since 1950s. *Plant and Soil* 168–169, 633–638.

Kerstiens, G., Townend, J., Heath, J. and Mansfield, T.A. (1995) Effects of water and nutrient availability on physiological responses of woody species to elevated CO_2. *Forestry* 68, 303–315.

Kiehl, J.T. and Briegleb, B.P. (1993) The relative roles of sulfate aerosols and greenhouse gases in climate forcing. *Science* 260, 311–314.

Kirchgessner, M., Windisch, W., Müller, H.L. and Kreuzer, M. (1991a) Release of methane and of carbon dioxide by dairy cattle. *Agribiological Research* 44(2–3), 92–103.

Kirchgessner, M., Kreuzer, M., Müller, H.L. and Windisch, W. (1991b) Release of methane and of carbon dioxide by the pig. *Agribiological Research* 44(2–3), 103–113.

Kirschbaum, M., Fischlin, A.S., Cannell, M.G.R., Cruz, R.V.O., Galinski, W. and Cramer, W.P. (1996) Climate change impacts on forests. In: Watson, R.T., Zinyowera, M.C. and Moss, R.H. (eds) *Climate Change 1995: Impacts, Adaptations and Mitigation of Climate Change: Scientific–Technical Analyses.* Cambridge University Press, Cambridge, pp. 95–129.

Kohlmaier, G.H., Häger, C., Würth, G., Lüdeke, M.K.B., Ramge, P., Badeck, F.W., Kindermann, J. and Lang, T. (1995) Effects of the age class distributions of the temperate and boreal forests on the global CO_2-source sink function. *Tellus* 47B, 212–231.

Kollert, W. (1990) Die Erfassung von Warnströmen des Holzmarktes in der Wirtschaftsstatistik der Bundesrepublik Deutschland. Dissertation, Universität München.

Kolstad, C. (1993) Mitigating climate change impacts: the conflicting effects of irreversibility in CO_2 accumulation and emission control investment. Paper presented to the International Workshop on the Integrative Assessment of Mitigation, Impacts and Adaptation to Climate Change, International Institute of Applied Systems Analysis, Laxenburg, October 1993.

Komen, M.H.C. and Peerlings, J.H.M. (1996) *Economy-wide Effects of Reducing Livestock Numbers in The Netherlands.* Department of Agricultural Economics and Policy, Wageningen Agricultural University, Wageningen.

Krankina, O. and Dixon, R. (1994) Forest management options to conserve and sequester terrestrial carbon in the Russian Federation. *World Resource Review* 6, 88–101.

Krankina, O.N., Harmon, M.E. and Winjum, J.K. (1996) Carbon storage and sequestration in the Russian forest sector. *Ambio* 25, 284–288.

Kreps, D.M. (1990) *A Course in Microeconomics Theory.* Harvester Wheatsheaf, New York.

Krjukov, N.V. (1987) *Main Guidelines for Forestry.* Oblast Organization and Development, Vologda, Moscow.

Kroth, W., Kollert, W. and Filippi, M. (1991) *Analyse und Quantifizierung der Holzverwendung im Bauwesen.* Untersuchung im Auftrag des BML, Lehrstuhl für Forstpolitik und Forstliche Betriebswirschafslehre der Ludwig-Maximilians-Universität München, Munich.

Labudda, V. (1995) *Entwicklung regionaler naturraumbezogener Leitbilder für eine Erstaufforstungskonzeption.* Technische Universität Dresden, Institut für Forstökonomie und Forsteinrichtung, Professur Forstpolitik, Dresden.

Lanly, J.P. (1982) *Tropical Forest Resources.* FAO Forestry Paper 30, FAO, Rome.

Lappalainen, E. and Hänninen, P. (1993) *Peat Reserves in Finland* (in Finnish, summary in English). Report of Investigation No. 117, Geological Survey of Finland, Espoo, Finland.

Leemans, R. and Solomon, A. (1993) Modeling the potential change in yield and distribution of the earth's crops under a warmed climate. *Climate Research* 3, 76–96.

Liljelund, L.-E., Pettersson, B. and Zackrisson, O. (1992) Skogsbruk och biologisk mengfald. *Svensk Botanisk Tidskrift* 86, 227–232.

Lovelock, J. (1995) *Gaia: A New Look at Life on Earth*, 2nd edn. Oxford University Press, Oxford.

Lyon, K. (1981) Mining of the forest and the time path of the price of timber. *Journal of Environmental Economics and Management* 8, 330–345.

McGinn, S.M., Akinremi, O.O. and Barr, A.G. (1995) *The Impacts of Climate Variability on Agricultural Sustainability in Alberta.* Report to the Nat Christie Foundation, Agriculture and Agri-Food Canada Research Centre, Lethbridge, Alberta.

Maddison, D. (1993) *The Shadow Price of Greenhouse Gases and Aerosols.* Centre for Social and Economic Research on the Global Environment, University College London and University of East Anglia, Norwich.

Maier-Reimer, E. and Hasselmann, K. (1987) Transport and storage of CO_2 in the ocean: an inorganic ocean-circulation carbon cycle model. *Climatic Dynamics* 2, 63–90.

Makundi, W.R., Sathaye, J. and Ketoff, A. (1995) COPATH: a spreadsheet model for the estimation of carbon flows associated with the use of forest resources. *Biomass and Bioenergy* 8, 369–380.

Manne, A.S. and Richels, R.G. (1992) *Buying Greenhouse Insurance*. MIT Press, Cambridge, Massachusetts.

Marland, G. (1988) *The Prospect of Solving the CO_2 Problem Through Global Reforestation*. DOE/NBB-0082, US Department of Energy, Office of Energy Research, Washington, DC.

Marsch, M. (1995a) *Bearbeitete Erstaufforstungsanträge lt. § 10 SächsWaldG in den Jahren 1992 bis 1994*. Referat 63 (unpublished), Saxon Ministry for Agriculture, Nutrition and Forestry, Dresden.

Marsch, M. (1995b) *Mehrung der Waldflächen in Sachsen – Ziele, Probleme und Wege*. Referat 63 (unpublished), Ministry for Agriculture, Nutrition and Forestry, Dresden.

Masera, O.R., Bellon, M.R. and Segura, G. (1995) Forest management options for sequestering carbon in Mexico. *Biomass and Bioenergy* 8, 357–367.

Mattsson, L. (1993) The non-timber value of nothern Swedish forests. *Scandinavian Journal of Forest Research* 5, 426–434.

Melillo, J., McGuire, A., Kicklighter, D., Moore, B. III, Vorosmarty, C. and Schloss, A. (1993) Global change and terrestrial net primary production. *Nature* 363, 234–240.

Melillo, J.M., Prentice, I.C., Farquhar, G.D. and Schulze, E.-D. (1996) Terrestrial ecosystems: biotic feedbacks to climate. In: Houghton, J.T., Meiro Filho, L.G., Callander, B.A., Harris, N., Kattenberg, A. and Maskell, K. (eds) *Climate Change 1995: The Science of Climate Change*. Cambridge University Press, Cambridge, pp. 445–481.

Metalnikov, A.P. (1994) *Analysis of Possibility of Ratification by Russia UN Framework Convention on Global Climate Change: Report of Workshop, 9–11 January 1994, Southampton, Bermuda*. Analytical Report by Federal Agency on Hydrometeorology and Criteria for Joint Implementation under Framework Convention on Climate Change, Woods Hole Research Center, Woods Hole, Massachusetts.

Ministero del'Ambiente, Italia (1995) *Prima Comunicazione Nazionale dell'Italia alla Convenzione-quadro sui Cambiamenti Climatici*. Ministero dell'Ambiente, Rome.

Ministry of Agriculture and Fisheries (NZ) (1994) *Situation and Outlook for New Zealand Agriculture*. MAF, Wellington.

Ministry of Agriculture, Fisheries and Food (MAFF) (1991) *Code of Good Agricultural Practice for the Protection of Water*. HMSO, London.

Ministry of Agriculture, Fisheries and Food (MAFF) (1992) *The Farm Woodland Premium Scheme: Rules and Procedures*. Ministry of Agriculture, Fisheries and Food, London.

Ministry of Agriculture, Fisheries and Food (MAFF) (1994a) *MAFF Statistics: Agricultural and Horticultural Census, 1 June 1993, United Kingdom and England*. Government Statistical Service, London.

Ministry of Agriculture, Fisheries and Food (MAFF) (1994b) *MAFF Statistics: Straw Disposal Survey 1993 – England and Wales*. Government Statistical Service, London.

Ministry of Environment Protection and Natural Resources (1995) *About Environmental Performance in Russian Federation in 1994*. State Report, Ministry of Environment Protection and Natural Resources, Moscow.

Mitchell, J.F.B., Johns, T.C., Gregory, J.M. and Tett, S.F.B. (1995) Climate response to increasing levels of greenhouse gases and sulphate aerosols. *Nature* 376, 501–504.

MLC (1993) *Beef Yearbook*. Meat and Livestock Commission, Milton Keynes.

Moeller, D. (1984) Estimation of global man-made sulphur emissions. *Atmospheric Environment* 18, 19–27.

Mohren, G.M.J. and Klein-Goldewijk, C.G.M. (1990) *CO_2FIX: Model Listing*. Report 614, Research Institute for Forestry and Urban Ecology, Wageningen Agricultural University, Wageningen, the Netherlands.

Monserud, R. and Leemans, R. (1992) Comparing global vegetation maps with the kappa statistic. *Ecological Modelling* 62, 275–293.

Montoya, G., L., S., de Jong, B., Nelson, K., Farías, P., Taylor, J.H. and Tipper, R. (1995) *Desarrollo forestal sustentable: Captura de carbono en las zonas tzeltal y tojolabal del estado de Chiapas*. Cuadernos de Trabajo 4, INE, CIES, San Cristóbal de las Casas, Chiapas, Mexico.

Moretto, M. (1995) Controllo ottimo stocastico, processi browniani regolati e optimal stopping. *Rivista Internazionale di Scienze Economiche e Commerciali* 42, 93–124.

Mosier, A.R. (1993) State of knowledge about nitrous oxide emissions from agricultural fields. *Mitteilungen der Deutschen Bodenkundlichen Gesellschaft* 69, 201–208.

Mosier, A.R., Schimel, D., Valentine, D., Bronson, K. and Parton, W. (1991) Methane and nitrous oxide fluxes in native, fertilised and cultivated grasslands. *Nature* 350, 330–332.

Moulton, R. and Richards, K. (1990) *Costs of Sequestrating Carbon Through Tree Planting and Forest Management in the United States*. General Technical Report WO-58, US Department of Agriculture Forest Service, Washington, DC.

Nabuurs, G.J. and Mohren, G.M.J. (1993) *Carbon Fixation Through Forestation Activities: A Study of the Carbon Sequestering Potential of Selected Forest Types*. IBN Research Report 93/4, Face Foundation and Institute for Forestry and Nature Research IBN-DLO, Wageningen, the Netherlands.

Nabuurs, G.J. and Mohren, G.M.J. (1995) Modelling analysis of potential carbon sequestration in selected forest types. *Canadian Journal of Forest Research* 25, 1157–1172.

National Audit Office (1986) *Forestry in Great Britain*. PIEDA for the National Audit Office, London.

Netherlands Central Bureau of Statistics (1992) *National Accounts 1992*. Central Statistical Bureau, Voorburg, the Netherlands.

Netherlands Central Bureau of Statistics (1993) *De Produktiestructuur van de Nederlandse Volkshuishouding, deel XIX Input-output Tabellen en Aanbod- en Grebuiktabellen 1988–1990*. Central Statistical Bureau, Voorburg, the Netherlands.

Netherlands Central Planning Bureau (1992) *Economische gevolgen op lange termijn van heffingen op energie*. Werkdocument 43, Netherlands Central Planning Bureau, The Hague.

Netherlands Central Planning Bureau (1993) *Effecten van een kleinverbruikersheffing op energie bij lage en hoge prijsniveaus*. Werkdocument 64, Netherlands Central Planning Bureau, The Hague.

Nicholls, N., Gruza, G.U., Jouzel, J., Karl, T.R., Ogallo, L.A. and Parker, D.E. (1996) Observed climate variability and change. In: Houghton, J.T., Jenkins,

G.J. and Ephraums, J.J. (eds) *Climatic Change 1995: The Science of Climate Change*. Cambridge University Press, Cambridge, pp. 137–192.

Nicholson, R.J. and Brewer, A.J. (1994) Unpublished report to MAFF. ADAS, Anstey Hall and ADAS, Maidstone.

Nielson, R. (1995) A model for predicting continental-scale vegetation distribution and water balance. *Ecological Applications* 5, 362–385.

Nielson, R., King, A. and Koerper, G. (1992) Toward a rule-based biome model. *Landscape Ecology* 7, 27–43.

Niesslein, E. (1992) *Privatwaldbewirtschaftung und Zusammenschulßwesen in den neuen Bundesländern*. Bundesministerium fur Ernehrung, Landwirtschaft und Forsten, Bonn.

Nilsson, S. and Schopfhauser, W. (1995) The carbon-sequestration potential of a global afforestation program. *Climatic Change* 30, 267–293.

Nordhaus, W.D. (1991) To slow or not to slow: the economics of the greenhouse effect. *Economic Journal* 101, 920–937.

Nordhaus, W.D. (1992) The optimal transition path for controlling greenhouse gases. *Science* 258, 1315–1319.

Nordhaus, W.D. (1993a) Reflections on the economics of climate change. *Journal of Economic Perspectives* 7(4), 11–25.

Nordhaus, W.D. (1993b) Rolling the 'DICE': an optimal transition path for controlling greenhouse gases. *Resource and Energy Economics* 15, 27–50.

Nordhaus, W.D. (1994) *Managing the Global Commons. The Economics of Climate Change*. MIT Press, Cambridge, Massachusetts.

OECD (1991) *Estimation of Greenhouse Gas Emissions and Sinks*. IPCC and OECD, Paris.

OECD (1994a) *Agricultural Policy Reform: New Approaches – The Role of Direct Income Payments*. OECD Publications, Paris.

OECD (1994b) *Climate Change Policy Initiatives: 1994 Updates*. OECD/IEA: Paris.

OECD (1995a) *Forestry, Agriculture and the Environment*. OECD Documents, Paris.

OECD (1995b) *Sustainable Agriculture – Concepts, Issues and Policies in OECD Countries*. OECD Publications, Paris.

OECD (1996) *Agricultural Policies, Markets and Trade, Monitoring and Evaluation 1996*. OECD, Paris.

Olsen, T.E. and Stensland, G. (1992) On optimal timing of investment when cost components are additive and follow geometric diffusions. *Journal of Dynamics and Control* 16, 39–51.

Olson, J., Watts, J. and Allison, L. (1983) *Carbon in Live Vegetation of Major World Ecosystems*. ORNL-5862, Oak Ridge National Laboratory, Oak Ridge, Tennessee.

O'Riordan, T. and Jäger, J. (eds) (1996) *Politics of Climate Change: A European Perspective*. Routledge, London.

Östlund, L. (1993) Exploitation and structural changes in the North Swedish boreal forests 1800–1992. Dissertations in Forest Vegetation Ecology 4, Swedish University of Agricultural Sciences, Umeå, Sweden.

Overpeck, J., Bartlein, P. and Webb, T. III. (1991) Potential magnitude of future vegetation in eastern North America: comparison with the past. *Science* 254, 692–695.

Papen, H., Hellmann, B., Papke, H. and Renneberg, H. (1993) Emission of N-oxides from acid irrigated and limed soils of a coniferous forest in Bavaria.

In: Oremland, R.S. (ed.) *Biogeochemistry of Global Change: Radiatively Active Trace Gases.* Chapman and Hall, New York, pp. 245–260.
Pareto W. (1896) (Italian edn, 1949) *Corso di Economia Politica.* Einaudi, Turin.
Parry, I.W.H. (1995) Pollution taxes and revenue recycling. *Journal of Environmental Economics and Management* 29, S64–S77.
Parton, W.J., Cole, C.V., Stewart, J.W.B., Ojima, D.S. and Schimel, D.S. (1989) Simulating regional patterns of soil C, N and P dynamics in the US central grasslands region. In: Clarholm, M. and Bergstrom, C. (eds) *Ecology of Arable Land.* Kluwer, Dordrecht, pp. 99–108.
Pearce, D.W. (1991a) Assessing the returns to the economy and society. In: Forestry Commission (ed.) *Forestry Expansion: a Study of Technical, Economic and Ecological Factors.* Forestry Commission, Edinburgh.
Pearce, D.W. (1991b) The role of carbon taxes in adjusting to global warming. *Economic Journal* 101, 938–948.
Pearce, D.W. (1994) The environment: assessing the social rate of return from investment in temperate zone forestry. In: Layard, R. and Glaister, S. (eds) *Cost–Benefit Analysis.* Cambridge University Press, Cambridge, pp. 464–490.
Pearce, D.W., Cline, W.R., Achanta, A.N., Frakhauser, S., Pachauri, R.K., Tol, R.S.J. and Vellinga, P. (1996) The social costs of climate change: greenhouse damage and the benefits of control. In: Bruce, J., Lee, H. and Haites, E. (eds) *Climate Change 1995: Economic and Social Dimensions of Climate Change.* Cambridge University Press, Cambridge, pp. 183–224.
Pearse, P.H. (1995) Farm forestry, agricultural policy reform and the environment: a summary and assessment of the workshop. In: OECD (ed.) *Forestry, Agriculture and the Environment.* OECD, Paris, pp. 25–44.
Peck, S.C. and Teisberg, T.J. (1992) CETA: a model for carbon emissions trajectory assessment. *Energy Journal* 13, 55–77.
Peck, S.C. and Teisberg, T.J. (1993a) CO_2 emission controls: comparing policy instruments. *Energy Policy* 21, 222–230.
Peck, S.C. and Teisberg, T.J. (1993b) Global warming uncertainties and the value of information: an analysis using CETA. *Resource and Energy Economics* 15, 71–97.
Peerlings, J.H.M. (1993) An Applied General Equilibrium Model for Dutch Agribusiness Policy Analysis. Academic thesis, Wageningen Agricultural University, Wageningen.
Petersen, L. (1993) Bodenschutz und property rights im Agrarsektor der USA: Die Mechanik von 'Compliance Policies'. *Gesellschaftliche Forderungen an die Landwirtschaft. Schriften der Gesellschaft für Wirtschafts – und Sozialwissenschaften des Landbaues e. V.* Münster-Hiltrup 30, pp. 259–269.
Petrov, A.P. (1994) *Forms and Property Rights for Russian Forests in Market Economy.* Economic and Legislative Aspects of Forest Management in Russia. Proceedings of an International Conference, Pushkino, Moscow, pp. 26–33.
Philipp, W. (1987) *Die Aufforstung als Beitrag zur Lösung des Überschußproblems in der Landwirtschaft Bayerns Nr 84.* Schriftenreihe der Forstwirtschaftlichen Fakultät der Universität München und der Bayerischen Forstilchen Versuchs – und Forschungsanstalten.
Pimentel, D., Dazhong, W. and Giamietro, M. (1990) Technological changes in energy use in US agricultural production. *Agroecology: Researching the Ecological*

Basis for Sustainable Agriculture, Ecological Studies: Analysis and Synthesis 78, 305–321.

Pinard, M.A. (1994) Reduced-impact logging project. *ITTO Tropical Forest Update* 4, 11–12.

Pinard, M.A., Putz, F.E., Tay, J. and Sullivan, T.E. (1995) Creating timber harvest guidelines for a reduced-impact logging project in Malaysia. *Journal of Forestry* 93(10), 41–45.

Pingoud, K., Savolainen, I. and Seppälä, H. (1995) Greenhouse impact of the Finnish forest sector including forest products and landfill management. *Ambio* 25, 318–326.

Plochmann, R. and Thoroe, C. (1991) Förderung der Erstaufforstung. *Schriftenreihe des Bundesministers für Ernährung Landwirtschaft und Forsten* 397, 119 pp.

Pontryagin, L., Boltyanskii, V., Gamkrelidze, R. and Mischenko, E. (1962) *The Mathematical Theory of Optimal Processes*. Wiley, New York.

Povellato, A. (1996) The implementation of Regulation 2078/92 in Italy. Paper presented at the Fifth Meeting of the European Forum on Nature Conservation and Pastoralism, Cogne, Italy, 18–21 September.

Prentice, I., Cramer, W., Harrison, S., Leemans, R., Monserud, R. and Solomon, A. (1992) A global biome model based on plant physiology and dominance, soil properties and climate. *Journal of Biogeography* 19, 117–134.

Price, C. (1993) *Time, Discounting and Value*. Blackwell, Oxford.

Price, C. (1995) Emissions, concentrations and disappearing CO_2. *Resource and Economics* 17, 87–97.

Price, C. (1996) Long time horizons, low discount rates and moderate investment criteria. *Project Appraisal* 11, 265–268.

Price, C. and Willis, R. (1993) Time, discounting and the valuation of forestry's carbon fluxes. *Commonwealth Forestry Review* 72, 265–271.

Price, C., Christensen, J.B. and Humphreys, S.K. (1986) Elasticities of demand for recreation site and for recreation experience. *Environment and Planning A* 18, 1259–1263.

Read, P. (1990) Global warming: why Mrs Thatcher should be more ambitious. *National Westminster Bank Review* November, 22–33.

Reilly, J.M. (1992) Climate change and the trace-gas index. In: Reilly, J.M. and Anderson, M. (eds) *Economic Issues in Global Climatic Change*. Westview Press, Boulder, Colorado, pp. 72–88.

Reinhardt, G.A. (1993) *Energie- und CO_2-Bilanzierung nachwachsender Rohstoffe: theoretische Grundlagen und Fallstudie Raps*. Verlag Vieweg, Braunschweig.

Richards, J.F. (1990) Land transformation. In: Turner, B.L., Clark, W.C., Kates, R.W., Richards, J.F., Mathews, J.T. and Meyer, W.B. (eds) *The Earth as Transformed by Human Action*. Cambridge University Press, Cambridge, pp. 163–178.

Richards, K.R., Alig, R., Kinsman, J.D., Palo, M. and Sohngen, B. (1995) Consideration of country and forestry/land-use characteristics in choosing forestry instruments to achieve climatic mitigation goals. In: Sampson, N.R., Sedjo, R.A. and Wisniewski, J. (eds) *Economics of Carbon Sequestration in Forestry*. Forest Policy Center, Washington, DC, pp. 23–36.

Richter, R. (1990) Sichtweise und Fragestellungen der Neuen Institutionenökonomik. *Zeitschrift für Wirtschafts – und Sozialwissenschaften* 110, 571–591.

Rodhe, H. and Svensson, B. (1994) *Impact on the Greenhouse Effect of Peat Mining and Combustion*. Report No. 4369, Swedish National Environment Protection Board, Stockholm, Sweden.

Rodhe, H., Hedlund, T., Eriksson, H., Jonsson, P., Klemendtsson, L. and Nilsson, M. (1995) *Sveriges växthusgasbudget*. Report No. 1/95, Klimatdelegationen.

Roemer-Mähler, J. (1992) Welche Hemmschwellen sind bei der Erzeugung und Verwertung nachwachsender Rohstoffe in der Bundesrepublik Deutschland zu erkennen und wie bzw. in welchem Bereich sind die mittelfristig überwinder? *Biomasseerzeugung zur direkten energetischen Nutang – agarpolitische, ökologische und ökonomische Möglichteiten und Grenzen – Arbeitsunterlagen*. DLG, Frankfurt, pp. 147–154.

Rosenzweig, C. and Parry, M.L. (1993) Potential impacts of climate change on world food supply: a summary of a recent international study. In: Kaiser, H.M. and Drennen, T.E. (eds) *Agricultural Dimensions of Global Climate Change*. St Lucie Press, Delray Beach, Florida, pp. 87–116.

Rosenzweig, C. and Parry, M.L. (1994) Potential impact of climate change on world food supply. *Nature* 367, 133–138.

Rotherham, T. (1996) Forest management certification: objectives, international background and the Canadian program. *Forestry Chronicle* 72, 247–252.

Sächsisches Staatsministerium fuer Landwirtschaft, Ernährung und Forsten (ed.) (1992) *Waldgesetz für den Freistaat Sachsen vom 10 April 1992*. Dresden.

Sächsisches Staatsministerium fuer Umwelt und Landesentwicklung (ed.) (1994) *Landesentwicklungsplan Sachsen*. Dresden.

Safley, L.M., Casada, M.E., Woodbury, J.W. and Roos, K.F. (1992) *Global Methane Emissions from Livestock and Poultry Manure*. Global Change Division, US Environmental Protection Agency, Washington, DC.

SAGyP (1994) *Evaluacion financeiera de modelos foresstales para producción de Madera y Fijarción de carbono en regiones seleccionadas*. Mimeo, Proyecto Forest AR, Secretaria de Agricultura, Ganadería y Pesca, República Argentina.

Sampson, N. and Hamilton, T. (1992) Can trees really help fight global warming? *American Forests* 98, 13–16.

Sampson, R.N., Moll, G.A. and Keilbaso, J.J. (1992) Opportunities to increase urban forests and the potential impacts on carbon storage and conservation. In: Sampson, N. (ed.) *Forests and Global Change*. American Forests, Washington, DC, pp. 51–67.

Santer, B.D., Taylor, K.E., Wigley, T.M.L., Johns, T.C., Jones, P.D., Karoly, D.J., Mitchell, J.F.B., Oort, A.H., Penner, J.E., Ramaswamy, V., Scharzkopf, M.D., Stouffer, R.J. and Tett, S. (1996) A search for human influences on the thermal structure of the atmosphere. *Nature* 382, 39–46.

SARH (1994) *Tropical Forest Action Plan of Mexico TFAP*, Vols I and II. Secretary of Agriculture and Hydraulic Resources, Undersecretary of Forests and Wildlife, Mexico City.

Saunders, C.M. (1994) *Agricultural Policy: an Update*. Working Paper 6, Centre for Rural Economy, University of Newcastle upon Tyne.

Schanzenbächer, B. (1994) Ermittlung externer ökologischer Effekte der Landwirtschaft und ökonomische und ökologische Auswirkungen von Maßnahmen zu deren Internalisierung. Dissertation, Universität Hohenheim.

Schelling, T.C. (1992) Some economics of global warming. *American Economic Review* 82, 1–14.

Schiffman, P.M. and Johnson, W.C. (1991) Phytomass and detrital carbon storage during forest regrowth in the southwestern US Piedmont. *Canadian Journal of Forest Research* 19, 97–98.

Schöpfer, W. (1993) Eine Schätzung des Nutzungspotentials der Wälder Baden-Württembergs. *Forst und Holz* 48, 148–155.

Scottish Office Environment Department (1994) *Our Forests: The Way Ahead.* HMSO, Edinburgh.

Sedjo, R. and Lyon, K. (1990) *The Long Term Adequacy of the World Timber Supply.* Resource for the Future, Washington, DC.

Sedjo, R.A. and Solomon, A.J. (1989) Climate and forests. In: Rosenberg, N.S., Easterling, W.E., Crosson, P.R. and Darmstadter, J. (eds) *Greenhouse Warming: Abatement and Adaptation.* Resources for the Future, Washington, DC, pp. 105–119.

Sedjo, R.A., Wisniewski, J., Sample, A.V. and Kinsman, J.D. (1995) The economics of managing carbon via forestry: assessment of existing studies. *Environmental and Resource Economics* 6, 139–165.

SEMARNAP (1996) *Cambio Climatico de Mexico: Reto Voluntario y Registro de Acciones.* Secretaria de Medio Ambiente, Recursos Naturales y Pesca, Mexico City.

Semenov, M.A. and Evans, L.G. (1996) Non-linearity in climate change and assessment of agricultural risks. *Proceedings of the International Congress on Environment and Climate*, Rome, Italy, 4–8 March.

Seppälä, H. (1994) Would radical carbon dioxide fees restructure future wood markets? *The Globalization of Wood: Supply, Processes, Products and Markets.* Forest Products Society Conference, Madison, Wisconsin, pp. 205–207.

Shoven, J.B. and Whalley, J. (1992) *Applying General Equilibrium.* Cambridge University Press, Cambridge.

Siegenthaler, U. and Sarmiento, J.L. (1993) Atmospheric carbon dioxide and the ocean. *Nature* 365, 119–125.

Simons, B., Burrell, A., Oskam, A., Peerlings, J. and Slangen, L. (1994) CO_2 *Emissions of Dutch Agriculture and Agribusiness: Method and Analysis.* Wageningen Economic Papers 1996–1, Wageningen Agricultural University, Wageningen.

Slesser, M. and Wallace, F. (1982) *Energy Consumption per Tonne of Competing Agricultural Products Available to the EC.* Report No. 85, Commision of the European Communities, Office for Official Publications of the European Communities, Luxemburg.

Sohngen, B. (1995) Integrating ecology and economics: the economic impacts of climate change on timber markets in the United States. PhD Dissertation, Yale School of Forestry and Environmental Studies, New Haven, Connecticut.

Sohngen, B. and Mendelsohn, R. (1996) *Integrating Ecology and Economics: The Timber Market Impacts of Climate Change on US Forests.* Report Prepared for the EPRI Climate Change Impacts Program, Yale School of Forestry and Environmental Studies, New Haven.

Solomon, A.M., Ravindranath, N.H., Stewart, R., Weber, M. and Nilsson, S. (1996) Wood production under changing climate and land use. In: Watson, R.T., Zinyowera, M.C. and Moss, R.H. (eds) *Climate Change 1995: Impacts, Adaptations and Mitigation of Climate Change: Scientific–Technical Analyses.* Cambridge University Press, Cambridge, pp. 489–510.

Statistisches Landesamt des Freistaates Sachsen (ed.) (1994) *Statistisches Jahrbuch Sachsen*. Kamenz.
Stout, B.A., and Nehring, R.F. (1988) Agricultural energy: 1988 and the future. *Phi Kappa Phi Journal* 68, 32–36.
Subak, S. (1996) The science and politics of national greenhouse gas inventories. In: O'Riordan, T. and Jäger, J. (eds) *Politics of Climate Change: A European Perspective*. Routledge, London, pp. 51–64.
Subak, S., Raskin, P. and Von Hippel, D. (1993) National greenhouse gas emissions: current anthropogenic sources and sinks. *Climatic Change* 25, 15–58.
Sverdrup, H., Warfvinge, P. and Nihlgerd, B. (1994) Assessment of soil acidification on forest growth in Sweden. *Water, Air and Soil Pollution* 77, 1–36.
Swart, R.J., Bouwman, L., Olivier, J. and van den Born, G.J. (1993) Inventory of greenhouse gas emissions in the Netherlands. *Ambio* 22, 518–523.
Swedish National Board of Forestry (1993) *Wood Balances 1992*. Report No. 2, National Board of Forestry, Stockholm, Sweden.
Swedish National Environment Protection Board (1992) *Åtgärder mot klimatförändringar*. Report No. 4120, Swedish National Environment Protection Board, Stockholm, Sweden.
Swisher, J.N. (1991) Cost and performance of CO_2 storage in forestry projects. *Biomass and Bioenergy* 1, 317–328.
Swisher, J.N. and Masters, G. (1992) A mechanism to reconcile equity and efficiency in global climate protection: international carbon emission offsets. *Ambio* 21, 154–159.
Tahvonen, O. (1994) *Net National Emissions, CO_2 Taxation and the Role of Forestry*. Research Papers 490, Finnish Forest Research Institute, Helsinki.
Tangermann, S. (1996) Implementation of the Uruguay Round Agreement on agriculture: issues and prospects. *Journal of Agricultural Economics* 47, 315–337.
Taylor, K.E. and Penner, J.E. (1994) Response of the climate system to atmospheric aerosols and greenhouse gases. *Nature* 369, 734–737.
Tchebakova, N., Monserud, R., Leemans, R. and Golovanov, S. (1993) A global vegetation model based on the climatological approach of Budyko. *Journal of Biogeography* 20, 129–144.
Titus, J. (1992) The cost of climate change to the United States. In: Majumdar, S.K., Kalkstein, L.S., Yarnal, B., Miller, E.W. and Rosenfeld, L.M. (eds) *Global Climate Change: Implications Challenges and Mitigation Measures*. Pennsylvania Academy of Science, Pennsylvania.
Tobey, J., Reilly, J. and Kane, S. (1992) Economic implications of global climate change for world agriculture. *Journal of Agricultural and Resource Economics* 17, 195–204.
Tol, R.S.J. (1995) The damage costs of climate change: toward more comprehensive calculations. *Environmental and Resource Economics* 5, 353–374.
Tol, R.S.J. (1996) The damage costs of climate change: towards a dynamic representation. *Ecological Economics* 19, 67–90.
Trexler, M.C. (1991) *Minding the Carbon Store: Weighing US Forestry Strategies to Slow Global Warming*. World Resources Institute, Washington, DC.
Trexler, M.C., Haugen, C.A. and Loewen, L.A. (1992) Global warming mitigation through forestry options in the tropics. In: Sampson, N. (ed.) *Forests and Global Change*. American Forests, Washington, DC, pp. 73–96.

Trunk, W. and Zeddies, J. (1996) Ökonomische Beurteilung von Strategien zur Vermeidung von Schadgasemissionen bei der Milcherzeugung. *Agrarwirtschaft* 245, 111–120.

Turner, D.P., Koerper, G.J., Harmon, M. and Lee, J.J. (1995) Carbon sequestration by forests of the United States: current status and projections to the year 2000. *Tellus* 41B, 232–239.

Turner, R.K., Pearce, D. and Bateman, I. (1994) *Environmental Economics: An Elementary Introduction*. Harvester Wheatsheaf, Hemel Hempstead.

Tyler, A. (1994) *The Influence of Site Quality on the Productivity and Flora of a Range of Tree Species in Scotland*. Report to the Scottish Forestry Trust, Edinburgh.

UK Department of Environment (DoE) (1994) *Climate Change: The UK Programme. United Kingdom's Report under the Framework Convention on Climate Change*. HMSO, London.

UNESCO/ICEC (1996) International Congress on Environment/Climate. *Book of Abstracts*. Tipografia Villaggio del Fanciullo, Opicina, Trieste, Italy.

United Nations (UN) (1993) *Handbook of National Accounting*. Integrated Environmental and Economic Accounting Series F, No. 61, Department for Economic and Social Information and Policy Analysis, United Nations, New York.

United Nations Commission on Trade and Development (UNCTAD) (1992) *Combatting Global Warming: Study on a Global System of Tradeable Carbon Emission Entitlements*. United Nations, New York.

US EPA (1993) *Options for Reducing Methane Emissions Internationally: Report to Congress*. Global Change Division, Office of Air and Radiation, Washington, DC.

van Amstel, A.R. (1993) *International IPCC Workshop: Methane and Nitrous Oxide: Methods in National Emissions Inventories and Options for Control*. International IPCC Workshop Proceedings, National Institute for Public Health and Environmental Protection, RIVM, the Netherlands.

van der Hamsvoort, C.P.C.M. and Latacz-Lohmann, U. (1996) *Auctions as a Mechanism for Allocating Conservation Contracts among Farmers*. Agricultural Economics Research Institute, The Hague, and Wye College, University of London.

van Kooten, G.C. (1995) Climatic change and Canada's boreal forests: socio-economic issues and implications for land use. *Canadian Journal of Agricultural Economics* 43, 133–148.

van Kooten, G.C., Arthur, L.M. and Wilson, W.R. (1992) Potential to sequester carbon in Canadian forests: some economic considerations. *Canadian Public Policy* 18, 127–138.

van Kooten, G.C., Thompson, W.A. and Vertinsky, I. (1993) The economics of reforestation in British Columbia when benefits of CO_2 reduction are taken into account. In: Adamowicz, W.L., White, W. and Phillips, W.E. (eds) *Forestry and the Environment: Economic Perspectives*. CAB International, Wallingford, pp. 227–244.

van Kooten, G.C., Binkley, C.S. and Delcourt, G. (1995) Effect of carbon taxes and subsidies on optimal forest rotation age and supply of carbon services. *American Journal of Agricultural Economics* 77, 365–374.

van Kooten, G.C., Grainger, A., Ley, E., Marland, G. and Solberg, B. (1997) Conceptual issues related to carbon sequestration: uncertainty and time. In: Sedjo, R.A., Sampson, R.N. and Wisniewski, J. (eds) *The Economics of Terrestrial Carbon Sequestration*. CRC Press, Boca Raton.

van Nevel, C.J. and Demeyer, D.I. (1996) Control of rumen methanogenesis. *Environmental Monitoring and Assessment* 42, 73–97.
van Soest, A. (1993) Structural models of family labour supply, a discrete choice approach. *Journal of Human Resources* 30, 63–88.
Vinson, T.S., Kolchugina, T.P. and Andrasko, K.O. (1996) Greenhouse gas mitigation options in the forest sector of Russia: national and project level assessments. *Environmental Management* 20, S111–S118.
Volz, K.R., Böswald, K. and Dinkelaker, F. (1996) *Forstpolitik – Entwicklungen und Perspektiven. Gutachten im Rahmen des Projektes.* Nachhaltige Land- und Forstwirtschaft-Expertisen, Veröffentilchungen der Akademie für Technikfolgenabschätzung, Springer Verlag, Berlin.
von Alvensleben, R. (1995) Naturschutz im Licht der Standorttheorie. *Agrarwirtschaft* 44, 230–236.
von Neumann, J. (1955) Can we survive technology? In: *The Fabulous Future.* Dutton, New York.
Watson, R.T., Meira Filho, L.G., Sanhueza, E., and Janetos, A. (1992) Greenhouse gases: sources and sinks. In: Houghton, J.T., Callander, B.A. and Varney, S.K. (eds) *Climate Change 1992: The Supplementary Report to the IPCC Scientific Assessment.* Cambridge University Press, Cambridge, pp. 29–46.
Watson, R.T., Zinyowera, M.C. and Moss, R.H. (eds) (1996) *Climate Change 1995: Impacts Adaptations and Mitigation of Climate Change: Scientific–Technical Analyses. Contribution of Working Group II to the Second Assessment Report of the IPCC.* Cambridge University Press, Cambridge.
Welsch, H. (1995a) *The Carbon Tax Game: Differential Tax Recycling in a Two-region General Equilibrium Model of the European Community.* Nota Di Lavoro 36.95, Fondazione Eni Enrico Mattei, Milan.
Welsch, H. (1995b) *Joint versus Unilateral Carbon/Energy Taxation in a Two-region General Equilibrium Model for the European Community.* Nota Di Lavoro 37.95, Fondazione Eni Enrico Mattei, Milan.
Westman, C.J., Fritze, H., Helmisaari, H.S., Ilvesniemi, H., Jauhianinen, M., Kitunen, V., Lehto, J., Mecke, M., Pietikäinen, J. and Smolander, A. (1994) Carbon in boreal soil. In: Kanninen, M. and Heikinheimo, P. (eds) *The Finnish Resesarch Program on Climate Change, Second Progress Report.* Publications of the Academy of Finland, Helsinki, pp. 212–214.
Weyant, J.P. (1993) Costs of reducing global carbon emissions. *Journal of Economic Perspectives* 7(4), 27–46.
Whitby, M. (1995) Transaction costs and property rights: the omitted variables? In: Albisu, L.M and Romeros, C. (eds) *Environmental and Land Use Issues: An Economic Perspective.* Wissenschaftverlag Vauk, Kiel, pp. 3–14.
Whitby, M.C. (ed.) (1996) *The European Environment and CAP Reform: Policies and Prospects for Conservation.* CAB International, Wallingford.
Whitby M.C. (1997) Agriculture and environment after CAP reform. In: Ferro, O. (ed.) *What Future for the CAP? Perspectives and Expectations for the Common Agricultural Policy of the European Union.* Wissenschaftverlag Vauk, Kiel (in press).
Wiersum, K.F. and Ketner, P. (1989) Reforestation, a feasible contribution to reducing the atmospheric carbon dioxide content. In: Okken, P.A., Swart, R.J. and Zwerver, S. (eds) *Climate and Energy: The Feasibility of Controlling CO_2 Emissions.* Kluwer, Dordrecht.

Wigley, T.M.L. (1995) Global mean temperature and sea-level consequences of greenhouse-gas concentration stabilization. *Geophysical Research Letters* 22, 45–48.

Wigley, T.M.L. and Raper, S.C.B. (1992) Implications for climate and sea level of revised IPCC emission scenarios. *Nature* 357, 293–300.

Wilhelmsson, E. (1988) *The Realism of Models for Regional Estimations of Long Term Potential Cut*. Report 48, Department of Forest Survey, Stockholm, Sweden.

Williams, A.E. (1994) *Methane Emissions*. Report of a Working Group Appointed by the WATT Committee on Energy, WATT Committee Report No. 28, London.

Willis, K.G. (1991) Recreational value of the Forestry Commission estate in Great Britain. *Scottish Journal of Political Economy* 38, 58–75.

Winjum, J.K., Dixon, R.K. and Schroeder, P.E. (1992) Estimating the global potential of forest and agroforest management practices to sequester carbon. *Water, Air and Soil Pollution* 64, 213–227.

Winter, M. (1996) *Rural Politics: Policies for Agriculture, Forestry and the Environment*. Routledge, London.

Wofsy, S.C., Munger, J.E., Bakwin, P.S., Daube, B.C. and Moore, T.R. (1993) Net CO_2 uptake by northern woodlands. *Science* 260, 1314–1317.

Woods Hole Research Center (1994) *Criteria for Joint Implementation under the Framework Convention on Climate Change*. Report of Workshop, 9–11 January 1994, Southampton, Massachusetts.

Woodward, F.I., Smith, T.M. and Emanuel, W.R. (1995) A global land primary productivity and phytogeography model. *Global Biogeochemical Cycles* 9, 471–490.

World Bank (1994) *World Tables 1994*. World Bank, Washington, DC.

World Resource Institute (1994) *World Resources 1994–95*. Oxford University Press, Oxford.

Yearbook of Forest Statistics, Finland (1995) *Official Statistics of Finland, Agriculture and Forestry 1995*. Finnish Forest Research Institute, Jyväskylä.

Yohe, G.W., Neumann, J.E., Marshall, P.B. and Ameden, H. (1996) The economic cost of greenhouse induced sea level rise for developed property in the United States. *Climatic Change* 32, 387–410.

Zamagni, S. (1994) Global environmental change, rationality and ethics. In: Campiglio, L., Pireschi, L., Siniscalco, D. and Treves, T. (eds) *The Environment after Rio: International Law and Economics*. Graham and Trotman, London.

Zecca, A. and Brusa, R.S. (1990) *The Missing Part of the Greenhouse Effect*. Internal Report, University of Trento, Trento, Italy.

Zecca, A. and Brusa, R.S. (1991) The missing part of the greenhouse effect. *Nuovo Cimento* 14C, 523–532.

Zecca, A. and Brusa, R.S. (1992) SO_2 cooling and detection of the greenhouse effect. *Nuovo Cimento* 15C, 481–483.

Zeelenberg, C., Huigen, R.D., Kooiman, P., van de Stadt, H. and Keller, W.J. (1991) *Tax Incidence in the Netherlands: Accounting and Simulations*. Statistische Onderzoekingen M42, Netherlands Central Bureau of Statistics, The Hague.

Zhang, Z.X. (1996) *Macroeconomic Effects of CO_2 Emission Limits: A Computable General Equilibrium Analysis for China*. Wageningen Economic Papers 1996-1, Wageningen Agricultural University, Wageningen.

Index

Adams 8, 59, 268, 286
Adger 7, 9, 27, 113, 297, 312
afforestation 9, 38, 43–45, 46, 74, 82, 125,
  199–213, 249, 297, 255–256
  of agricultural lands 10, 17, 44, 113,
    120, 124, 188, 193, 197,
    235–236, 237
  of non-forest land 190, 227, 234
  social benefit of 45, 188, 297
Agenda 21 1, 37
agricultural intensification 16, 24, 25, 146
agricultural policy reform 3, 18, 23–34
  see also European Union agriculture
    reforms; Common Agricultural
    Policy, reform
agricultural production and climate impacts
  17, 60
agricultural support, targeting of 26
agricultural systems 8, 129
agricultural yield 60, 202
agri-environmental arena 298
agroforestry 45, 111, 256, 269–284
  carbon-sequestration potential 282
  systems 276
  Western Europe 45
Ahrens 210, 211
air pollutants on forest production 242

Allison 191
Alriksson 249
ammonia (NH_4) 159, 162–163, 164,
  165, 166
animal husbandry 298
annual increment of growth 246
Anderson 70, 194
Andrasko 257, 269
Anuchin 286
Anz 188
applied general equilibrium (AGE) 171,
  173–176, 181
  Armington assumption 175
  Constant Elasticity of Transformation
    (CET) 175, 176
Apps 12, 38
Arable Area Payments Scheme 16, 130
arable crops 298
ARABLE model 130, 135
Argentina 19, 255–268
Armstrong Brown 129, 135
Arrow 13, 70
Arthur 119
Assmann 231, 232
auctions 306
Australia 24, 25, 258
  Landcare Programme 33

Austria 20
  fertilizer levies 31
  Joint Implementation 271
  National Communications 14
  UAA 16
available moisture 12, 120
Ayres 67, 194

back stop technologies 70
Barrow 74
Bateman 195
beef cattle, methane emissions 130–140, 145
beef, greenhouse-gas intensity 147, 155–156
beef production 147, 150
  systems in Europe 145–158
beef special premium (BSP) 134, 138
Begon 251
Belgium 4, 20, 124
Berg 243
Bergman 173
Bernthal 45
berries and mushrooms 239, 242
'best guess estimates' 61, 67, 296
Beuermann 6
biodiversity 12, 188, 190, 195, 197, 217, 239, 243–244, 283, 297
Biodiversity Convention 1
biofuels 32, 40, 125
  see also wood-based energy
biogas 159, 166–167, 169
biogeochemical cycle models 110
biomass
  below-ground 122, 123
  burning 8, 9, 11, 24, 32, 39–40, 145
Bishop 305
Bohm 303
Bonan 56
Bonny 155
boreal forests 12, 43, 119, 123, 126, 216, 286
  geographical distribution 56
Borman 230
Böswald 229, 230, 231, 235
'bottom up' land-use policy 308
Bouwman 162

Bovenberg 172, 173, 182
bovine spongiform encephalopathy (BSE) 18
Brabänder 199, 208
Bradley 132
Bramryd 233
Brazee 110
Brazil, beef 147
Breuss 173
Bromley 204, 205, 307
Broome 84
Brouwer 16, 125, 126, 307
Brown 10, 12, 15, 47, 113, 115
Burschel 228, 229, 230, 232, 233, 234, 237
Burton 18
business-as-usual 1, 90, 114

Cairns 282
Canada
  agricultural policies 125
  climate change impacts on 117–119
  land conversion 32
  PSE 25
  surpluses 28
  wheat yields 115
Cannell 10, 14, 74, 76, 191, 192, 193, 194, 230
Capros 173
carbon
  credit 266, 287, 288, 289–294, 298
    association 293–294
    system 286
  emission
    permits 306
    social costs 69, 70, 202
  fixing estimates 10, 73, 191
  fluxes 8–9, 10, 13, 71, 72–74, 79, 84, 103, 113, 123, 129, 130, 216, 218, 244, 274–275, 276–278, 282
  reservoirs 217–219
  retention 35, 122, 191–196
    of farm woodlands 187–197
    valuation of 194, 196

sequestration 10, 11, 15, 28–29, 35, 38, 45, 47, 113, 123, 141, 142, 143, 144, 160, 189, 191, 194, 225, 239, 244–252, 255, 259, 267–269, 276–280, 289, 296, 311
    value of 12, 38, 202, 234, 235–236, 257, 262, 266, 280
    tax 2, 85, 113, 122, 167, 215, 248, 252, 302
    as shadow price of carbon 194
carbon dioxide (CO_2)
    credit 292
    emissions, marginal social costs of 67
    equivalents 3, 8, 143, 160, 165, 167
    fertilization, effects of increased 38, 114, 115, 116
    tax 167–169, 224
Carraro 182, 303
Century model 132, 136, 141, 142, 143
Charlson 51
Chichilnisky 89
China
    agroforestry 45
    food supply 115
Cicerone 8
Ciesla 35, 38, 308
clear-felled 191, 231
Cleveland 155
climate change
    on agricultural areas 2, 111
    developed country impact 61
    developing country impact 61
    dynamic issues 103
    ecological adjustments 103, 108
    economic costs 59–70, 85, 103, 104, 106, 248
    global scope 103
    human adaptation 103–112
    impacts 59, 60–69
        environmental 35, 56, 103, 112
        on Europe 117–119
        on food production 28, 115, 116
        health 60
    land-use change 59–70, 113, 117–124
    losers 59, 69
    mitigation 3, 18, 57, 119, 306
    non-linear 68
    potential benefits from 89
    projections 1, 35, 71–87
    spatial issues 103
    winners 59, 69
Cline 13, 59, 64, 66, 67, 75, 82, 84, 85, 103, 248
coal 15, 115
    mining 9
Coase 203
codes of practice 32, 137, 301
'Cohesion States' 3
Cole 132
Colman 305
'command and control' 301
    regulatory measures 92
commercial planting 191
Common Agricultural Policy (CAP) 13, 124, 188, 199–213, 307
    reform 15–18, 26, 27, 125, 130, 202, 209, 304, 305, 309
        impact on net emissions of greenhouse gases 17
        livestock 16
Conrad 173
construction 270
conventional logging 74
CORINAIR 13, 14, 21–22
cost–benefit 37, 69, 71
    analysis 38, 188, 195
cost effectiveness 306
    analysis 38
cost efficient analysis 196
cost function 12, 256
cost of agricultural support 195
Cox 94, 100
Coxworth 113
Crabtree 120, 122, 187, 191, 196
Cramer 105
crop–land–forest 122–124
crop insurance 125
crop response models 116
cross-compliance 30, 304–305
Crutzen 11, 137
CSFR (former) 20
Curtin 113
Czech Republic, National Communications 14
Czerkwaski 146

dairy 28, 132–134, 136–138, 139, 143, 163, 166
damage 60–66, 68, 90, 95, 122, 202
    costs 65, 248
    profiles 82
Danilov-Daniljan 286, 287
D'Arge 85
Darwin 111, 114, 115, 116, 117, 118, 127
Dasgupta 240
De Alessi 205
De Benedictis 303
deforestation 8, 11, 27, 35, 38–39, 113
de Haan 181
De Jong 270, 272, 283
Delcourt 126
de Melo 175
Demeyer 146
Demmel 236
Denmark 4, 20, 124
depopulation 126
Detwiler 270
Dewar 74, 192
de Wit 172
dietary interventions 146
Directive 2080/92 199, 208
discount rate 13, 66, 74, 76, 79
discounted damage 66, 71
discounting 13, 70, 72, 73, 82, 83–86, 257, 296, 310
    climate change effects 71
distortionary tax system 172
Dixit 100
Dixon 10, 12, 257, 270, 287
Dobris Assessment 21
Dosi 100
double dividend 172, 182, 302
Dowlatabadi 66
Downing 60
Dudek 286, 292, 293, 294
Dykstra 41
dynamic models 105–110
    climate 106
    economic 103, 106, 108–110, 111
    ecosystem 106–108
dynamic issues 111

economic costs of impacts 66, 296
economic damage 64
economic efficiency 85
economic institutions 114
economic instruments 29–30, 113–128
economic value of carbon stock changes 247–249
Edmonds 202
efficiency and equity 225
Eggleston 137
Eichner 162
Ekins 66
Eliasson 239, 241, 247, 252
Emanuel 105
emissions
    from agriculture 13–15, 17, 297
    reduction 3, 285
    trading 286
endangered species 42, 243
energy
    crops 32, 44
    demand 61
    embodied in agriculture 150–154
    inputs 150
    policy 289
    saving strategy 289
    tax 177, 181, 182, 302
        Dutch 171–185
        unilateral 172, 180
        European Union 171
    wood-based 32, 44, 224
England (yield-nitrogen formulae) 130
Enquete Commission 24, 228
enteric fermentation 9, 145, 146, 147–149, 154
environmental accounting 240, 247
environmental indicators 240
environmental taxation, at EU level 302
Environmentally Sensitive Areas (ESAs) 134, 141–142, 143, 144
Eriksson 245, 249
Estonia 20
eucalyptus 258
European Union (EU) 1–22, 25, 26, 31, 34, 117, 118, 123, 126, 156, 194, 199, 301

Agreement on Agriculture 17
agriculture reforms 3
'block vote' 2
change in CO_2 emissions 1990–1994 6
environmental policies 309
FCCC ratification 2
heterogeneous policy 2–7
'monitoring mechanism' 15
National Communications of 15
extreme weather events 60

Fankhauser 59, 64, 65, 67, 68, 194
Farm Woodland Premium Scheme (FWPS) 134, 142–143, 189, 190, 191, 192, 193, 195
feed-supplement 146
Ferro 309
fertilizer 16, 130–132, 135, 144, 150, 155, 160, 163, 167, 179, 180
    manufacture 134, 136, 140, 143, 146
    regulatory instruments 29, 31, 125
    use 135–136, 138, 140, 141, 142, 144
        N_2O emissions 7, 9, 24, 27, 158, 162, 245
fertilization effects, atmospheric 12
Finland 10, 20, 215, 216–226
    fertilizer levies 31
    National Communications 14
    national forest inventory 217
fire control 217
fiscal instruments 187, 304
Fischer 90
Foerster 229, 230
Food and Agriculture Organization (FAO) 39, 46, 308
foreign debt 215, 217
forest
    carbon fixing 42, 43–44, 72–74, 80–82, 86, 191–194
    illegal grazing 11
    increasing area 10, 111, 112
    industry products 215
    investments 311
    litter 14
    management 10, 12, 37, 42, 76, 106, 191, 200, 215–226, 256–257

    mitigation options 35–47, 256–257, 270
    mixed-species 263
    natural 10, 42, 44
    planting 43–45, 189–191
    practices 10
    products 42, 118, 219
    substitution of natural 297
    taiga 111
forestry
    community 282
    conversion of grazing and arable land to 123, 155
    Europe 10–13
    fires 11, 35, 36, 42, 46
    indicators for sustainable 37
    large-scale projects 255–268
    pests 11, 35, 36, 42, 46
    policy, UK 17, 18, 187–189
    recreation and countryside protection 12, 188, 189
    sector 36–38, 60, 216–221, 237
fossil-fuel
    emissions 36, 47, 52, 141, 151, 152, 153, 161
    total 10
    energy 7, 24, 166
    sources 249
    substitution 10, 11, 32, 40, 160, 166, 167, 215, 225, 227, 231–232
    tax on 225
France 20
    tax reliefs 302
    UAA 16
Franz 229
Framework Convention on Climate Change (FCCC) 1, 129, 171, 255, 271, 285, 286, 299
    National Communications 3, 19–22
    see also European Union (EU), FCCC ratification
Froud 142
Frühwald 232, 233, 238
Furubotn 204
Future Agricultural Resources Model (FARM) 115–116, 117, 119

Gaia hypothesis 310
Galinski 10
game hunting 242
García 272
Gaskins 173
gas production 9
Gay 270
General Agreement on Tariffs and Trade (GATT) 3, 15, 25, 125
general equilibrium (GE) model 115, 224
German forests 227–238
Germany 3, 4, 10, 20, 199
    agri-environmental funds 16
    emissions 6
    farming systems 159–169
    Joint Implementation 271
Giampietro 147
Gibbs 134, 137
global
    afforestation and CO_2 sequestration 44
    average temperatures 51, 52, 54, 55, 66, 114
    circulation models (GCM) 103, 105, 115, 119
      projections 51, 104, 116, 117
    damage 104, 195
    scope 104, 111
    warming potential 43, 46, 49–57, 60, 82, 157, 160
Goulder 172, 173, 303
Grabherr 56
Grainger 42, 43, 44
grassland 8, 28, 44, 111, 113, 125, 141, 144, 163–167, 164
grassland–crop production margin 121–122
    use 163
'green consumerism' 307
green investments 89–101
greenhouse effect, runaway 68
greenhouse-gas
    cycles 296
    emissions, mitigation summarized 7–13
    land-use-related 295
    minimizing strategy 35, 145
Greece 3, 4, 20, 39
    UAA 16

'greening' of agriculture and forestry policy 18
Gregory 188
Grossman 240
gross national product (GNP) 61, 240
gross world product 65, 82, 85, 116, 117
Grubb 4, 15, 66, 152

Haigh 3, 6
Hanley 195
Hansen 307
Harmon 10
Harold 28
Harrison 100
Health 10
Henrichsmeyer 203, 206
Heyer 162
Hodge 299
Hoel 172
Hoen 224
Holdridge 105
Holling 114
Holtz–Eakin 240
Hope 66
Horlacher 163
horticulture 180, 182, 183
Houghton, J.T. 49, 50, 51, 80, 81, 103, 104, 105, 106, 295
Houghton, R.A. 8, 44, 122, 123, 257, 269
Hourcade 12
Howes 31
Hultkrantz 239, 241, 243, 252
human adaptation 102–112, 114, 126–127
Hungary 20

Iceland, PSE 25
industrial wood 262, 264
institutional
    arrangements 285
    change 37, 205
    constraint 44, 64 200, 212
    obstacles 210–212, 213
      to afforestation 188, 203–207
    rigidity 125
instruments, market-based 299
insurance industry 60

Intergovernmental Panel on Climate
    Change (IPCC) 10, 14, 51,
    53, 54, 55, 56, 60, 61, 68, 104,
    133, 134, 137, 255, 295, 296,
    302, 308
  forestry 47
  scenario of reversing deforestation 114
  Second Assessment reports 12, 32
  Summary for Policymakers 50
  Working Groups II and III 59, 61
Intergovernmental Panel on Forests (IPF)
    37
International Tropical Timber Organization
    (ITTO) 37
intensive systems 154
Ireland 3, 4, 20
irreversibility 12, 89
Italy 4, 10, 20
  National Communications 13

Jarvis 137
Jenkinson model 132, 141
Jepma 12
Joint Implementation 43, 271, 306
Jones 52
Jorgenson 173, 182
Josling 3

Kaiser 24, 120
Karaban 286
Karjalainen 10, 221, 222, 232
Karl 55
Kats 308
Kauppi 10, 11, 13, 21, 22, 217
Kerstiens 12
Kiehl 55
Kirchgessner 162
Kirschbaum 12, 56
Kohlmaier 230
Kollert 230
Kolstad 90
Komen 173, 174, 176
Krankina 257, 287
Kreps 187
Krjukov 293
Kroth 233

Labudda 210
landfills 9, 15, 219, 221
  wastes 220
land tenure 310
land-use
  historical trends 7
  relative global contribution 8
Lanly 39
Lappalainen 219
least cost approach 197
Leemans 111
Less Favoured Areas 126, 190, 193,
    212
lichen 239, 243
Liljelund 243
linear programming models 161, 162
livestock
  maximum stocking rates 32
  production 27, 115, 124, 145–157
  quotas 188
  systems 145, 146
  units 138, 163
livestock pasture risk diversifying strategy
    125
location-rent theory 201
Lovelock 310
Luxemburg 4, 20
Lyon 110

McGinn 127
Maddison 67
Maier-Reimer 77
maize, average yields of 115
Makundi 282
Malaysia, reduced impact logging in 41
management agreements 305, 306
Manne 90
manure 9, 16, 124, 141, 145, 146,
    149–150, 154, 158, 162–163,
    166
Marland 269
marginal damage costs 194
market failure 114
Marsch 208
Masera 270
Mattsson 242
mean annual increment (MAI) 274

mean global surface temperature 49
Mediterranean forests 11
Melillo 10, 12, 105
Mesopotamia 255, 257, 258, 260, 261, 263, 267
Metalnikov 288
methane (CH_4) 14, 15, 18, 68, 80, 82, 129, 143–144
    arable sector 130–132, 135–136, 140–141, 144
    and BSE 17–18
    conversion factors 137
    enteric fermentation 7, 134, 140, 145, 146, 147–149
    from coal 15
    from fossil fuel energy 7
    from landfills 15, 219
    livestock 17, 24, 27, 28, 32, 79, 134, 136–140, 141, 145–157, 158, 160, 162, 163–166, 168
        abatement strategies 146
    stored slurry 7, 137–138, 149–150, 162
    total emissions in EU 9
Mexico 19, 269–284
    Chiapas 269, 272–273, 277, 279, 281
    costs of carbon sequestration 270, 272, 274, 280
    ejido 272
        defined 271
    land reform 270
    Land Tenure Law 271
migration 61, 106
milk
    production quotas 132
    yield 159, 165
minimum tillage 301
missing sinks 298
Mitchell 51
mitigation strategies, benefits of 71
Moeller 51, 53
Mohren 274
monetary
    measurement of people's preferences 63
    quantification of all impacts 59, 66
Monserud 105

montane forests, geographical distribution 56
Montoya 272
Montreal Protocol 37, 114
Moretto 100
Mosier 131, 162
Moulton 268

Nabuurs 274, 278
National Communications 3, 13–15, 19–22, 129
    political process 14
'natural' fluxes 9
Nepal 39
net national emissions 11, 223
net national income 239
net national product (NNP) 239, 240, 241
net primary productivity (NPP) 10, 105, 108
Netherlands 3, 5, 10, 20, 31, 124, 171–185
    Joint Implementation 271
New Zealand
    afforestation 28
    livestock 27
    PSE 25
Nicaragua 39
Nicholls 54
Nicholson 137
Nielson 105, 107
Niesslein 207
Nilsson 82
Nitrate Sensitive Areas (NSAs) 134, 141, 144
nitrogen cycle 160, 161
nitrous oxide (N_2O) emissions 24, 145, 146, 162, 163, 164
    see also fertilizer use
non-market and environmental goods 188, 200, 240, 241, 242
non-market impacts 61, 69, 72
non-timber
    forest products 12, 42
    services 12
non-wood products 42
Nordhaus 13, 59, 64, 66, 67, 70, 78, 79, 80, 82, 103, 194, 248

'no regret' 12
　　policies 295, 308, 311
North American Free Trade Agreement 271
Norway 5, 20
　　fertilizer levies 31
nutrient, availability 12

observed global temperature increase, partitioning 54
oceanic sinks 71, 76, 79, 80, 86
oceanic uptake 74–79, 81, 86
Olsen 100
Olson 111
opportunity cost of beef 152
Organization for Economic Cooperation and Development (OECD) 23–34, 65
　　Emissions Inventory Guidelines 14, 140
O'Riordan 2
Östlund 250
Overpeck 107

Papen 159
paper 232
Pareto 310
Parry 172
participatory methods 271
Parton 132
pasture, risk-diversifying strategy 125
Patagonia 255, 258, 263, 264, 265, 266, 267
Pearce 13, 15, 59, 61, 62, 65, 67, 73, 84, 172, 194
Pearse 125, 188
peatland 219, 297
　　deep 192
　　soils 74, 220
Peck 67, 70, 194
Peerlings 175
pesticides 160
Petersen 204
Petrov 292
Philipp 234
'Pigouvian subsidy' 95–96
Pimentel 24

Pinard 41, 74
Pingoud 218, 220, 221, 232
plantation 81, 106, 109, 269
　　on agricultural land 190, 287
　　establishment 43, 110, 257, 259, 261, 264
　　forests 10, 257, 258, 297, 304
　　large-scale 255, 259
　　short-rotation 74, 255, 269
　　timber 197
Plochmann 210
Poland 10, 20
policy failures 200
policy instrument 16, 29–33, 92, 99, 187–197, 302–305, 308
　　juridical 299–301
　　mandatory 295, 300, 301
　　voluntary 295, 300
policy-makers, inertia of 298
policy objectives, forestry 188, 196
'polluter pays' principle 302
pollution
　　abatement 90
　　tax 172
　　threshold 91, 96
Pontryagin 110
Portugal 2, 3, 20
　　National Communications 13, 14
positive-feedback loop 56, 82
Povellato 305
precautionary principle 69
Prentice 105
Price 72, 74, 78, 84, 194
principal-agent 187
Producer Subsidy Equivalent (PSE) 25
　　defined 34
　　and fertilizer use 28
producers' surplus 305
profitability-related constraint 199
property rights 2, 198, 199, 200, 203–205, 207, 211, 297, 299

Read 75
reafforestation 292
recreation 42, 200
redistribution between generations 310
reduced impact logging (RIL) 41, 74–75

reforestation projects 70
Regulations 124, 127, 130, 172, 204, 297, 301, 309
    2092/91 307
    2078/92 16, 304, 305
    2079/92 16
    2080/92 10, 16, 188, 235, 297, 304, 305
    2081/92 307
    2082/92 307
    and legislation 30, 173
Reilly 80
Reinhardt 161
renewable energy technologies 224
rent function, forestry 123
resource accounting, forest 241–244
rice 9
    average yields of 115
    methane 8, 24, 32, 79
Richards 7
Richter 204, 205
Rodhe 245
Roemar-Mähler 234
Rosenzweig 59, 114, 115, 116, 127
rotations 122
    lengthening of period 233
    short 269
Rotherham 37
ruminant 9, 32, 159
rural communities 270, 308
Russia 19, 20, 117, 285
    average yields of 115
    institutional carbon credit 285–294
Russian forest sector 287

Safley 150
Sahelian Africa, livestock systems 146
Sampson 44, 46
Santer 51
satellite account 239, 241
Saunders 16
sawn timber 225
Saxon 209
Saxony 199, 207–209
Scandinavian countries, forestry research 296
Schazenbächer 161

Schelling 75
Schiffman 143
Schöpfer 230
Scotland 193
sea-levels, rise in 1, 60, 68, 82, 89
Sedjo 43, 104, 109, 113, 257, 269
Semenov 56
Seppälä 224
set-aside 16, 17, 31–32, 125, 130, 132, 136, 168, 188, 204, 209, 213
shadow prices of carbon 69, 70, 194
shopping 172
Shoven 176
Siegenthaler 76
Simons 181
Slesser 147, 150, 151
social accounting matrix (SAM) 173–174, 176, 184, 185
Sohngen 104, 106, 109, 110, 112
soil
    carbon 10, 28, 31, 73, 74, 113, 123, 124, 126, 135, 140, 228, 230, 249, 256, 261, 269
    sequestration, set-aside field 136
    sink 132, 219
    erosion 113
    methane sink 131–132, 134, 135 138, 141, 142, 144
    moisture, available 60, 119
    nutrients 239
    organic matter 228, 274
Solomon 11
soybeans, average yield of 115
small island states, climate impacts 60
Spain 3, 5, 20
    National Communications 14
spatial issues 111
species extinction 60, 89
standards and licences 301
steady-state models 104–105
stewardship 305
stocking-density 134, 139, 142
Stout 155
straw-burning 140–141, 144, 301
stumpage price 122
Subak 14, 15, 145
subsidies 113, 114, 122, 123, 124, 125, 176, 204

sulphur dioxide (SO$_2$) 50–55
    cooling 51–52, 54, 296
    doubling 50
sustainable development 33, 37, 47, 69, 126
Sverdrup 242
Swart 181
Sweden 5, 10, 20, 239–253
    average standing-timber volume 252
    emission tax in 248
    fertilizer levies 31
    National Communication 14
Swisher 270
Switzerland 5, 20
SWOPSIM, trade model 114
Sydney model 133, 136
system of national accounts (SNA) 240

Tahvonen 224
Tangermann 3, 17
taxes 26, 31, 85, 92, 124, 125, 171–185, 302
    greenhouse-gas 223–225
    see also carbon dioxide (CO$_2$), tax; carbon, tax; energy, tax; environmental taxation, at EU level; fossil-fuel, tax on; pollution, tax
tax
    land conversion 122
    market-based 299
    unilateral 182
Taylor 51
Tchebakova 105
technical change 115, 169, 236
    climate-induced 116
technological change 32–33, 38, 85, 94, 95, 97, 98, 99, 100, 182, 203
    optimal private timing 91
technological escape 303
technological innovation 303, 311
technological optimism 85
technology 37, 90, 92, 93, 94, 95, 96, 99, 100, 270, 288
temperate
    European forests 12
    forest 11, 43
    grasslands 122

tenancy contracts 211
terrestrial sinks 71, 78, 79, 80, 86
timber
    markets 108, 109, 112, 216
    global 104
    prices 224
Titus 59, 64
Tobey 114
Tol 59, 64, 65, 66
tourism, climate impacts 60
transaction costs 203, 205–207
transfers 288
transport 9, 51, 60, 124, 129
tree planting subsidies 125
Trexler 45
Tropical Forestry Action Plan 271
Trunk 160, 161, 164, 165, 166
Turner 13, 310
Tyler 191

uncertainties 38, 46, 49, 50, 89, 94, 97, 106, 296
    forestry fluxes 10
unemployment 172, 215, 217, 225
UK 5, 10, 20
    agricultural policy 129–144
    arable areas 144
    beef imports ban 17
    livestock 27
        systems 123, 132
    National Communications 14, 129
    N$_2$O fluxes 129
    forestry 9
    UAA 16
United Nations Commission on Sustainable Development (CSD) 37
United Nations Conference on Environment and Development (UNCED) 1, 37
USA
    available forestry area 44
    climate impacts 59
    land conversion 31
    PSE 25
    surpluses 28
    timber markets 112

Uruguay Round 17, 25
urban tree 45, 46
utilized agricultural area (UAA) 203, 204

van Amstel 137
van der Hamsvoort 306
van Kooten 84, 119, 120, 123, 126, 194, 268
van Nevel 146
van Soest 176
Vinson 289
voluntary approaches (VAs) 30, 33
Volz 229
von Alvensleben 201
von Neumann 85

waste storage, livestock 134, 149–150, 156, 162
water pollution 124
  impacts 11
water supply 12, 116
watershed protection 12
Watson 38, 59, 60, 145
Welsch 172
Westman 219
wetlands 28, 61, 192
  conversion of 113, 125
  methane 8, 79
  soil processes 79
Weyant 173
Whitby 16, 18, 305, 307
Wiersum 143

Wigley 50, 51, 82
wildfire management 39
Wilhelmsson 243, 244
Williams 131, 137
willingness to accept (WTA) 61, 63, 196, 306
willingness to pay (WTP) 61, 63, 248–249, 252, 307
  for safety 63
Willis 195
Winjum 199
Winter 18
Wofsy 11
wood-based
  energy 32, 44, 224
  products 122, 215, 216, 219, 223, 259, 262
wood-burning cooking stoves 40, 270
wood utilization 230–232, 233, 237
Woodward 105
World Commission on Environment and Development 37
World Trade Organization (WTO) *see* General Agreement on Tariffs and Trade (GATT)

Yohe 59

Zamagni 307, 310
Zecca 50, 51, 52, 54, 55
Zeelenberg 176
Zhang 173